Car / Griesebner / Strobl (Eds.)

Geospatial Crossroads @ GI_Forum '10

Adrijana Car / Gerald Griesebner / Josef Strobl
(Eds.)

Geospatial Crossroads @ GI_Forum '10

Proceedings of the
Geoinformatics Forum Salzburg

All explanations, data, results etc. contained in this book have been made by the authors to the best of their knowledge and have been approved with care. However, some errors could not be excluded. For this reason the explanations etc. are given without any obligations or guarantee by the authors, editors and publisher. They cannot take over any responsibility for eventual erroneous contents.

Bibliographic information published by the Deutsche Nationalbibliothek
The Deutsche Nationalbibliothek lists this publication in the Deutsche Nationalbibliografie.
Detailed bibliographic data are available in the Internet at http://dnb.d-nb.de.

ISBN 978-3-87907-496-9

Bismarckstr. 33, 10625 Berlin, Germany
www.vde-verlag.de
www.wichmann-verlag.de

Printing: H. Heenemann GmbH & Co., Berlin, Germany

Foreword by the Editors

Advancing the Dialog

The beautiful world cultural heritage city of Salzburg is the setting for the fourth Geoinformatics Forum (GI_Forum, www.gi-forum.org) Symposium from July 6-9, 2010 – jointly organised by the University of Salzburg Centre for Geoinformatics (Z_GIS) and the Institute for GIScience at the Austrian Academy of Sciences (OeAW-GIScience). The interdisciplinary GI_Forum has become an annual event for a vibrant GI community from academia, industry, and government to advance the dialogue on progress in GIScience and to explore new research directions. The symposium is co-located with the highly regarded annual German language conference on Applied Geoinformatics (AGIT, www.agit.at).

The proceedings of the GI_Forum'10 consist of papers on emerging topics and research outcomes related to Geoinformatics methodology. These topics range from advanced spatial analysis, knowledge extraction and geovisualization, to standards and spatial data infrastructures, mobile services, and dynamic modelling and simulation. GI_Forum'10 also attracted contributions pertaining to the following specific topics:

- *Digital Cities and Urban Sustainability*
  Digital Cities are meant to provide a collaborative platform based on modelling of existing and planned urban infrastructure. This concept supports applications of a wide scale range, from the whole city, a city quarter, or a group of buildings down to a single building. A number of contributions focus on these issues ranging from concepts to applications.

- *Global Change: Monitoring and Modelling*
  "Global change" – a short formula for a multitude of anticipated shifts in societal and environmental domains due to strong global drivers such as climate change – calls for monitoring and modelling techniques to better understand regional implications and potential dynamics of such changes in a geospatial manner. International programmes and visions (GEO, GMES, SEIS, …) envisage unified systems based on quality standards for data, products and services to establish optimized observation and forecasting capacity within Europe, and globally. Again this year contributions in two discussion sessions focused on this challenging topic offering local and regional solutions to a rather global problem ranging from LULC changes to natural hazards and rapid geospatial reporting.

- *Vulnerability: Spatial Assessment and Analysis*
  A one day workshop dealt with different developed and currently investigated methodologies to spatially assess vulnerability. The workshop specifically addressed the issue of vulnerability assessment, independent from conceptual discussions. The participants reviewed and discussed different methods of GIScience employed to assess, quantify and represent vulnerability as integrated spatial phenomena. Current achievements and future research challenges were identified and formulated in a plenary session. The contributions will be published in separate proceedings later this year.

- *Learning with Geoinformation*
  Attracted contributions discussing various issues related to geoinformation in education. This year a special focus was put on the concept of 'Spatial Citizenship' and the role of GI-based learning in developing the relevant competences. The bilingual conference on Learning with Geoinformation hosted such contributions in respective sessions shared with GI_Forum. These contributions are published in separate proceedings.

The programme committee has selected over 40 contributions to be presented as publications and oral presentation, in a discussion session, or as poster presentations at the GI_Forum'10. We thank all authors who submitted their contributions to the peer review process! 32 of these contributions qualified for publication in this book based on the review outcomes. Roughly two thirds of them were accepted based on a full paper review; they profoundly contribute to further development in the field of GIScience and Technology. At the GI_Forum'08 we introduced the thematic *discussion session* format to facilitate the exchange of experience and further enhance discussion among the presenters and audience, which proved particularly fruitful for presenting early ideas. Due to excellent feedback by both presenters and audience we decided to again organise a number of sessions in this manner. Summaries of the respective contributions are available as extended abstracts in these proceedings.

The quality of all the contributions greatly depends on critical and constructive feedback from the GI_Forum programme committee members – an internationally acknowledged team of experts from academia and industry. We thank all of them for this well done, challenging and time consuming task!

Our very special "thank you" goes to Gerold Olbrich from Wichmann Verlag for yet another highly professional cooperation that resulted in a high quality book published in time for the GI_Forum'10.

The GI_Forum programme is significantly enriched by four keynote speakers; Athina Trakas, Open Geospatial Consortium, Inc.; Arup Dasgupta, GIS Development; Ed Parsons, Google, USA; and Laxmi Ramasubramanian, Hunter College at City University New York.

The organisation of this GI_Forum as well as of the annual AGIT, AGIT EXPO and accompanying events would be impossible without the Z_GIS team members who work all year to make everything happen. Bernhard Zagel supported by Dagmar Baumgartner is in charge of the overall event organisation. Gerald Griesebner and Bernhard Bretz are responsible for the smooth flow of the review process from paper submission to the published proceedings. Petra Jenewein and Anna Karnassioti are involved with the operational side of GI_Forum ensuring answered emails and successful symposium registrations. To them and all the remaining team members we owe a big thank you!

Annual events of and around AGIT and GI_Forum also present a showcase of Salzburg's unique scientific environment in the field of Geoinformatics and GIScience: basic and applied research activities at Z_GIS and OeAW-GIScience and the research studio "iSpace" and complement academic study programs at a graduate and postgraduate level (recently revised MSc "Applied Geoinformatics" started in winter semester 2009, UNIGIS postgraduate online distance learning programs and an innovative doctoral study programme). In such an environment the AGIT / GI_Forum events stimulate creative, new activities resulting in regional, national and international initiatives, new projects, seminars, and research. For more information visit www.zgis.at, www.oeaw-giscience.org, or www.giscience-research.org/, an entry to a world wide GIS community network.

Geospatial crossroads @ GI_Forum is a great opportunity for advancing the dialogue, interconnecting people, ideas and institutions. We hope you will find this book a revealing and motivating reading and hope to see you at the GI_Forum 2011!

Adrijana Car *Gerald Griesebner* *Josef Strobl*

Table of Contents

Foreword by the Editors V

Albrecht, F. and Moser, J.:
Potential of 3D City Models for Municipalities – The User-Oriented Case Study of Salzburg 1

Amelunxen, C.:
On the Suitability of Volunteered Geographic Information for the Purpose of Geocoding 11

Anantsuksomsri, S.:
The Spatial Analysis on the Impacts of Mass Transit Improvements on Residential Land Values – the Case Study of Bangkok Metropolitan Area, Thailand 18

Barbosa, M., Dias, L., Seoane, J. and Buratto, M.:
GIS Prioritization Model Based on Forest Fire Hotspots and Land Susceptibility 22

Brunner, D., Egger, G. and Leitner, M.:
Statistical Analysis for Discovering Distribution Patterns of Succession Types and Abiotic Factors Along Rivers 26

Chima, C. and Trodd, N.:
Comparison of DMC Nigeriasat-1 SLIM, SPOT-5 HRG and Landsat 7 ETM+ Data for Urban Land Cover Analysis 36

Dasgupta, A.:
Taking Geospatial Applications to the Grassroots 40

Eisank, C. and Drăguţ, L.:
Detecting Characteristic Scales of Slope Gradient 48

Epitropou, V., Karatzas, K. and Bassoukos, A.:
A Method for the Inverse Reconstruction of Environmental Data Applicable at the Chemical Weather Portal 58

Erlacher, C., Anders, K.-H. and Gröchenig, S.:
VestiGO! – More Than an Adaptable Location-Based Mobile Game 69

Van Gasselt, S. and Nass, A.:
Challenges in Planetary Mapping – Application of Data Models and Geo-Information-Systems for Planetary Mapping 79

Helbich, M. and Brunauer, W.:
Mixed Geographically Weighted Regression for Hedonic House Price Modelling in Austria 87

Hochmair, H. H.:
Spatial Association of Geotagged Photos with Scenic Locations 91

Hoechstetter, S., Krüger, T., Goldberg, V., Kurbjuhn, C., Hennersdorf, J. and Lehmann, I.:
Using Geospatial Data for Assessing Thermal Stress in Cities 101

Karampourniotis, I. and Paraschakis, I.:
Using Mobile GIS for Water Networks' Management – The Open Source Approach ... 111

Klinger, G.:
Enabling INSPIRE for Aeronautical Information Management 121

Krüger, T.:
Algorithms for Detecting and Extracting Dikes from Digital Terrain Models 130

Manohar, S. S. and Kiechle, G.:
Performance Evaluation of Two Solution Methods for Time Dependent Bi-objective Shortest Path Problems 140

Marjanović, M.:
Landslide Susceptibility Mapping with Support Vector Machine Algorithm 150

Möller, M., Bertermann, D. and Roßner, P.
THERMOMAP – Mapping Subsurface Thermal Potential for Selected Test Sites in the EC 160

Nass, A., Van Gasselt, S., Jaumann, R. and Asche, H.:
GIS in Planetary Geology – Standardised Moduls for Planetary Mapping 164

Nduwamungu, J.:
Applied GIS in Forest Mapping and Inventory in Rwanda 168

Nedkov, S.
Modeling Flood Hazard Due to Climate Change in Small Mountainous Catchments.... 172

Olang, L. O., Kundu, P., Bauer, T. and Fürst, J.:
Assessing Spatio-Temporal Land Cover Changes Within the Nyando River Basin of Kenya Using Landsat Satellite Data Aided by Community Based Mapping – A Case Study 176

Pickle, E.:
GeoNode – A New Approach to Developing SDI 184

Smreček, R.:
Measuring the Stem Breast Diameter Using A Terrestrial Laser Scanner 188

Steinnocher, K., Aubrecht, C. and Köstl, M.:
Geocoded Address Point Data and Its Potential for Spatial Modeling of Urban Functional Parameters – A Case Study 198

Tiede, D., Hoffmann, C., Füreder, P., Hölbling, D. and Lang, S.:
Automated Damage Assessment for Rapid Geospatial Reporting – First Experiences from the Haiti Earthquake 2010 207

Tontisirin, N. and Anantsuksomsri, S.:
Residential Location Decisions – A Case of the Bangkok Metropolitan Area 211

Vatseva, R. and Koulov, B.:
Land Use/Cover Changes and Sustainability of Tourism Development in the Bulgarian Black Sea Coastal Zone 215

Wallentin, G. and Car, A.:
Spatio-Temporal Uncertainty in Individual Based Tree Line Modelling 219

Zalavari, P., Klug, H. and Weinke, E.:
Three Steps Towards Spatially Explicit Climate Change Analysis 229

Neis, P., Singler, P. and Zipf, A.:
Collaborative Mapping and Emergency Routing for Disaster Logistics – Case Studies from the Haiti Earthquake and the UN Portal for Afrika 239

Acknowledgement 249

Index of Authors 251

Potential of 3D City Models for Municipalities – The User-Oriented Case Study of Salzburg

Florian ALBRECHT and Julia MOSER

The GI_Forum Program Committee accepted this paper as reviewed full paper.

Abstract

A current trend can be observed that cities want to extend their 2D GIS implementations to the third dimension. There already exist 3D city models for a large number of municipalities, mainly with a focus on visualization for marketing and tourism purposes. With the availability of city models a lot of new applications become obvious that can be supported by advancing this technology. The presented research conducted within the *Digital Cities* project focused on adapting city modelling technology for a selection of these applications in the City of Salzburg, Austria. For an optimal integration into the working processes a user-centred approach was implemented. Workshops with the city departments of urban planning and facility management were conducted to gather information about the tasks in present workflows and the intended support that a digital city can provide to them. The collected information was structured and analysed into specific requirements to the various components of the digital cities environment. The resulting requirements reflected the users' needs and were well-founded by expert statements. Therefore, they were optimally suited for the development of applications supporting experts in the discussed disciplines of a municipality in their every day work.

1 Introduction

Digital 3D City Models have achieved a high presence during the last years. Many cities like Berlin[1], Köln[2] and Paris[3] already implemented a 3D city model with attractive visualizations of the urban infrastructure. The requirements and prospects of 3D city models are different. High level visualizations aim mainly at adequate online presentations of cities and the main emphasis is on touristic aspects. According to the increasing number of applications in conjunction with high costs for data acquisition the development of standards like CityGML (KOLBE et.al. 2005) and XPlanung (BENNER et.al. 2009) was established (STROBL 2009). Beyond that current 3D city model projects focus more on the integration of such models in working processes of communities. The REFINA 3D project (ROSS et. al. 2008) deals with the utilization of virtual 3D city models in the context of urban land management. Other research discusses the benefits of 3D city models in facility management (BLEIFUSS 2009). AG 3D STADTMODELLE (2004) published a guideline for

[1] http://www.3d-stadtmodell-berlin.de
[2] http://www.3dstadtmodellkoeln.de/
[3] http://v3d.pagesjaunes.fr/paris/

3D city models that includes various applications and benefits for municipalities. These approaches commonly pursue the objective that integrated 3D city models can support working processes and decision making processes within communities. This study was accomplished within the research project *Digital Cities*[4] that Autodesk conducted with Z_GIS, Centre for Geoinformatics, as research partner and the City of Salzburg, Austria, as a "pilot city" (cf. Fig. 1). The project focused on supporting a defined subset of workflows in selected organisational units of a municipality by an integrated environment for 3D city models. The workflows were mainly related to urban planning and facility management. *Digital Cities* placed special emphasis on the demands of the designated users. Therefore, the to-be supported workflows were discussed in workshops with members of the relevant departments of the City of Salzburg. The users could express their needs that then were analysed and structured into specific requirements for a *digital city*. The requirements covered all the components of the *digital city* working environment, such as the software, the data models, or the infrastructure. Within the "*Digital Cities*" project prototypical software was implemented concurrently to the requirements gathering process and will be developed further.

Fig. 1:
Salzburg – Pilot city of Digital Cities (Courtesy of Autodesk)

2 Methods

2.1 Performing Workshops to identify user needs

With the case study of the city of Salzburg the requirements for a *digital cities* environment were analysed. They aimed at the essential tasks that this platform needs to be able to perform in order to have value for the city of Salzburg. Therefore it is focused on tasks that have specific requirements to certain components of the environment. These critical components are for instance analysis capabilities, the data model for the city data, or access for involved stakeholders. Other components that are concerned with less critical functionality (e.g. the appearance of the graphical user interface or other aspects of software usability) are not considered here.

[4] http://www.autodesk.com/digitalcities

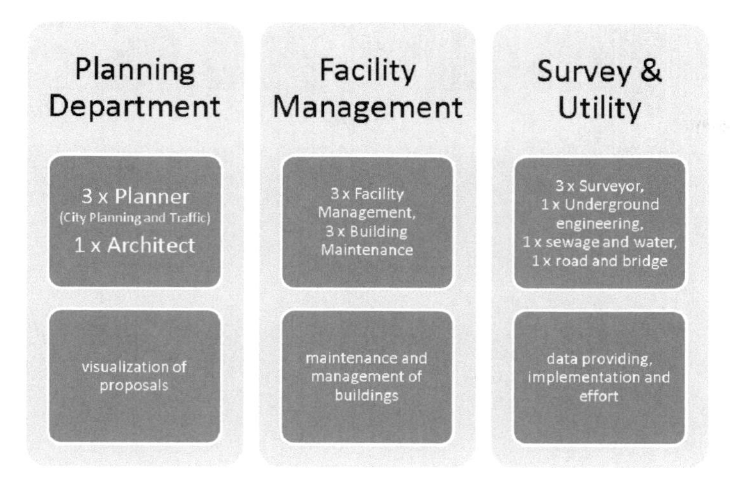

Fig. 2: Workshops with the city of Salzburg

The basis for the requirements analysis was the information mainly provided by workshops and interviews with experts of the city of Salzburg as the prospected users of the to-be developed technology (cf. Fig. 2).

The workshops were carried out with selected departments. Fig. 2 illustrates the departments, number of participants, the represented professions of the respective workshop participants and the main workshop objectives. Additional information was provided by the internet presence of the involved city departments. Workshops and interviews with selected experts have been added for getting more detailed information and for making already collected information more precise.

2.2 Structuring requirements

The gathered expert information was analyzed and structured in such a way that requirements for specific components of the *digital cities* environment could be extracted (cf. Fig. 1). Therefore, tasks were identified and described. Complete workflows have been investigated as well as single work steps. Each of these tasks was described in a general way by an expressive title and by the assignment to a field of expertise. A condensed description was added in the manner of a use case scenario. Scenarios are a tool from the application development community (CARROLL 1995). They describe the 'who', 'what', and 'why' of actions taken with a system. So, apart from the system interaction the motivation and the user needs are included (ibid.). Scenarios are a somewhat unstructured format of information, like an image containing implicit information that can be made explicit by image interpretation if needed. Providing a narrative description of the tasks may confirm explicit statements and can make them more precise. It may also contain information that is difficult to structure in a different way.

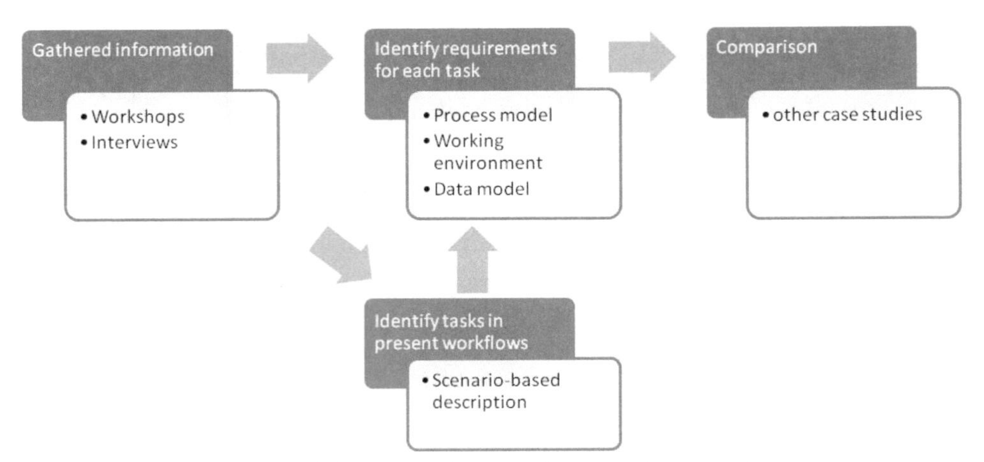

Fig. 1: Structuring expert information into requirements for *digital cities*

Each of the tasks had its own requirements to *digital cities*. Several perspectives on the tasks are considered to identify the requirements of the specific system components that represent the environment. These perspectives focus (a) on the process model, (b) on the working environment, and (c) on the supporting data models of each task.

a) For acquiring information about the process model the workflows are evaluated. Fig. 2 shows an example of a workflow where relevant information, such as the involved stakeholders and the channels of communication, were directly visible. Other information is implicitly included such as the type of exchanged information products.

When regarding the example in Fig. 2, the facility management requests approval for a building study from the building preservation authority based on visualizations of the study. The specified purpose of this task is another requirement for the *digital city* environment as it relates to design and quality of the information products.

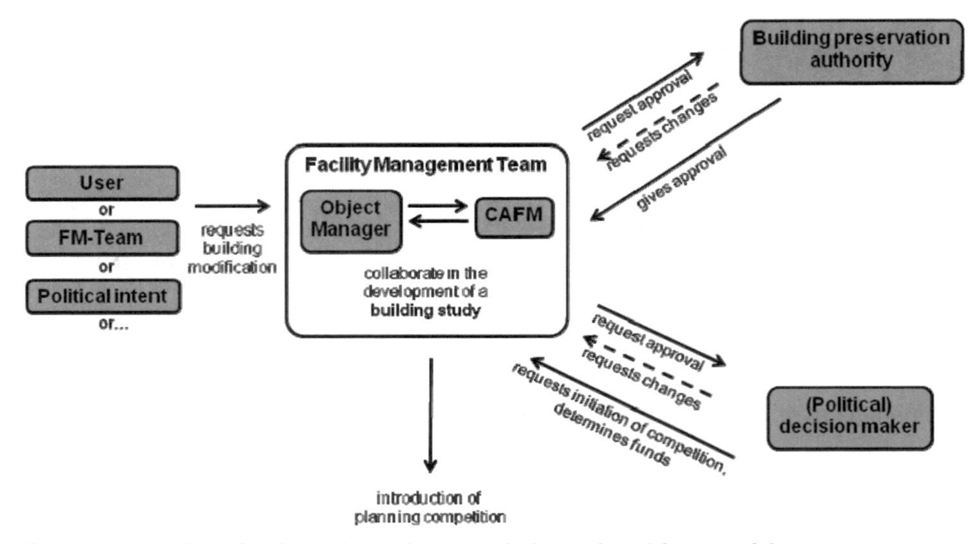

Fig. 2: Workflow for the task *Evaluation of Planned Building Modification*

b) The working environment perspective adds information about requirements on a more detailed level that already has been identified on a general level within the process model perspective. Two aspects of the working environment are distinguished here. One of them is the information exchange between stakeholders. This has to be facilitated by supporting a corresponding infrastructure and by the management of access to the working environment. The other aspect is concerned with the functionality that needs to be available, for example tools for manipulating the city model data, for analysis and for simulation. Each task can require different functionality for 3D city modelling. Therefore, two requirements categories are defined within the working environment perspective, the first one for infrastructure and access management, the second one for provided functionality.

c) To complete the view on the *digital cities* environment a perspective on the conceptual data model was appended. The data model provides the structure for all the information about the city infrastructure that needs to be represented within the *digital cities* environment. The requirements of the data model perspective define how the underlying city model has to be built in order to support the tasks. As the working environment before, the data model is related to the process model. The data model makes the requirements to the information products and their purpose more specific. They focus especially on the kind of represented objects, e.g. houses, rooms, or doors. The associated attributes of each object and eventually linked documents are a second requirement that specifies the represented information about an object more detailed. Though data quality aspects would complete the requirements they have not been included at this point in time. The initial workshops were not configured to yield reliable information about data quality. Data quality statements derived from the specified requirements of each task were not substantial without a validation by the user.

3 Results

From the workshops with the city of Salzburg tasks were defined mainly of the field of planning and the field of facility management (cf. Fig. 2). The survey & utility workshop helped largely with additional requirements and played only a minor role in identifying new tasks. For each task a structured outline of all the requirements are displayed in Tables 1 and 2. The outputs of Tab. 1 are requirements related to the process model perspective.

The first application area is visual communication of planned development. There are several related workflows and tasks that address different scales, from building modifications to the planning of whole city districts. These applications employ the perspective view on the proposed development in its environment via visualizations of and navigation in the city model. By this the *digital cities* environment represents the communication basis for decision making processes that involve stakeholders of different areas of expertise. The second application area is the management of geographic objects that profit from being represented in three dimensions and that are distributed over a wide area of the city. Workflows especially of the Salzburg city facility management are connected to this application area. In both application areas planned modifications and the geographic objects, respectively, are embedded in their surroundings and can be analyzed by their spatial relations to other geographic objects.

Table 1: Characterization of the tasks and requirements related to the process model perspective

General task information		**Process model of the task**		
Task [Application field]	Short Description	Stakeholder/ involved actors	Kind of exchanged information/data (e.g. Reports, Visualizations, Models)	Purpose of exchanged information/ data
Export of a surrounding area model [City planning]	The architect gets a selection of the 3D city model (model subset of surrounding area) in order to work on the proposal and implement models	Planner, Architect	Subset of the digital 3D city model: in case of single building projects the immediate surrounding is selected, for projects of urban development a larger area is selected.	The selection of the 3D model is the basis for the architect to design, visualize and review new projects
Upload of a building model (proposal) to participate in the competition [Architect]	An architect participates in the competition with his uploaded proposal (a 3D model). The proposal is now ready for review	Architect	3D building model with components (depending on the announcement)	Representing a new building in a surrounding area of the 3D city model
Comparison of proposals with the master plan [City planning]	During the competition the architects have uploaded their proposals. The planning department validates the proposal with the master plan. The system checks the proposals against con-straints of the zoning maps/plans and the land use plan.	Planner, (Architect)	Input 3D model of the architect, conflict model, conflict report	Controlling if the proposal agree with the master plan > decision support
Use of simulations (sun and shadow simulation, visibility, noise simulation, environmental simulation) [City planning]	Planners are able to perform simulations of planned and existing buildings/areas. Visualized simulations are helpful for decision making during competitions and are used as a communication tool to the public	Planner, Architect	Results presented in thematic maps, figures or mathematical outputs	Visualization and presentation of simulation results for decision making
3D Plotting [City planning]	3D plotting directly out of the 3D model for a presentation of a proposal	Planner	Selection of the digital 3D city model, plastic model	Representing projects about buildings/ areas

Evaluation of planned building modification [Facility Management]	Building inspections and user demands show the need for a building modification. First models and visualizations of different options are done by the CAFM-constructor for estimating costs and identifying options that the Building Preservation Authority (BPA) approves of. These results are the basis for an architectural competition.	CAFM-Constructor, Architect, Building Preservation Authority	Reports on changes through modification, visualizations, 3D model views, digital models	Estimating costs, decision support
Monitoring of building state [Facility Management]	Building inspection re-ports are entered into a database for all managed buildings. The stored in-formation of different levels of detail can be displayed and accessed in the 3D city model (for display, linked to the database input mask).	Owner, Operator	Reports, thematic 3D model views	Exploration and evaluation of the managed buildings
Gathering information for the need of air conditioning [Facility Management]	User reports are collected about constricted usage of specific rooms due to high temperature levels in summer. The requirements for air conditioning are estimated based on room volume and existing ventilation installations	Owner, CAFM Constructor	User reports via e-mail, reports from site inspection; analysis report, visualizations	Decision support
Accessing all information about the city buildings through a city model [Facility Management]	The facility management employees use digital cities as a tool to organize and access all the information, including plans and documents, about the buildings that they manage. Depending on the available detail some basic GIS analyses will be performed.	Owner, CAFM Constructor	Reports, thematic 3D model views, visualizations, analysis reports	Decision support

On the basis of the characterization of the tasks (cf. Table 1), Table 2 illustrates requirements related to the working environment and to the data model.

Table 2: Requirements related to the working environment and to the data model

General task information	Requirements of working environment		Requirements of the data model	
Task [Application field]	**Infrastructure and access**	**Analyses – Functionality of the software**	**Represented objects**	**Attributes and linked data**
Export of a surrounding area model [City planning]	Internet (restricted), Personal computer	Selection of a building or urban area, implementation of analyses, presentation of results	Footprint, building line, infrastructure of the environment block models	Building heights and use of buildings of the surrounding area, zoning plans
Upload of a building model (proposal) to participate in the competition [Architect]	Internet (restricted)	Indication of the area in which the building model will be uploaded	Depends on available model and the announcement of the competition	Zoning plans
Comparison of proposals with the master plan [City planning]	Personal computer	Comparison of attributes, geo-metrical comparison of areas & volumes, Summary of conflicts compiled in a report	Mass models (building shape & roof shape)	Proposal data (use of buildings, height, floors, etc.)
Use of simulations (sun and shadow, visibility, noise, environmental simulation) [City planning]	Results are available only internally and for architects, maps are also available for the general public	Selection of a building city area, implementation of analyses, presentation of results	Buildings or area of buildings added with simulation results in form of cartographic methods	Topography, building height, at noise simulation: land use (differences in the reflection of noise, e.g. vegetation)
3D Plotting [City planning]	Internal, personal computer, 3D Plotter	Tool for the selection of a subset with a polygon of interest (e.g. artificially defined), export function (3D model to printer)	Depends on available model	Not applicable
Evaluation of planned building modification [Facility Management]	Model accessible through intranet and internet	Modelling/Import of building modification models, bookmarks with description, screen-shot, video, line-of-sight-analysis	Objects of building shell, surfaces (walls, roof, windows, doors, chimney,…), eventually rooms	Association of each object to "existing", "plan1", "plan2", "planX"; room area

Monitoring of building state [Facility Management]	Intranet, connection to building inspection Database required	Display of thematic content, query, search, editing thematic content with Linked input mask of building inspection database	Buildings as points or block models (identifiable as one object), relevant Building parts ("roof", "fascade", "windows", "doors, gates", "rooms", "building technique", "fire control assets")	Building state ("red", "yellow", "green"), objects of all levels of detail, Quality description for detailed objects, link to input mask, eventually other attributes and links
Gathering information for the need of air conditioning [Facility Management]	Intranet	Query data, network analysis, produce maps/reports	Rooms, windows, doors, ventilation system components, roof, facade	Room volume, isolation of walls/roof, aggregated room temperatures, number of days with constricted usage, ventilation system capacity
Accessing all information about the city buildings through a city model [Facility Management]	Intranet, connection to corresponding databases required	Display, query, search, editing of thematic content, basic thematic visualizations	3 levels of detail: (1) Building as a point or block model (identifiable as one object), (2) building parts (building divided into thematically defined block objects, e.g. floors, or according to age of building part), (3) single rooms, doors, windows, cable structure, pipes and ventilation ...	Owner information, building energy consumption, occupancy, age, building state, rented or not, ... ; linked documents: rental contracts, 2D drawings, profiles

4 Conclusion and Outlook

For many municipalities virtual city modelling is a new approach to manage cities and to facilitate participation of the public in planning processes. A *digital cities* environment can help cities like Salzburg to visualize, analyze and communicate proposed changes to their urban environment. Therefore, user-oriented requirements for a *digital cities* environment were collected and specified for workflows applied in the city of Salzburg. The extracted information, especially about tasks of urban planning, is applicable for other cities with similarly arranged workflows. They appear in cities with an equal size, number of inhabitants, historical building structure or level of development. This was confirmed by the findings of other workshops within the *digital cities* initiative that Autodesk conducted with the second pilot city, Vancouver, Canada.

As a tool for more transparent communication and better decision making *Digital Cities* is expected to improve the quality of the planning process. Although the planning department does not see savings in work time, this probably will occur in the facility management, where *Digital Cities* will fulfil the role of an integrated information tool. Facility management staff anticipates a shift of the workload towards the preparation and maintenance of the underlying information. They even expect changes in the organizational structure within the municipality. Therefore, employing a *digital cities* environment within municipalities will have consequences in government and government processes. By enabling this new approach cities will move from isolated tasks of individual departments to an integrated tool for urban planning and managing.

Acknowledgements

The research was financed through the *Digital Cities* project. The authors thank the project partners (Autodesk, City of Salzburg) and the reviewers for their constructive discussion.

References

AG 3D STADTMODELLE DES AK KOMMUNALES VERMESSUNGS- UND LIEGENSCHAFTSWESEN DES STÄDTETAGES NORDRHEIN-WESTFALEN (2004), Orientierungshilfe 3D Stadtmodelle. Städtetag Nordrhein-Westfalen.

BENNER, J. & KRAUSE, K. U. (2009), XPlanung – ein standardisiertes Datenformat zum verlustfreien Datenaustausch. Planerin – Fachzeitschrift für Stadt-, Regional- und Landesplanung, 5, pp. 20-22.

BLEIFUSS, R. (2009), 3D Innenraummodellierung: Entwicklung und Test einer CityGML-Erweiterung für das Facility Management. Diplomarbeit, TU München.

CARROLL, J. M. (1995), Introduction: The scenario perspective on system development. In CARROLL, J. M. (Ed.), Scenario-based design – Envisioning work and technology in system development. New York, Wiley & Sons, pp. 1-17.

KOLBE, T. H., GRÖGER, G. & PLÜMER, L. (2005), CityGML – Interoperable Access to 3D City Models. Proceedings of the First International Symposium on Geo-Information for Disaster Management (Delft, Netherlands, March 21–23, 2005). GI4DM, ed. by OOSTEROM, P. VAN, ZLATANOVA, S. & FENDEL, E. M. Springer.

ROSS, L., KLEINSCHMIT, B., DÖLLNER J. & KEGEL, A. (2008), Entwicklung von Flächeninformationssystemen auf Basis virtueller 3D-Stadtmodelle. 13th International Conference on Urban Planning, Regional Development and Information Society (REAL CORP, ed. by SCHRENK, M., POPOVICH, V. V., ENGELKE, D. & ELISEI, P.). CORP – Competence Center of Urban and Regional Planning, pp. 565-570.

STROBL, J. (2009), Digital City Salzburg. – Beitrag zum 14. Münchner Fortbildungsseminar Geoinformationssysteme. München.

On the Suitability of Volunteered Geographic Information for the Purpose of Geocoding

Christof AMELUNXEN

The GI_Forum Program Committee accepted this paper as reviewed full paper.

Abstract

The automated process of assigning geographic coordinates to textual descriptions of a place, generally referred to as geocoding, plays an important role in various fields of geographic information technologies, ranging from the analysis of health records or crime incidents to location based services like route planning applications. However, since the collection and maintenance of appropriate spatial data is the traditional domain of official surveying offices and commercial companies, there are only very few publicly available geocoding services which can be used free of charge, and those which exist are usually limited to a specific country or even smaller units. Furthermore, no freely available geocoding service offering house number level precision had so far been implemented based on volunteered geographic data. The goal of the work summarized in this paper (originating from the author's MSc thesis) was thus to explore the suitability of volunteered geographic information for the purpose of geocoding.

1 Introduction

Until recently, the generation, maintenance and distribution of geographic information had been, with only very few exceptions, solely the domain of either official land surveying offices or commercial companies. This was presumably mainly due to the immense costs related to the actual surveying and maintenance and the lack of possibilities to effectively share and distribute the collected spatial data. However, this has recently changed, for the two following reasons (based on suggestions by GOODCHILD (2007b):

1. The dramatically reduced costs along with the enhanced usability of modern satellite navigation handheld devices have enabled a mass of people to collect geographic data with ease of use and in precision levels which had formerly been simply beyond reach for private persons.
2. The progress of the internet from a formerly "read-only media" to the "web 2.0" participatory approach has made collaborative efforts to generate and share content of various kinds very common.

The OpenStreetMap (OSM) project has been selected as the data source for this research as it provides an impressively extensive database originating from collaborative volunteered effort and the exponential growth of the project data since its start in 2004 is very promising. Its primarily goal is to generate a free map of the world through volunteered effort. Nevertheless, although the generation of maps still is the focus of the project, the collected spatial data is made publicly available and may be used for other purposes as well[1]. OpenRouteService (ORS) (http://www.openrouteservice.org/) e.g. is an example of a project that has successfully implemented a routing service based on OpenStreetMap data.

The definition and usage of the term geocoding varies in scientific literature. Some authors limit the scope of input data to postal addresses (BAKSHI et. al. 2004, BEHR et. al. 2008, CAYO & TALBOT 2003) whereas others widen the scope to include named places (DAVIS et. al. 2003) or even arbitrary textual representations of a place (GOLDBERG 2008, POULIQUEN et. al. 2004). The focus of the work presented in this paper was to explore the suitability of OpenStreetMap data for the purpose of geocoding, simplified as the conversion of textual address information into point coordinates and vice versa. If a working geocoding service could successfully be build based on OpenStreetMap data, this would be a substantial advance in the improvement and progression of a wide range of projects, based in the field of volunteered geographic information.

A major objective of the work was further to evaluate the possibilities to compensate for incomplete data. Because house number data in OpenStreetMap is still rare and inhomogeneously distributed (some areas are almost completely mapped whereas others contain no house number data at all) one specific challenge of this work was to find out, whether the location of a house number along a given street may be effectively approximated by probability based approaches. In other words, the question to be answered was: "Is it possible to effectively approximate the position of a house number along a given street in the absence of real house number data?". The term "effectively" in this case was meant in the sense of "better than simply returning the centerpoint of the street".

2 Approach

The first task was thus to analyze the data provided by the project and to develop an appropriate process to transform the data in a format usable for geocoding purposes. The next task was the actual design and implementation of the geocoding application.

The geocoding application has been integrated into the OpenRouteService (NEIS 2008, NEIS & ZIPF 2008a, NEIS & ZIPF 2008b) project, providing a framework compliant to the OpenGIS Location Service (OGC 2008) specification.

At first, the general suitability of the OpenStreetMap data for geocoding purposes was evaluated with respect to its data model, relational integrity and completeness. Based on this analysis the proposed data model for the geocoder's reference dataset was designed and an appropriate data transformation and integration processes were developed following concepts presented by HAN & KAMBER (2006) and RAHM & DO (2000).

This has been followed by the definition and analysis of the use cases to be provided by the geocoding service. The actual processing of the geocoding use cases was then designed following standard geocoding practices as described by GOLDBERG (2008), DAVIS et. al. (2003), BORKAR et. al. (2001) and CHRISTEN & CHURCHES (2005).

---

[1] GOODCHILD (2007a) proposed the term "Volunteered Geographic Information (VGI)" for this type of geographic data, which is generated by collaborative volunteered effort.

The treatment of incomplete house number data received special attention. In order to compensate for missing house number data in OpenStreetMap, probability based approaches were developed in order to effectively approximate house number locations. The approximation is based on the following parameters:

1. Average distance between two house numbers
2. Average offset of the first house from the beginning of the street
3. Direction of the street

This approach is limited to the sequential alternating house numbering system that is used in most parts of Europe (FARVACQUE-VITKOVIC et. al. 2005). Given these parameters are known, the position of a house number may be approximated as shown in figure 1.

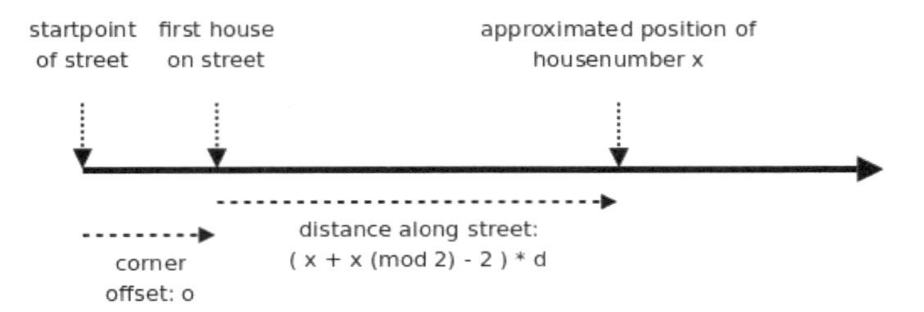

Fig. 1: Calculating approximate house number positions along a street

The first two of these parameters were determined by spatial analyses of known reference datasets including house numbers. In order to guess the direction of a street, official guidelines on house numbering available for the study area – the federal state of North-rhine-Westfalia, Germany – were consulted (STÄDTETAG NRW 1979), resulting in the following hypotheses:

1. If only one of the end points of a street is connected to another street, the street starts at this point (dead-end street approach).
2. If a street proceeds away from the city center in a radial direction, the street starts at the end point closest to the city center (city center approach).
3. If a street is at one end connected to a street of a higher rank than at the other end point, the street starts at the end point connected to the higher level street (street rank approach).

3 Results

The quality of the geocoder, implemented according to the concepts and guidelines developed before, was measured using the standard key figures match rate and positional accuracy as described by CAYO & TALBOT (2003) and additionally by comparing the positional accuracy measured to a commercial geocoding service provided by Google™. The match rate, defined as the percentage of requests returning a correct match, was 96% at the municipal level requests (sample size n = 333), 83% at the street level requests (n =

1000), and 5% (n = 1000) at the house number level requests for randomly chosen addresses within the study area.

When considering a match rate of 85% to be the minimum acceptable rate necessary to reliably detect spatial patterns in address datasets as proposed by RATCLIFFE (2003), it has to be concluded that the achieved match rate at the street and house number level is not yet sufficient for detailed spatial analysis purposes.

The average positional error for house number level requests, determined by comparing the results to the real positions of the buildings as provided by the surveying office for the study area, was measured differentiating the availability of house number positions in the OpenStreetMap data (see table 1).

Table 1: Geocoding accuracy of ORS geocoder depending on the house number data availability in OSM

Location Method	Sample Size	Mean Positional Error
Exact house number match	13933	11m
Interpolation between known house number positions	890	31m
No house number data available	255073	142m

These figures must be considered as non-suitable for fine-scale spatial analyses of address datasets unless house number data is available. ZANDBERGEN (2007) e.g. demonstrates that even a medium error of 41 meters with a 90th percentile of just 100 meters can significantly bias analysis results as shown on the example analysis of traffic-related air pollution affecting school children (using a sample of 104,865 addresses). The average positional accuracy achieved when interpolation between two known house number positions was possible, is nevertheless significantly better than the medium error of 41m measured by Zandbergen for 104,865 sample addresses located in Orange County, Floria; these addresses were geocoded using official street centerline and parcel data of the Property Appraisers Office of the Orange County.

The measured medium positional error of merely 11m for exact house number matches can be considered as an extraordinary accuracy. Literature research revealed no case study presenting a geocoding service providing accuracy figures even close (CAYO & TALBOT 2003, DEARWENT et. al. 2001, GOLDBERG et. al. 2008, GRUBESIC & MURRAY 2004, KRIEGER et. al. 2001, MAZUMDAR et. al. 2008, RATCLIFFE 2001, WHITSEL et. al. 2004).

A comparison with the accuracy provided by the geocoding service offered by Google™ (see: http://code.google.com/apis/maps/documentation/geocoding/) shows that whenever house number data was available, the positional error was significantly lower than Google's (see table 2 and figure 2) and about equal when interpolation between two known house numbers was possible. Yet for the case when no house number data was available, the average positional accuracy proved significantly worse than the one provided by Google.

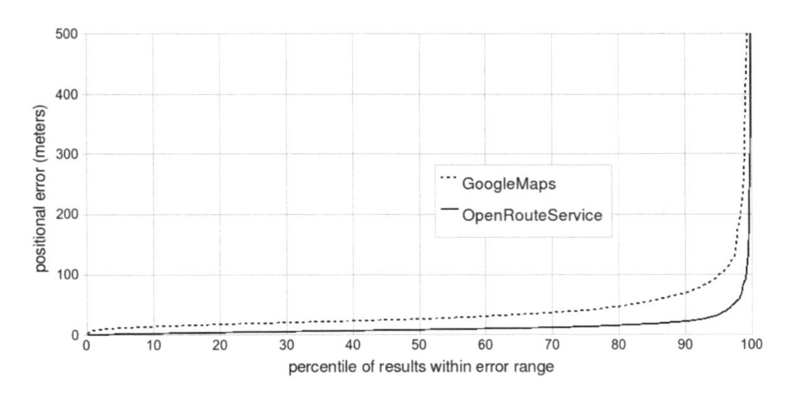

Fig. 2: Positional errors of OpenRouteService and GoogleMaps geocoder when house number data is available in OSM

It was further found that it is indeed possible to effectively approximate house number locations by two different probability based approaches to guess street directions (see table 3). It was also found that the effectiveness of these approaches, although showing a significant overall average improvement, depends heavily on the study area. It was further found that these improvements are still insufficient to generate accuracy levels comparable to cases where actual house number data is available.

Table 2: Comparing the positional accuracy of ORS and Google geocoder depending on house number availability in OSM

Location Method	Sample Size	Mean Error ORS	Mean Error Google
Exact house number match	13283	11m	32m
Interpolation between known house number positions	853	31m	32m
No house number data available	54889	142m	34m

Table 3: Effectiveness of probability based house number approximations

Approach to guess street direction	Sample Size	Improvement of positional accuracy
Dead-end street approach	23656	29%
City center approach	10423	28%
Street rank approach	17242	0%

4 Conclusion and Outlook

The research presented can serve as a proof of concept for the use of volunteered spatial data as a reference dataset for geocoding services. It could be shown that it is indeed possible to build a working geocoder based on volunteered geographic information. The inherent inconsistencies presented in the OpenStreetMap data however required substantial concessions in terms of referential integrity. Furthermore, the positional accuracy to be expected strongly depends on the availability of house number data, although means to partially compensate incomplete data have successfully been developed.

The result of this work is already used operationally as the geocoding engine for various research projects, of which OpenRouteService (http://www.openrouteservice) and OSM-3D (http://www.osm-3d.org/) presumably are the most prominent.

The recent development of the OpenStreetMap project is very promising, too. The amount of house number locations stored in the OpenStreetMap database for the area of Germany has almost doubled during the implementation phase of the research presented. Starting with 172,000 house numbers at the end of December 2008 the amount increased to more than 330,000 house numbers at the end of April 2009. At the time of writing this paper (February 2010) there were around 600,000 house numbers in the database for the area of Germany and about 4.5 million for the scope of Europe.

References

BAKSHI, R., KNOBLOK, C. A. & THAKKAR, S. (2004), Exploiting online sources to accurately geocode addresses. Proc. 12th annual ACM international workshop on Geographic information systems. Washington DC, USA, pp. 194-203.

BEHR, F-J., RIMAYANTI, A. & LI, H. (2008), Opengeocoding.org – A free, participatory, community oriented geocoding service. Technical report, Department of Geomatics, Computer Science and Mathematics, University of Applied Sciences Stuttgart, Stuttgart, Germany.

BORKAR, V. R., DESHMUKH, K. & SARAWAGI, S. (2001), Automatic segmentation of text into structured records. Proc. SIGMOD Conference, Santa Barbara, California.

CAYO, M. R.. & TALBOT, T. O. (2003), Positional error in automated geocoding of residential addresses. International Journal of Health Geographics, 2 (10), December 2003.

CHRISTEN, P. & CHURCHES, T. (2005), A probabilistic deduplication, record linkage and geocoding system. Proc. ARC Health Data Mining workshop, Adelaide, Australia, April 2005.

DAVIS, C., FONSECA, F. & BORGES, K. (2003), A flexible addressing system for approximate geocoding. Brazilian Symposium on GeoInformatics, 2003.

DEARWENT, S. M., JACOBS, R. R. & HALBERT, J. B. (2001), Locational uncertainty in georeferencing public health datasets. Journal of Exposure Analysis and Environmental Epidemiology, 11, pp. 329-334.

FARVACQUE-VITKOVIC, C., GODIN, L., LEROUX, H., VERDET, F. & CHAVEZ, R. (2005), Street Addressing and the Management of Cities. The International Bank for Reconstruction and Development/The World Bank, Washington, DC, USA.

GOLDBERG, D. (2008), A Geocoding Best Practices Guide. University of Southern California, GIS Research Laboratory.

GOLDBERG, D., WILSON, J., KNOBLOCK, C., RITZ, B. & COCKBURN, M. (2008), An effective and efficient approach for manually improving geocoded data. International Journal of Health Geographics, 7 (1), p. 60.

GOODCHILD, M. F. (2007a), Citizens as sensors: the world of volunteered geography. GeoJournal, 69 (4), pp. 211-221.

GOODCHILD, M. F. (2007b), Citizens as sensors: Web 2.0 and the volunteering of geographic information. Geofocus, 7, pp. 8-10.

GRUBESIC, T. H. & MURRAY, A. T. (2004), Assessing positional uncertainty in geocoded data. Proceedings of the 24th Urban Data Management Symposium, Chioggia, Italy, 2004.

HAN, J. & KAMBER, M. (2006), Data Mining: Concepts and Techniques. The Morgan Kaufmann Series in Data Management. Diane Cerra, San Francisco, USA, 2nd edition.

HARRIS, K. (1999), Mapping Crime: Principle and Practice. Diane Pub Co.

KRIEGER, N., WATERMAN, P., LEMIEUX, K., ZIERLER, S. & HOGAN, J. (2001), On the wrong side of the tracts? Evaluating the accuracy of geocoding in public health research. American Journal of Public Health, 91 (7), pp. 1114-1116.

MAZUMDAR, S., RUSHTON, G., SMITH, B., ZIMMERMAN, D. & DONHAM, K. (2008), Geocoding accuracy and the recovery of relationships between environmental exposures and health. International Journal of Health Geographics, 7 (1), p.13.

NEIS, P. (2008), Location Based Services mit Openstreetmap Daten. Master's thesis, Fachhochschule Mainz, Fachbereich I.

NEIS, P. & ZIPF, A. (2008a), Zur Kopplung von Opensource, OpenLS und Openstreetmaps in openrouteservice.org. Proc. AGIT, Salzburg, Austria, 2008.

NEIS, P. & ZIPF, A. (2008b), Openrouteservice.org is three times open: Combining opensource, openls and openstreetmaps. Proc. GISRUK 2008 conference, Manchester, April 2008. UNIGIS UK.

OGC (2008), OpenGIS Location Service (OpenLS) implementation specification: Core Services, Sept. 2008.

POULIQUEN, B., STEINBERGER, R., IGNAT, C. & DE GROEVE, T. (2004), Geographical information recognition and visualization in texts written in various languages. SAC '04: Proc. 2004 ACM symposium on Applied computing. New York, USA, pp. 1051-1058.

RAHM, E. & DO, H. H. (2000), Data cleaning: Problems and current approaches. IEEE Data Engineering Bulletin, 23, p. 2000.

RATCLIFFE, J. H. (2001), On the accuracy of tiger-type geocoded address data in relation to cadastral and census areal units. Geographical Information Science, 15, pp. 473-485.

RUSTHON, G., ARMSTRONG, M. P., GITTLER, J., GREENE, B. R., PAVLIK, C. E., WEST, M. M. & ZIMMERMAN, D. L. (2006), Geocoding in cancer research: A review. American Journal of Preventive Medicine, 30, pp. 16-24.

STÄDTETAG NRW (1979), Richtlinien für die Nummerierung von Gebäuden oder bebauten Grundstücken.

WHITSEL, E. A., ROSE, K. M., WOOD, J. L., HENLEY, A. C., LIAO, D. & HEISS, G. (2004), Accuracy and repeatability of commercial geocoding. American Journal of Epidemiology, 160 (10), pp. 1023-1029.

ZANDBERGEN, P. (2007), Influence of geocoding quality on environmental exposure assessment of children living near high traffic roads. BMC Public Health, 7 (1), p. 37.

The Spatial Analysis on the Impacts of Mass Transit Improvements on Residential Land Values – the Case Study of Bangkok Metropolitan Area, Thailand

Sutee ANANTSUKSOMSRI

Abstract

For more than a decade, rapid public transport systems have alleviated traffic congestion, facilitated economic agglomeration, and increased land values in Bangkok, Thailand. Although there have been some investigations of the effects of public transit on Bangkok residential land use and development, very few spatial analysis of this relationship has been examined. In this study, hedonic price and spatial econometric models, together with the Geographic Information System (GIS) and data on real estate development, are used to analyze the effects of mass transit systems on land values in the Bangkok metropolitan area. The study will explore changes in urban structure and provides some guidance on policy implementation such as land value taxation.

1 Introduction

Infrastructure improvement is one of major factors leading to socioeconomic growth in developing countries. However, the impact of public transit improvement on land use and land price has not been well examined in these countries. An understanding of this relationship is necessary for formulating efficient urban planning policies. In this paper, the hedonic price models with spatial econometrics techniques are developed to investigate the effects of existing and proposed public transit systems on residential land prices and the changes of urban structure in the metropolitan area of Greater Bangkok.

Bangkok's urban structure has been dominated by the expanding network of arterial roads and ring roads since the 1960s. Urbanized areas have grown without effective control from the government, while housing development in the city has been led mainly by real estate developers. Consequently, the urban area of Bangkok has expanded along road networks, resulting in strip developments, especially in the outskirt of the city. Since the transportation policies in the past concentrated mostly on construction of new roads, rather than on traffic management, traffic congestion has become one of the most crucial problems of the city.

In order to ease the chronic traffic congestion in Bangkok, a rapid transit project was initiated in the early 1990s. However, due to political interference and economic problems, the processes of planning and construction of the rail system were very slow. Not until the late 1990s, the first mass transit system in Bangkok, BTS Skytrain, was introduced. Soon later, MRT Subway started their operations in 2004. These mass transit systems have not only alleviated traffic problems, but also assisted the economic development and shaped the urban structure of Bangkok.

Since 1960s, Bangkok urban form has been characterized by automobile-oriented transportation and urban sprawl toward its periphery. Since the emergence of the mass transit in the early 2000's, Bangkok urban development has intensified, especially in the inner areas where the mass transit is accessible. A great number of new residential developments, for example, can be seen along mass transit routes. Like the impact of road improvement, the impact of the rapid transit system may affect values of lands in the city, resulting in great disparities in land prices between land parcels located on and away from mass transit stations.

To examine the impact of mass transits on property values, hedonic regression analysis has been widely used. For the case of Bangkok, CHARLERMPONG (2008) studies relationships of BTS stations and property prices in the inner city and finds that a property price gradient decreases at approximately US$ 10 per sq.m. of livable areas for an additional meter away from BTS stations, and also indicates that price elasticity is -0.09. A real estate in Bangkok, however, has market segmentation (WISAWEISUAN 2001), so a land price seems to be a better indicator than a property price for evaluating value changes in residential land use.

This analysis of current housing and real estate developments will reveal the evidence of changes in residential land value and pattern of urban structure with the impacts of existing and proposed rapid transit systems in Bangkok. The result of this study can also be used as guidance for urban policies such as urban development plan and land taxation. In addition, this research will be one of the first attempts to examine the spatial relations of the public transit and residential development in the full coverage of the Greater Bangkok metropolitan area using the spatial analysis methods.

2 Methodology and Data

Spatial association and correlation are considered in examining the impacts of mass transit on land value. Three modelling approaches are used in this study: 1) the ordinary least square (OLS) model, 2) the spatial lag model, and 3) the spatial errors model. The dependent variable in the hedonic regression is a land price of observed land parcels. The independent variables are housing and population densities of sub-districts in the metropolitan area, as well as distances from existing and proposed rapid transit stations, from central business district (CBD), and from other urban developments such as arterial roads, top schools, universities, hospitals, shopping centres and parks. These distance variables are calculated using ArcGIS.

The study area of the Greater Bangkok metropolitan region includes Bangkok, Nonthaburi, Pathumthani, Nakornprathom, Samutprakarn and Samutsakorn. The data set of 939 on-market single-family residential development projects was collected in 2009 by the Real Estate Information Centre (REIC). The prices of land range from 5,000 Bahts (US$ 150) to 275,000 Bahts (US$ 8,450) per sq.wah (1 sq.wah = 4 sq.m.). A land price, which is the dependent variable in this study, is an average land price of the sold housing units in each housing project. After deleting the data with missing value, 796 observations are included in the final data set. Together with geographical data of the Greater Bangkok from the department of city planning, Bangkok Metropolitan Administrative, the land price data set is used to examine relationships of land parcels and urban development.

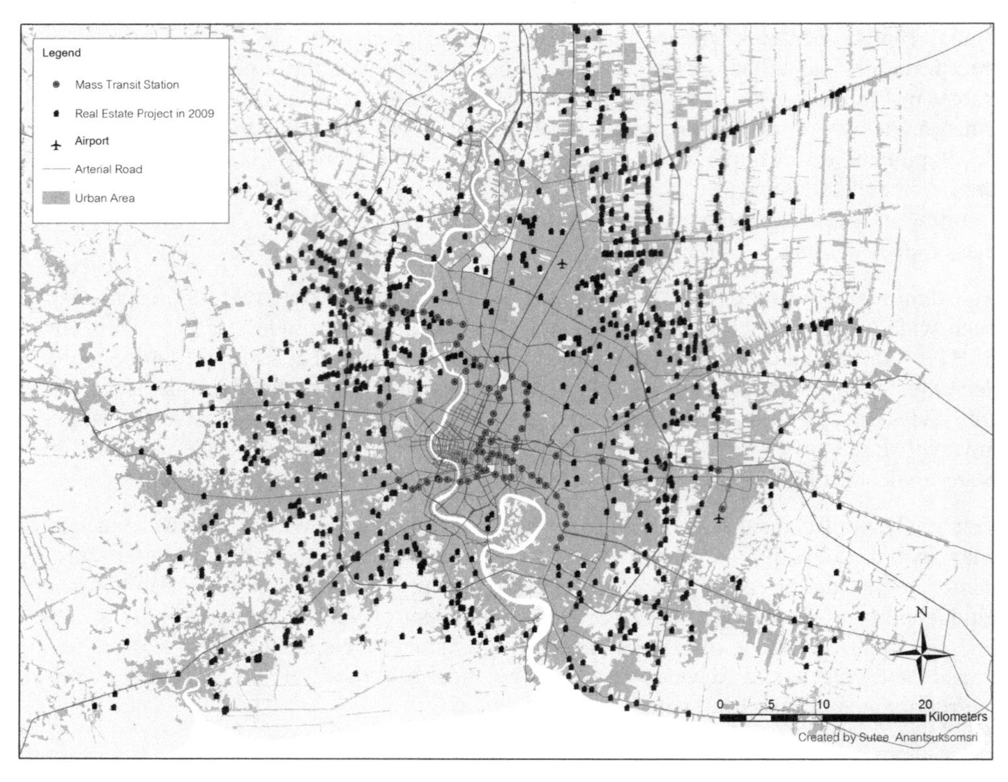

Fig. 1: Single-family residential development projects in Bangkok and its vicinity, 2009

Figure 1 depicts locations of single-family real estate development projects in 2009 in the Greater Bangkok. As can be seen, most of the residential projects are clustered along the outer ring road in the eastern and western parts of the metropolitan area. Consequently, after completion of the ring road during last ten years, several real estate investors have acquired developable agricultural land along the ring road. Availabilities and prices of lands seem to play an important role in housing development patterns reflecting the current urban structure of Bangkok. New developments in Figure1 indicate that the proposed route of MRT subway route are one of the major factors leading to high density clusters of residential developments in the western part of the city.

3 Results

The estimates of linear and log-linear hedonic regressions of residential land price are generated using the techniques of OLS, spatial lag, and spatial error models. The results of the linear regression imply land price gradient, while the results of the log-linear regression presents elasticity of independent variables on land price. The results from three models are similar in trends and significance. The results also suggest that physical attributes play an important role in determining land prices since most of the distance variables, especially the distance to nearest mass transit station, are significant.

The Lagrange multipliers of the spatial lag model show that spatial autocorrelation is significant, and suggest that the OLS model may overestimate the impact of the distance to nearest station in the linear regression since the spatial clustering of locations and prices of residential projects is not taken into account. These spatial effects also reflect a fact that the real estate market in Bangkok is mostly market-driven.

The results of linear regression's coefficient of distance to station from the spatial lag model is at approximately -0.32, imply that a price per sq.wah of land decreases at roughly 0.32 Baht (US$ 0.01) for an additional meter away from a rapid transit station, ceteris paribus. A price per sq.wah of land parcel located one kilometre away from a rapid transit station, for example, will be 320 Bahts (US$ 10) per sq.wah lower than the price of an identical land parcel located adjacent to the station.

As for the log-linear regression, the coefficients of distance to rapid transit station from the OLS and spatial error models, which are roughly at average of -0.04, suggest that a price per sq.wah of land discounts of approximately 0.04% for a percent increase in the distance away from a station, ceteris paribus. For instance, a land parcel with a price of 32,000 Bahts (US$ 1,000) per sq.wah located 100 meters away from the station will be valued at 13 Bahts (US$ 0.4) higher than an identical parcel located 101 meters away from the station, which has a price of 31,987 Bahts (US$ 999.6).

4 Conclusion and Outlook

The results from the study in this paper suggest that the mass transit improvement systems have been contributing to an increase in values of residential land price. In fact, land tax law legislation may soon be imposed. Most of current appraised values of land in Bangkok and its vicinity, however, do not reflect market values, especially the areas with proposed mass transit routes. The results from this study may be applied in land value taxation and other value capture policies.

In this study, only land prices of residential development projects located outside the inner-ring road are used to represent land values. As for further studies, the spatial analysis of the resale-home market can provide the clearer picture of residential price and land use. Furthermore, an understanding of the impact of mass transit improvements on other types of land use such as commercial and industrial is also crucial for the urban development of Bangkok.

References

CHALERMPONG, S. (2007), Rail Transit and Residential Land Use in Developing Countries: Hedonic Study of Residential Property Prices in Bangkok, Thailand. Transportation Research Record: Journal of the Transportation Research Board, 2038 (1), pp. 111-19.

DOWALL, D. E. (1992), A Second Look at the Bangkok Land and Housing Market. Urban Studies, 29 (1), pp. 25-37.

WARD, M. D. & GLDITSCH, K. S. (2008), Spatial regression models. CA, Sage Publications.

WISAWEISUAN, N. (2001), A Case Study of Bangkok, Thailand. Ph.D. Dissertation, Department of Land Economy, University of Cambridge, UK.

GIS Prioritization Model Based on Forest Fire Hotspots and Land Susceptibility

Marcelo BARBOSA, Leonardo DIAS, José SEOANE and Mario BURATTO

Abstract

A computer application was developed whose main function is to interact hotspots (generated from satellites) with different risk areas in a fire susceptibility map, and classify them automatically into five alert levels represented respectively with five different symbols. This interaction function changes the symbology of the hotspots in real time depending on which level of susceptibility they occurred. This resulting map "Alert Level Map" helps forest managers in the prioritization of the hotspot to be verified in the field, thus improving the use of fire fighting resources. The hotspots are available continuously every four hours and a susceptibility map is produced daily through map algebra algorithm, which uses static (topography, vegetation and land-use) and dynamic (weather) variables. All processes run through automated geoprocessing routines and are displayed in a GIS software.

1 Introduction

The launching of the first Landsat satellite in 1972 enabled detection changes in forest areas from space. Since then, thermal images and short-wave infrared bands have been used to detect fires, allowing the mapping of burned vs. unburned areas through contrast of the thermal gradients. In Brazil, the National Institute for Space Research (INPE) monitors occurrences of fire in every Brazilian state through satellites (SETZER & PEREIRA 2004). The developed technology identifies and locates hotspots geographically, making them available on the internet as dots and sending them as a geographical files package if required.

Making the most from this technological advance for hotspot detection in Brazil, this work aims to contribute further to this theme, by integrating geographical information system resources as ESRI products and geoprocessing technologies that are readily available, in order to establish alert levels for the satellite hotspots through an effective logic, which identifies their real degree of danger.

2 Material and Data Processing

2.1 Study Area

We choose an area in Para State, Brazil. This area is located in the Amazon forest and is extremely vulnerable for forest fires and permanently monitored by INPE through satellites hotspots generation.

2.2 Hotspots Map

Upon request INPE started sending the hotspots of our study area directly to a FTP server on average every four hours. The georeferenced hotspots arrived in digital file packages in shapefile format as dots and we took all of them. A computer routine identifies the arrival of these files and loads them to a geographic database, making them available for visualization on the hotspot map in a GISystem. Each hotspot is represented graphically by a small green square, when located outside the study area, and by a yellow square, when inside it. Every file carries an attributes table with technical information on hotspot collection for further query and statistics.

2.3 Susceptibility Map

2.3.1 Thematic Maps

Considering ignition and spreading factors two categories of thematic maps were defined for the calculation of the susceptibility map: static and dynamic. The static maps provide information which is either stable or changes slowly through time. Theses maps, such as *land-use and topographic maps* can be produced once a year. Dynamic thematic maps in turn, such as *fire weather maps*, generate information that changes rapidly and this variation influences the calculation of susceptibility. Such maps can be produced once a day.

2.3.1.1 Landuse Map

A *Landuse Map* represents the different uses of the land surface, interpreted from satellite imagery, through vector drawings as dots, lines and polygons. It was obtained by supervised classification image processing software from CBERS (China-Brazil Earth Resources Satellite) images, with field validation and final editing through visual interpretation on the screen, using GIS software.

2.3.1.2 Topographic Maps

These are maps that depict hypsometric values. Two types of maps were generated in raster file format, with the representation of values through continuous surface – a *Slope Map* (representing the slope angles of the terrain between specified locations) and an *Aspect Map* (representing the directions of the slope faces). The basic input used was a mosaic of images from the Shuttle Radar Topography Mission (SRTM).

2.3.1.3 Fire Weather Maps

These are maps which depict values of weather variables. Seven types of maps were elaborated in raster file format, with the representation of values by continuous surface- *Maximum Temperature Map* and *Minimum Temperature Map* (representing the temperature variation in °C); *Rainfall Map* (representing rainfall values in mm); *Wind Speed Map* (representing the wind speed variation in m/s); *Wind Direction Map* (representing wind direction variation in degrees azimuth); *Maximum Relative Humidity Map* and *Minimum Relative Humidity Map* (representing the relative humidity variation in %). In order to produce these maps spline interpolation was used. The basic input for these maps production was georeferenced weather data, contracted from a weather company, which is delivered daily for our FTP server.

2.3.2 Susceptibility Map Calculation

2.3.2.1 Classification/Reclassification of Risk (risk index)

In order to obtain the reclassified thematic maps in raster format, to be used effectively in the spatial analysis using map algebra, each of the *ten thematic maps* (indicated above) went through three classification steps based on its own risk situation when applicable. The first, automatic classification of empiric basis, considers a number of distribution classes intervals of uniform values, from the highest to the lowest, as to facilitate its numeric clustering and to allow the first separation of areas (natural breaks). The second, manual classification, groups values or attributes, through knowledge basis, leading to identification of the areas with high and low risk. The third classification, or reclassification itself, allocates numeric values- risk index- to previously classified intervals, in order to rank them as a function of their potential to generate fires (AKPINAR & USUL 2005).

The resulting eight reclassified thematic maps (because 3 maps were gathered to empowered the risk – aspect x wind direction x wind speed) were stored in a geographic database. These maps are ready to enter the automated algorithm routine in order to be intercorrelated spatially and generate areas susceptible to forest fires.

2.3.2.2 Structuring Algorithm and Susceptibility Index Calculation

The weighting method was adopted for the algorithm applied to all eight reclassified thematic maps, using arithmetic operators. Five values for the weights – integers from 1 to 5 – were used according to the relevance of maps in the occurrence of fires; i.e., the theme that contributes the most to cause or spread of fires and its intensity is multiplied by a larger weight (CHRISTIANSEN 2005). The choice of weight values was based on forest fire experts discussions; other compositions were tested as well. The current composition of weights was the most appropriate, in relation to approximate reality. The algorithm is implemented in a GIS software as follows:

Susceptibility Index =
LandUse (reclass) * 5 + Slope (reclasss) * 2 + Aspect_WindDirectionSpeed (reclass) * 3 + MaxTemperature (reclass) * 4 + MinTemperature (reclass) * 2 + Rainfall (reclass) * 3 + MaxRelative Humidity * 2 + MinRelative Humidity * 4

The end of the described process results in the susceptibility map as a raster file which is converted into polygons. Everything is carried out through an automated geoprocessing routine in the application as well as uses a gray scale palette to create a legend for five class intervals, where the different tones indicate areas susceptible to fires.

3 Results

3.1 Alert Level Map

To generate the alert level map, a function that connects the hotspots with the risk areas indicated in the fire susceptibility map, classifies them in real time into five alert levels, discriminating them according to special symbols (crescent squares, from the lower to the higher values, respectively), as shown in Figure 1.

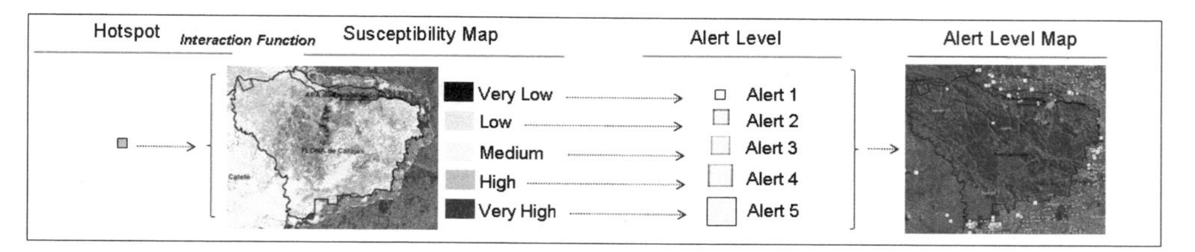

Fig. 1: Schematic representation of the interaction between hotspots and susceptibility map, resulting in different alert levels in their respective symbology

Hotspot locations are available every four hours during the day, that is, from 00:00 am to 11:59 pm of the same day; the application receives six packages of files with hotspot data per day and immediately makes them available for visualization in the GIS software. The susceptibility map is produced daily starting at 7:00 am, with the weather data arrival from the previous day. From 8:00 am, every day, the routine runs the interaction function, producing the different alert levels based on the susceptibility.

4 Conclusion

According to preliminary results, where we compared real forest fires location in the study area with the classified hotspots in the Alert Level Map, the accuracy was satisfactory and demonstrated that GIS routines are able to determine the relationship between a reality-based interpreted susceptibility map of the area and satellite-generated hotspots, highlighting the ones of highest hazard level through the alert classification, becoming an important tool to help decisions from the fire control centre, especially for large high risk regions. The methodology may be applied to any forested areas.

References

AKPINAR, E. & USUL, N. (2005), GIS in Forest Fire. ESRI 2005 International User Conference. – http://proceedings.esri.com/library/userconf/proc05/papers/pap1052.pdf.

CHRISTIANSEN, J. (2005), Calculating Wildfire Hazard Levels: Algebraic Raster. Construction using Spatial Analyst. ESRI 2005 International User Conference. – http://proceedings.esri.com/library/userconf/proc05/papers/pap2128.pdf.

SETZER, A. & PEREIRA, J. A. R. (2004), The operational fire alert system of Brazil: a detection and management tool using multiple satellites. Workshop on the Use of Space Technology for Disaster Management. Germany: UNOOSA-United Nations International.

Statistical Analysis for Discovering Distribution Patterns of Succession Types and Abiotic Factors Along Rivers

Daniela BRUNNER, Gregory EGGER and Michael LEITNER

The GI_Forum Program Committee accepted this paper as reviewed full paper.

Abstract

The aim of this study is to evaluate the correlation between abiotic factors and the succession phase along river reaches and to find typical differences of this correlation between rivers close to their natural state and river reaches under influence of dam operation. For the analysis 6 study sites situated in US and Canada were mapped and the abiotic factors flood inundation, geomorphic disturbance and distance to groundwater as well as the succession phases were captured and used for area bilances, cluster maps and Discriminant Analysis. The overall prediction for the succession phases based on Discriminant Analysis is between 65 and 95 percent. Most problems occurred by classifying pixels of secondary succession. Including the distance to the river in the analysis does not improve the result. Furthermore the evaluations showed that the impact of dam operation leads to less flood inundation, geomorphic disturbance and to a larger distance to the groundwater in general.

1 Introduction

Ecosystems along rivers are influenced by a lot of natural and human impacts. This leads to steadily changing vegetation along the river reaches. Each change (natural or human) in flow regime of a river leads to ecological responses, like loss of species, migration, tree mortality or altered plant cover types (POFF et. al 1997). The idea behind this study was, to observe the changes of succession phases by a small set of abiotic factors and to find out the correlation between these factors and the succession phase on the one hand and the change of the behavior of this correlation at river reaches under influence of dam operation.

In particular the aim of this study is to test the following hypotheses:

Hypothesis 1: For the distribution of succession phases at river reaches close to their natural stage, a typical pattern can be found:

- Braided reaches should present a high blending of young and old succession types
- Meandering reaches should show clear cold spots (aggregation of areas of a young succession phase) and hot spots (aggregation of areas with of old succession phases). Cold spots should always be situated at the inside curve of the river.

Hypothesis 2: The minimal distance of each pixel is depending on the succession phase and on the type of the river. There should be found the following pattern, which allows distinguishing between braided and meandering rivers:

- At braided reaches young succession phases (phases 1, 2, 3) should be situated near the river. The distances of the other succession phases are varying.
- At meandering river reaches only young succession phases should be found near the river. Older succession phases are farther.

At river reaches influenced by dam operations the dispersion of distances should decrease. Furthermore there should exist similarities in the distribution of the succession phases for rivers of the same type (meandering or braided).

Hypothesis 3: The parameter value of the abiotic factors is dependent on the succession phase. At rivers under influence of dam operations the parameter value should generally tend to be lower than it would be at natural river reaches.

2 Materials and Methods

2.1 Study sites

The whole study site includes six river reaches situated in Montana and Idaho (both US) and British Columbia, Canada. The river reaches extend over an area of between 7 to 58 hectares and show the following characteristics: Three of the six river reaches are meandering, the other three are braided. Two of the three meandering river reaches are close to their natural state, while the third has been under the influence of a dam operation. Similarly, two of the three braided river reaches are close to their natural state, while the third has been under the influence of a dam operation.

Table 1 summarizes the river characteristics and Figure 1 shows the location of all six river reaches included in this study.

Table 1: River reaches considered in the analysis

	Braided	**Meandering**
River reaches close to natural state	Elk River	Upper Kootenai River – Fenwick
	Flathead River	Upper Kootenai River – Wasa
River reaches under influence of dam operation	Lower Kootenai River (a)	Lower Kootenai River (b)

2.2 Field survey and mapping

All data for this research were sampled in August 2007. Plots were selected according to the “subjective sampling-approach”, described in MUELLER-DOMBOIS & ELLENBERG (1974). The plot size was about 5 – 10 meters, depending on the distribution of indicator species and most common vascular plants of a community.

Furthermore the study sites were mapped at a scale of 1:5.000 and the Lower Kootenai River at a scale of 1:10.000. The same date as within sample plots were recorded for all polygons, expect for the cover proportions of deciduous and coniferous tress, which were ocularly estimated within each plot as percentage of surface area. For this analysis mostly

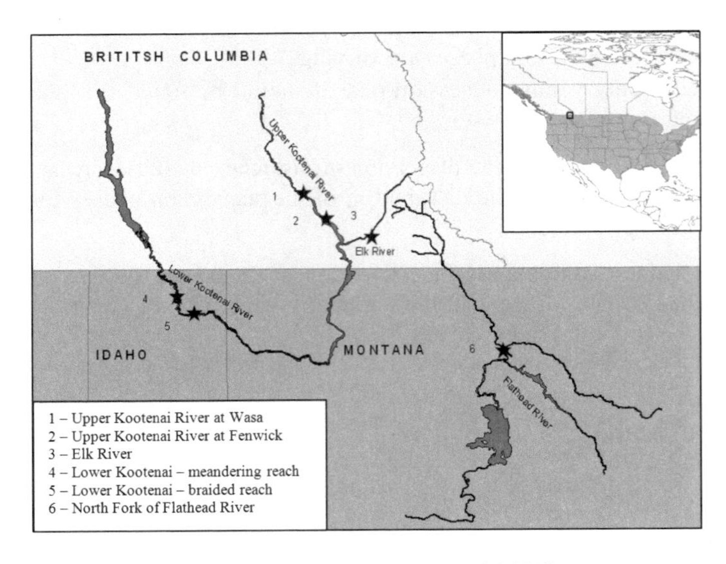

Fig. 1: Study sites situated in Idaho (US), Monatana (US), British Columbia (CA)

Table 2: Succession phases (EGGER et. al 2007)

ID	Succession phase	Duration in years
0	Water	–
1	Initial phase	0-1
2	Pioneer phase	1-3
3	Herb phase	3-5
4	Shrub phase	3-15
5	Early successional woodland phase – primary succession	15-55 (70)
6	Established forest phase – primary succession	55-110
7	Early successional woodland phase – secondary succession	25-55 (70)
8	Established forest phase – secondary succession	45-90
9	Mature stage	>110 (200-400)
10	Farmland, grassland	>200 (400)

the abiotic factors and the succession phases were used. The following section describes the succession phases (Table 2) as well as the classification of the abiotic conditions, including flood inundation (Table 3), geomorphic disturbance (Table 4), and distance to groundwater (Table 5). For more detail on the sampled data look at EGGER et. al 2007.

2.3 Discriminant function analysis

"Discriminant function analysis is used to determine which variables discriminate between two or more naturally occurring groups" (POULSEN & FRENCH 2010). Discriminant Analysis (DA) can be calculated on a base dataset with independent and dependent variables. The independent variables are the predictors and the dependent variable is the group variable. The process of Discriminant Analysis can be divided into two steps: 1. Testing significance of a set of discriminant functions, 2. Classification. The significance test is done by comparing two matrices (matrix of total variances and matrix of covariances) via multivariate F-Tests. This test decides whether the variables are statistically significant and which of the

variables have significantly different means across the groups (POULSEN & FRENCH 2010). After significance test the canonical functions are determined via a canonical correlation analysis. The maximal number of functions is equal to the number of variables or the degrees of freedom, depending on which one is smaller. As a result of Discriminant Analysis we get an optimal combination of variables, so that the first function explains most overall variation between groups, the second function second most etc. (POULSEN & FRENCH 2010).

Table 3: Classification of flood inundation

	Flood inundation
0	Non significant (water)
1	Very low/non: > 100 years
2	Low: 10-100 years
3	Moderate: 3-10 years
4	High: 1-3 years (bank full)
5	Very high: >1x/year (bank zone))

Table 4: Classification of geomorphic disturbance

	Geomorphic disturbance
0	Non significant (water)
1	Very low/non: >10 years
2	Low: weak intensity of erosion, local, 1-10 years
3	Moderate, 1-3 years
4	High: moderately intensive erosion, 1-3 years
5	Very high: intensive erosion, every year

Table 5: Class. of distance to groundwater

	Distance to groundwater
0	Non significant (water)
1	Very low (0.1 m)
2	Low (0.1-0.5 m)
3	Moderate (0.5-1.5 m)
4	High (1.5-3 m)
5	Very high (> 3m)

Eigenvalues

The eigenvalue (EV) reflects the ratio of importance of the dimensions which classify cases of the dependent variable. For each discriminant function there exists one eigenvalue. "*For two-group DA, there is one discriminant function and one eigenvalue, which accounts for 100% of the explained variance. If there is more than one discriminant function, the first will be the largest and most important, the second next most important in explanatory power, and so on.*" (GARSON 2008)

Wilks' Lambda

Wilks lambda is used to test the null hypothesis that the populations have identical means on the linear discriminant equatation. Wilks lambda is: $\Lambda = \frac{SS_{within\_groups}}{SS_{total}}$, so the smaller the Λ the more doubt cast upon that null hypothesis. (Where SS_{within_groups} differences within groups, SS_{total} and differences between groups + differences within groups) (POULSEN & FRENCH 2010).

The group variable was given by the succession phases and its range was from 0-11 for the first run of analysis and from 0-9 for the second and third run. As predictors or independent variables different combinations of 5 available variables (flood inundation, geomorphic disturbance, distance to groundwater, minimum distance to river, fire – as an indicator for succession phases of secondary succession) were used:

1. Flood inundation, geomorphic disturbance, distance to groundwater
2. Flood inundation, geomorphic disturbance, distance to groundwater, minimum distance to river
3. Flood inundation, geomorphic disturbance, distance to groundwater, fire

All variables were converted to a resolution of a 5 × 5 meter raster dataset. To each pixel the value of the 6 variables were assigned to get a 6 column dataset of all pixels.

2.4 Analysis of Local Indicators of Spatial Association (LISA)

The local univariate LISA in GeoDA is based on a revised local Moran statistics, proposed by ANSELIN (1995) and is able to answer these questions. Local Moran is defined as:

$$I_i(d) = \sum_{j-i}^{n} w_{ij} Z_j ,$$

Where the observations Zi and Zj are in standardized form (with mean of zero and variance of one). The spatial weight wij are in row-standardized form. So, Ii is a product of Zi and the average of the observations in the surrounding locations (BAO 1998).

The LISA Cluster Map generated on the base of Moran's I is is able to illustrate the local spatial autocorrelation functionality. In GeoDA, there are four different region types corresponding to the four possible types of local spatial association between a region and its neighbors. Regions with a high value (relative to the mean) surrounded by regions with high values are classified as "High-High", so called Hot Spots. On the opposite, regions with low values, surrounded by regions with low values are found as "Low-Low". Furthermore regions with low values, surrounded by regions with high values (LH) regions with high values, surrounded by regions with low values (HL) are called "Spatial Outliers. Regions HH and LL (respectively LH and HL) refer to positive (resp. negative) spatial autocorrelation indicating the spatial clustering of similar (resp. dissimilar) values (WANG & XIANG 2007).

For this specific analysis young succession types (ID 0-5, table 2) are interpreted as low values and old succession types (ID 5-9) are interpreted as high values.

Software settings and Input data
To evaluate spatial association GeoDA (Version 0.9.5, ANSELIN 2004) was used. First weights were calculated with the setting of queen neighborhood with first order of contiguity and a threshold of 0.01. Then the cluster map was created by using these weights for an univariate LISA on the succession phase variable.

The cluster map was calculated for each study site and shows "hot spots", "cold spots" as well as "spatial outliers" by analyzing the local clusters.

3 Results

3.1 Discriminant Analysis

The discriminant analysis all lead to models with 3 or 4 discrimination functions (depending on the number of independent variables included – the table only shows the values of the eigenvalue and Wilks' Lambda for the first three discriminant functions), due to the large size of samples they are all highly significant (0,000). If we consider Wilk's Lambda as an indicator of the quality of the functions, only the 1st functions of all models have a value < 0,1 (except Flathead River and Upper Kootenai River at Wasa – there Wilk's Lambda for the 2nd function is still below 0,1). The 1st functions of the models explain between 65% and 95% of Variance.

The first run of DA lead to the worst results of correct classified pixels. The results could be improved for 3 to 20 percent by removing the pixels classified as infrastructure and farmland or grassland. The correct classification for the third run rose again for 0 to 10 percent, but for one river reach (Flathead River) the result fell down for 2 percent. The following section discusses in detail the results of run 3 partially compared with the results of run 2.

At the river reaches under influence of dam operation the pixels of succession phase 9 are completely misclassified in run 2. In all cases 100 percent are classified as pixels of succession phase 5. Run 3 already classified between 50 and 65 percent of succession phase 9 correctly. The pixels of succession phase 8 were always misclassified, most often they were classified as succession phase 6 and furthermore as succession phase 9. The only exception was run 3 for the Upper Kootenai River at Wasa. Here 51 percent of the pixels of succession phase 8 were classified correctly.

Table 6: Results of Discriminant Analysis

	Elk	**Flat-head**	**Lower Koo-tenai River braided**	**Upper Koo-tenai River – Wasa**	**Upper Koo-tenai River – Fenwick**	**Lower Koo-tenai River meandering**
EV1	14.716	13.648	13.102	34.591	12.691	16.752
% of Variance EV1	83.6	65.2	92.2	84.1	77.4	89.3
EV2	2.693	6.642	0.712	3.876	3.146	1.683
% of Variance EV2	115.3	31.7	5	9.4	19.2	9
EV3	0.195	0.565	0.397	2.451	0.497	0.322
% of Variance EV3	1.1	2.7	2.8	6	3	1.7
EV4	0.008	0.080	0.004	0.237	0.063	0.13
% of Variance EV4	0	0.4	0	0.6	0.4	0.1
Wilks' Lambda 1	0.014	0.005	0.030	0.001	0.011	0.016
Wilks' Lambda 2	0.225	0.077	0.416	0.048	0.152	0.278
Wilks' Lambda 3	0.830	0.592	0.713	0.234	0.628	0.747
Wilks' Lambda 4	0.992	0.926	0.996	0.808	0.940	0.987
% correctly	74.1	80.2	82	76.9	86	91.3

EV = Eigenvalue, % correctly = Percentage of original grouped cases correctly classified

Table 7: Correct classified pixels for each succession phase for every study site

	Elk	Flathead	Lower Kootenai River braided	Upper Koo-tenai River – Wasa	Upper Koo-tenai River – Fenwick	Lower Kootenai River meandering
0	100.0	100.0	100.0	100.0	100.0	100.0
1	100.0	71.0	(n.a.)	52.7	100.0	100.0
2	76.2	100.0	71.5	77.9	100.0	100.0
3	0	(n.a.)	80.9	100.0	(n.a.)	95.4
4	58.5	87.1	98.9	100.0	31.0	25.6
5	45.5	91.6	76.6	68.4	75.4	93.0
6	3.2	81.3	56.6	73.8	(n.a.)	(n.a.)
7	(n.a.)	(n.a.)	0	53.1	90.4	(n.a.)
8	(n.a.)	0	(n.a.)	50.8	74.1	(n.a.)
9	99.6	10.4	49.7	84.2	(n.a.)	63.9

3.2 LISA

Hot spots

River reaches close to their natural state show clear hot spots for succession phases. At meandering river reaches (Upper Kootenai River at Wasa and Upper Kootenai River at Fenwick) we found one large hot spot for each project area. At Wasa the cluster covers about 43% of the project area and consists of polygons of succession phases early successional woodland phase and established forest phase, each of them is of secondary succession. At Fenwick the hot spot covers 35 % of the project area and also consists of polygons of succession phases early successional woodland phase and established forest phase, again secondary succession.

The braided river reaches of the Elk river and the Flathead river also show hot spots. For the Elk River two large-area hot spots could be found. They are composed by polygons of farmland and grassland bordering early successional woodland of primary succession and mature stage woodland. The sum of the hot spot area is about 40% of the whole project area. At the project area of the Flathead River the hot spot covers about 10% of the whole area. At project areas influenced by dam operation only at the braided reference of the Lower Kootenai River hot spots could be found (Fig. 3 and 6). The hot spots mainly consist out of areas of phase farmland and grassland and a minor part are regions of mature stage. The hot spots cover 25% of the whole project area.

Cold spots

At meandering river reaches influenced by dam operation no cold spots could be found. At all natural reference sites cold spots were displayed. At meandering river reaches the cold spots occur as large-area clusters whereas at braided river reaches there are shown more and smaller clusters.

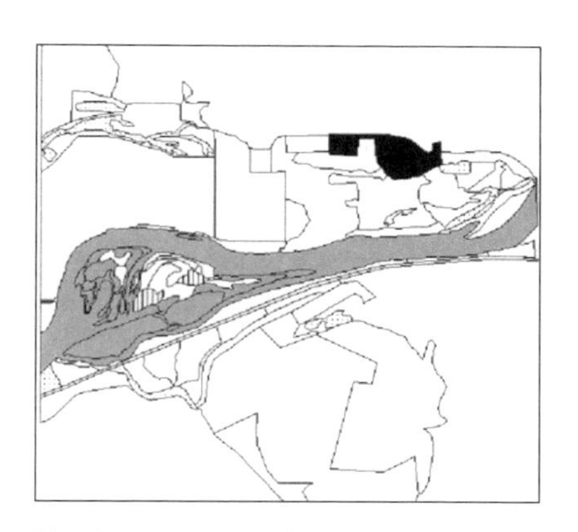

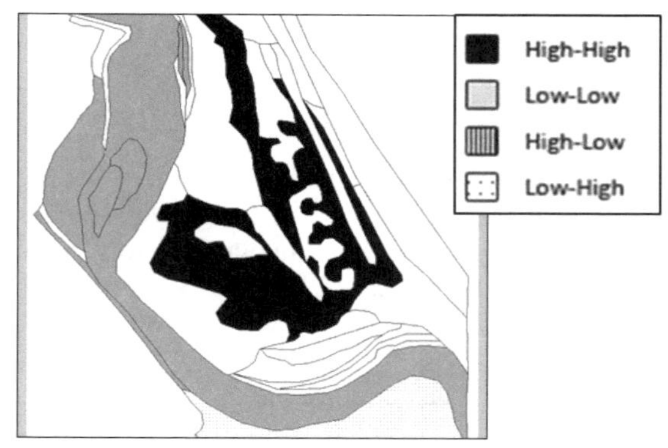

Fig. 2: LISA Cluster Map, Elk River, Variable: Successionphase
Fig. 3: LISA Cluster Map, Flathead River, Variable: Successionphase

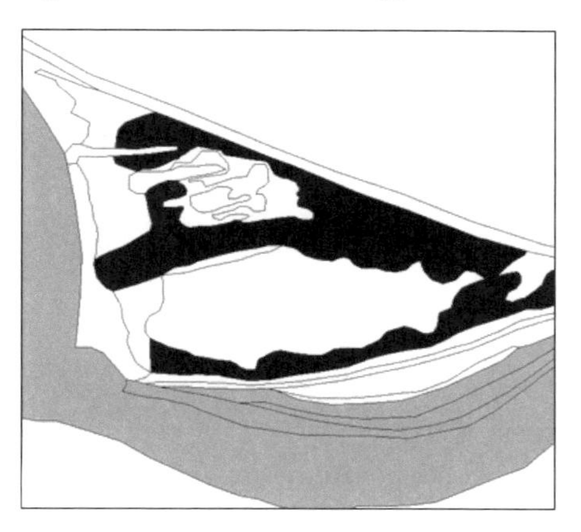

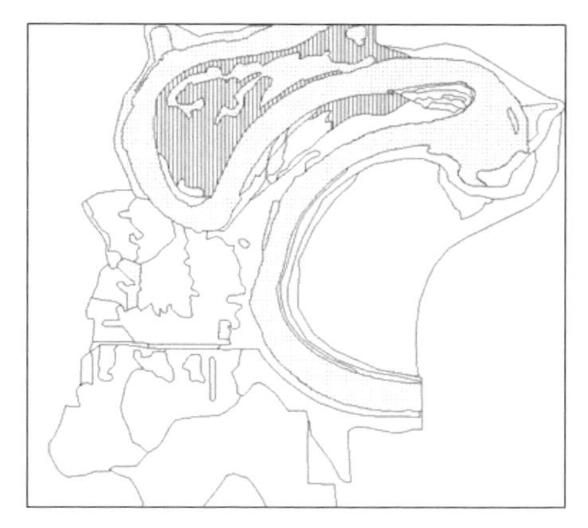

Fig. 4: LISA Cluster Map, Lower Kootenai (braided), Variable: Successionphase
Fig. 5: LISA Cluster Map, Upper Kootenai (Wasa), Variable: Successionphase

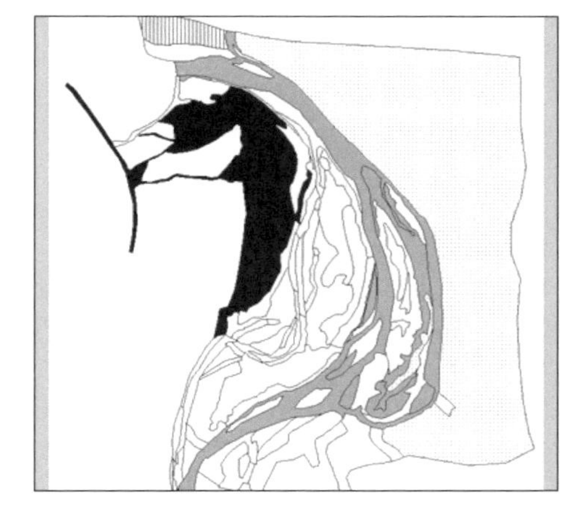

Fig. 6: LISA Cluster Map, Upper Kootenai (Fenwick) , Variable: Successionphase
Fig. 7: LISA Cluster Map, Lower Kootenai (meandering), Variable: Successionphase

3.3 Area balances

At rivers influenced by dam operation the percentage of the area assigned to flood inundation class 1 is significantly higher than at rivers close to their natural stage. In Figure 8 the proportion of areas assigned to the different flood inundation classes for the meandering rivers is visualized. At meandering river reaches close to their natural state there are no polygons assigned to flood inundation class 1. 34 and 44 percent of the area were assigned to flood inundation class 2. At the reach of the Lower Kootenai River under influence of dam operation 72 percent of the area are classified as flood inundation class 1.

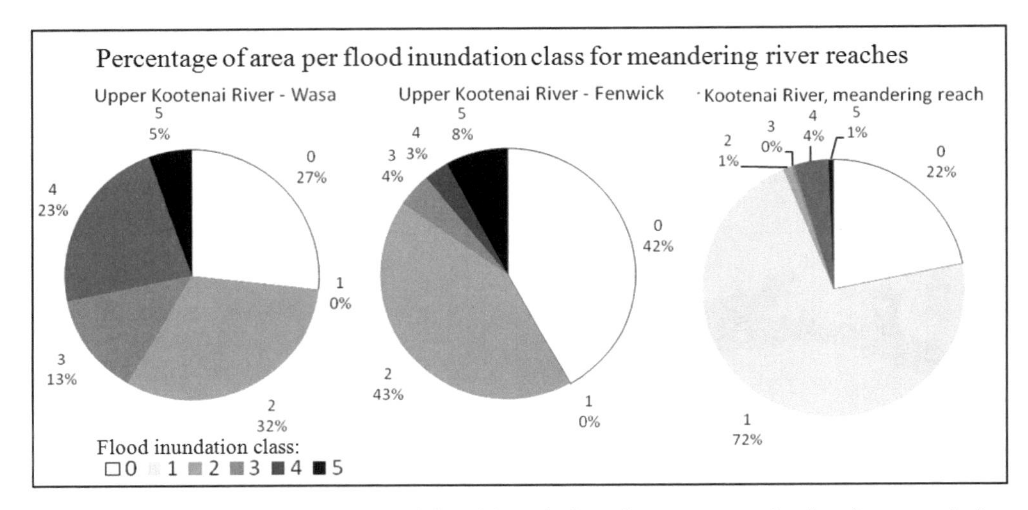

Fig. 8: Area balance, percentage of flood inundation classes per study site, for meandering river reaches

4 Discussion

The hypotheses number one and three presented in chapter 1 could not be refuted. The results indicate a strong correlation between the abiotic factors and the succession phases. In detail the results in respect of the hypotheses are as follows:

Hypothesis 1: At braided river reaches the cluster analysis showed a stronger blending of old and young succession types than at meandering river reaches. At meandering river reaches instead, the hot spots were always situated farther away from the river and the cold spots were found only directly at the river. At river reaches under influence of dam operation there were less cold or hot spots.

Hypotheses 2: The correlation between the distance to the river and the succession phase could not be supported. The results of the area balances do not show a clearly visible pattern. Also the result of discriminant analysis did not improve by regarding the Euclidean distance as a fourth predictor (Table 6).

Hypotheses 3: At river reaches under influence of dam operation the value of the abiotic factors at the study sites always tended to be at least 1 class lower. In Figure 8 this is visualized for the meandering rivers. For braided rivers the same pattern could be found.

The misclassification of pixels of succession phase 8 in DA can be explained by the secondary succession of these areas. If there would not have been any fire the pixels would probably be of succession phase 6, which is the phase the DA would suggest for the pixels.

Literature

ANSELIN, L. (2004), GeoDa 0.9.5-i Release Notes, Urbana: Spatial Analysis Laboratory and Center for Spatially Integrated Social Science, Department of Agriculture and Consumer Economics, University of Illinois, Urbana-Champaign.

BAO, S. (1998), Exploratory Spatial Data Analysis with Multilayer Information. – http://chinadatacenter.org/cdc/papers/workshop_emu.pdf (accessed on 22.03.2010).

FRIEL, C., Criminal Justice Center, Notes on Discriminant Analysis, Sam Houston State University. – http://www.shsu.edu/~icc_cmf/cj_742/Old%20Files/mod_7.doc (accessed on 22.03.2010).

EGGER, G., BENJANKAR, R., DAVIS, L. & JORDE, K. (2007), Simulated effects of dam operation and water diverstion on riparian vegetation of the Lower Boise River, Idoho, USA. In: Harmonizing the Demands of Art and Nature in Hydraulics, 32nd Congress of IAHR, Juli 1-6, 2007, Venice, Italy, Proceeding 1-14.

GARSON, D. (2008), Discriminant Function Analysis. – http://faculty.chass.ncsu.edu/garson/PA765/discrim.htm (accessed on 30.03.2010).

POFF, N. L., ALLAN, J. D., BAIN, M. B., KARR, J. R., PRESTEGAARD, K. L., RICHTER, B. D., SPARKS, R. E. & STROMBERG, J. C. (1997), The natural flow regime: A paradigm for river conservation and restoration. Bio-Science, 47 (11).

POULSEN, J. & FRENCH, A., Discriminant Function Analysis. – http://www.docstoc.com/docs/21097876/DISCRIMINANT-FUNCTION-ANALYSIS-%28DA%29 (accessed on 30.03.2010).

SPSS Inc. (2009), PASW Statistics Release 17.0.2, Chicago, IL: SPSS Inc.

STATSOFT, Electronic statistics textbook, Discriminant Function Analysis. – http://www.statsoft.com/textbook/discriminant-function-analysis/ (accessed on 22.03.2010).

TAPPEINER, U., TASSER, E. & TAPPEINER, G. (1998), Modelling vegetation patterns using natural and anthropogenic influence factors: Preliminary experience with a GIS based model applied to an Alpine area (1998). Ecological Modelling, 113 (1-3), pp. 225-237.

TASSER, E., TAPPEINER, U. (2002), Impact of land use changes on mountain vegetation. Applied Vegetation Science, 5, pp. 173-184.

WANG, X. & JIANG (2007), Analysis on the Spatial Distribution and Evolution of FDI in Mainland China. ICMSE 2007.

WUENSCH, K. J. (2008), Two group discriminant function analysis. – http://core.ecu.edu/psyc/wuenschk/MV/DFA/DFA2.doc (accessed on 22.03.2010).

Comparison of DMC Nigeriasat-1 SLIM, SPOT-5 HRG and Landsat 7 ETM+ Data for Urban Land Cover Analysis

Chris CHIMA and Nigel TRODD

1 Introduction

The usefulness of Nigeriasat-1 SLIM data for urban land cover mapping and analysis is ascertained by comparing it with SPOT-5 HRG and Landsat7 ETM+ data in per-pixel image classification and object based image analysis (OBIA). This work is part of a larger study to assess the use of multi-date/multi-sensor satellite remote sensing to monitor and model urban growth in African cities. Given the long term objective of combining information from multiple Earth observation satellites and the development and launch of new Earth observation missions, it is necessary to have a better understanding of their compatibility. Nigeriasat-1 is one of six satellites, each of which carries a Surrey Linear Imager sensor (SLIM), that form the Disaster Monitoring Constellation (DMC) (SURREY SATELLITE TECHNOLOGY LTD 2010). Launched in 2003, it captures data in the green, red and near infrared (NIR) bands at 32m ground resolution. Earlier attempts to classify Nigeriasat-1 data for urban land use analysis with single date pixel based classifiers have not been encouraging due to the limited number of spectral bands and low spatial resolution (OMOJOLA 2004, OYINLOYE et al. 2004).

Urban development in Africa can be planned or informal. The former is associated with land use zoning and, in general, more reliable demographic statistics. It follows that it may be possible to integrate ancillary data with satellite data to extract more useful information on land cover and land use in these situations (POHL & VAN GENDEREN 1998). Geodata fusion can be at pixel, feature and decision levels (LEUKERT et al. 2003) and this paper considers only the first two levels. At pixel level different sensor data are co-registered and combined for change detection; while at feature level, objects are extracted based on extent, shape and neighbourhood patterns in an image or from parcel boundaries in the ancillary data.

2 Study Area and Data Sets

Abuja, capital city of Nigeria, is located between latitude 8° 52' and 9° 07' North and longitude 7° 22' and 7° 32' East in the geographical centre of the country (Figure 1). The city was planned in 1976 and the area covered by the urban plan occupies approximately 800 square kilometres. According to the Nigerian census Abuja grew from a population of 107,067 in 1991 when it officially became the seat of government to 590,400 in 2006. The demographic data and 2006 urban plan provide the ancillary data.

Three cloud-free images were acquired. The scenes were captured by Landsat ETM+, Nigeriasat-1 SLIM and SPOT-5 HRG (Table 1). They were all acquired during the dry season to reduce the effects of seasonal variations in vegetal cover and solar illumination.

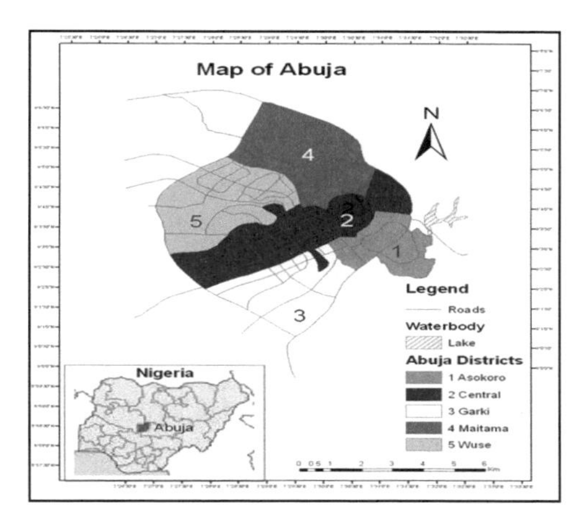

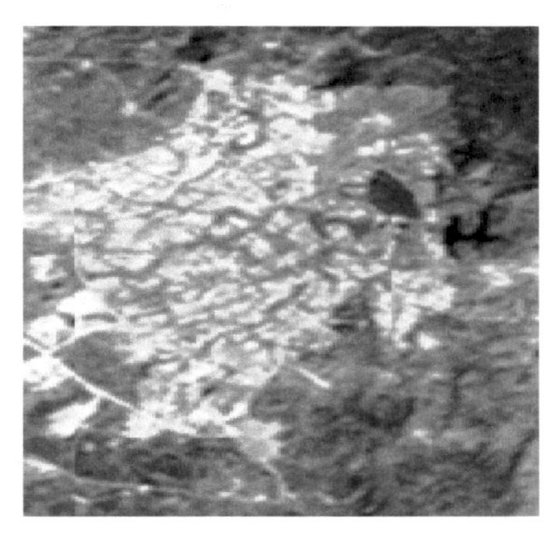

Fig. 1: Study area: Map of Abuja and equivalent area of Nigeriasat-1 SLIM image

Table 1: Image properties

Data	Date of acquisition	Image bands used	Image resolution	Scene identification	Local time of overpass
Landsat 7 ETM+	27 Dec 2001	Bands 1-5, 7	28.5m	LE71890542001361EDC01	09:39:04am
SPOT 5 HRG	26 Nov 2006	Bands 1-4	10m	KJ 74-332	10:06:54am
Nigeriasat-1 SLIM	4 Dec 2003	Bands 1-3	32m	Abuja and environs	09:24:49am

3 Method

A four category land cover schema, adapted from ANDERSON et al. (1976), was developed from field reconnaissance in 2008. Four land cover types were classified namely: built up, vegetation, water body and bare (soil) surface.

Each image was classified using a per-pixel maximum likelihood classifier and an OBIA approach. To enable direct comparison, similar training and accuracy assessment sites were prepared and imported into all classifiers. 150 samples were used in the training for both classifiers.

The OBIA approach employed an additional land parcel layer from the urban plan and the integrated data sets were segmented hierarchically into three levels using multi-resolution segmentation (DEFINIENS 2006). Areas outside the coverage of the urban plan were segmented based only on local homogeneity criteria without the spatial constraint of parcel boundaries shape file (DEFINIENS 2006).

Pixels and objects were allocated to one of the four land cover categories. Error matrices were compiled from 150 samples for per pixel and OBIA analysis and overall accuracy, overall kappa and conditional kappa values were computed (CONGALTON & GREEN 1999, FOODY 2002).

4 Results and Discussion

The overall accuracy of the per-pixel maximum likelihood classification of SLIM data was 63%, substantially less than the 88% and 89% achieved with ETM+ and HRG data respectively (Fig. 2). In the OBIA approach, the HRG / urban plan combination produced the highest overall accuracy at 97% whilst the equivalent classifications with ETM+ and SLIM data were 92% and 90% respectively. These accuracies were boosted by the very strong levels of agreement recorded for the vegetation and water categories. Built-up areas also had a conditional kappa in excess of 80% in every classification but only HRG data was able to discriminate land parcels of bare surfaces with an accuracy of more than 80%.

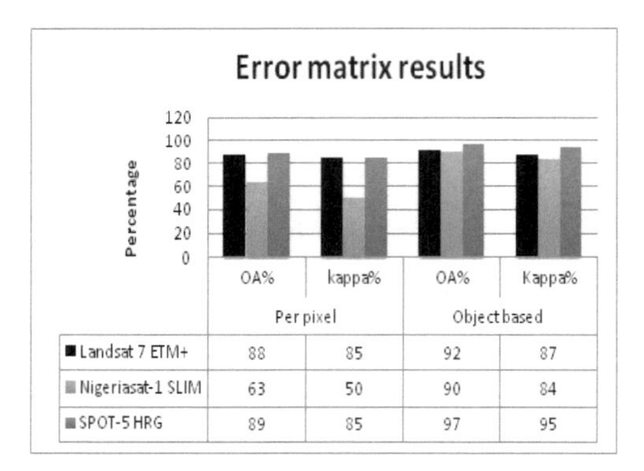

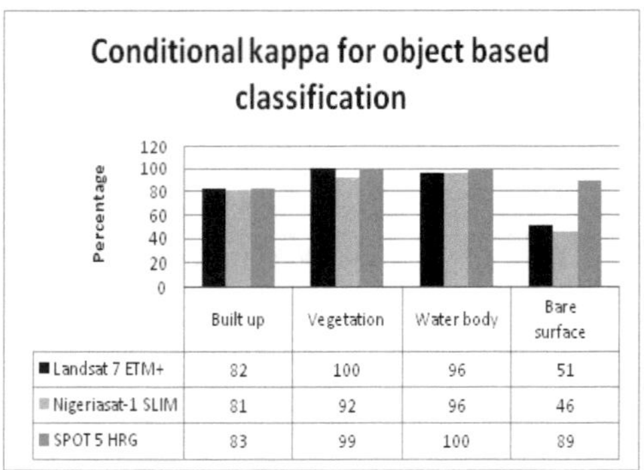

Fig. 2: Accuracy assessment results

Whilst HRG data were captured in the same year as the field survey the ETM+ data had been captured 5 years earlier. Despite this difference an overall kappa statistic of 85% was achieved for per-pixel maximum likelihood classifications of both ETM+ and HRG data. It was felt that the small number of categories contributed to these levels of agreement. The relatively poor result for SLIM data agreed with previous studies (OMOJOLA 2004, OYINLOYE et al. 2004) and raises further questions about the quality of individual SLIM DN values e.g. OGUNBADEWA (2009).

The ability of an OBIA approach to enable SLIM data to match the overall performance of ETM+ data in classifying land cover and to achieve similar results to HRG data in three of the four categories is, therefore, of considerable importance. The failure of SLIM and ETM+ data to accurately classify bare surfaces even with the assistance of land parcel boundaries, however, remains a concern, more so if the objective is to characterise urban land cover change as bare surfaces may represent an intermediate stage in the transition from vegetated to built up environments.

5 Conclusions

This research has demonstrated that conventional classificatory approaches based on the spectral properties of individual pixels of Nigeriasat-1 SLIM data are less likely to make a substantial contribution to urban land cover analysis than SPOT-5 HRG and Landsat7 ETM+ data. Data fusion techniques that provide a framework to process data at the object level open up the prospect of using SLIM data as part of a multi-date/multi-sensor study of land cover. Where land parcel boundary data are widely available, as found for several of the 'new' capital cities of African nations, SLIM data could be used to support an OBIA approach to urban monitoring and modelling. Further research in this direction should be encouraged given the growing archive of data from the DMC mission.

Acknowledgements. The authors wish to thank Nigerian Space Research Development Agency, SPOT OASIS Project, the Global Land Cover Facility and Abuja Geographic Information Systems for making the data available for this research.

References

ANDERSON, J. R., HARDY, H. E., ROACH, J. T. & WITMAR, R. E. (1976), A Land Use and Land Cover Classification System for Use with Remote Sensor Data. Geological Survey Professional Paper 964. Washington: United States Government Printing Office.

CONGALTON, R. G. & GREEN, K. (1999), Assessing the Accuracy of Remotely Sensed Data: Principles and Practices. Boca Raton, FL CRC/Lewis Press.

DEFINIENS (2006), Definiens Professional 5 User Guides. Munich, Definiens AG.

FOODY, G. M. (2002), Studies of Land Cover Classification Accuracy Assessment. Remote Sensing of Environment (80) 1, pp. 185-201.

LEUKERT, K., DARWISH, A., & REINHARDT, W. (2003), Urban Land Cover Classification: An Object Based Perspective. 2nd GRSS/ISPRS Joint Workshop on Remote Sensing and Data Fusion over Urban Areas. Berlin, 22-23 May 2003.

OGUNBADEWA, E. Y. (2008), The Characteristics of Nigerisat-1 and its Potential Application for Environmental Monitoring. African Skies 12 (October).

OMOJOLA, A (2004), Urban and Rural Land Use Mapping of Parts of Ekiti State using Nigeriasat-1 Imagery, Proceedings of National Workshop on Satellite Remote Sensing (NigeriaSat-1) and GIS: A Solution to sustainable National Development Challenges. Abuja, Nigeria, 15–17 June 2004.

POHL, C. & VAN GENDEREN, J. L. (1998), Multisensor image fusion in remote sensing: concepts, methods and applications. Int. Journal of Remote Sensing (19) 5, pp. 823-854.

OYINLOYE R. O., AGBO, B. F. & ALIYU, Z. O. (2004), Application of NigeriaSat-1 Data for Landuse/Landcover Mapping, Proceedings of National Workshop on Satellite Remote Sensing (NigeriaSat-1) and GIS: A Solution to sustainable National Development Challenges. Abuja, Nigeria, 15 – 17 June 2004.

SURREY SATELLITE TECHNOLOGY LTD. – http://www.sstl.co.uk/ (accessed on 23 January 2010).

Taking Geospatial Applications to the Grassroots

Arup DASGUPTA

Keynote at the GI_Forum 2010

Abstract

The effectiveness of geospatial systems for solving spatial problems is a given. These systems are being used for disaster management, environmental assessment and planning for sustainable development to name a few applications. However, these are all applications spearheaded by the government or by private industry for the government. The pressures on the planet's resources demands not just compliance to regulations and behavioural changes but also innovative applications. Such innovative applications to be inclusive must happen at the grassroots. This presents a dilemma. Geospatial technologies take a synoptic view, and suggest global solutions while the need for innovation at the grassroots requires solutions customised for the individual. Non Governmental Organisations have realised the need for such customised solutions that take into account the needs of the individuals. They have developed an interactive public-private partnership (P-P-P) model which has proved to be very effective over time. This paper explores the integration of the geospatial synoptic, global model with the P-P-P individual customisation model for the effective use of geospatial systems for meeting the information needs of individuals at the grassroots.

1 Introduction

Traditionally remote sensing and GIS applications are designed to cover large geographic extents. Remote sensing provides a synoptic view which highlights features which are masked in smaller segmented views. Coverage of disasters like floods and drought or the estimation of crops on a regional basis is best served by remotely sensed data. The handling of such data requires a digital spatial database management system exemplified by the ubiquitous GIS. By handling large amounts of data covering vast geographic spreads the high cost of software and systems can be spread out resulting in a low cost per unit area served. This approach is equivalent to the mass production model in manufacturing. Individual needs are not considered in the search for economy of production. In fact the individual needs to adjust to the mass production market. At best the market may segment the consumers into groups but within a group the product is uniform. The consumer has to decide which group fits his requirement the best. Similarly, in the large-scale model of geospatial applications we see that universal prescriptions are made for the entire community or segments of the community. Within each segment there is only one solution. Further, the users are government functionaries whose main aim is to implement government programmes. The end beneficiary is treated as a passive receptor.

Non Governmental Organisations (NGO) on the other hand, consider the individual needs and prepare their plan in a participative environment. Techniques like Rapid Participatory Rural Appraisal are used to get the information at the grass root level. Solutions are found iteratively with the full participation of the intended beneficiaries. The best solution is the one the beneficiaries prefer and not the one dictated by the straight forward application of geospatial technology. Further, most NGOs are put off by the high cost of geospatial systems and prefer to work with ground level data. Consequently their geographical area of coverage is small. They also feel that spatial heterogeneity, as postulated by Tobler's Law makes large scale development projects unviable as the same solution cannot be applied over vast swaths of heterogeneous geography and socio-economic conditions. Customisation of solutions must take into account the needs and aspirations of the intended beneficiaries. In the same geographical area the needs may vary based on socio-economic variables and individual preferences.

The use of geospatial technology and science for solving spatial problems is well understood. Access to the technology and applications which was restricted to a few hundred geospatial specialists or a few thousand resource managers is now available to hundreds of thousands of users through Web enablement. The scenario we are looking at goes beyond these numbers and addresses millions of potential users. Greater inter-operability between different hardware platforms and between different GIS software and the convergence of various technologies makes this goal achievable. Geospatial technologies are now device independent and can provide great value when associated with other domains. For example the integration of GIS with IT based e-governance has opened up many spatially enabled citizen services. GIS integrated with GPS and mobile communications systems provide location based and location aware services. The Internet provides a means of distributing geospatial data, services and applications. What is required is for the access to these types of services to become independent and customizable to individual needs through innovative solutions.

2 Developing a Strategy

The strategy for addressing the needs at the grass roots has to go beyond serving up maps or even applications. The system should provide an environment for decision making for the individual. For example, it could provide the kind of extension service a Village Level Worker (VLW) provides. A VLW acts as a facilitator between the agricultural research units and the farmers by helping to transfer laboratory results to the land and transferring back the success and failure stories.

Figure 1 outlines the approach needed to address this issue. This model assumes that the service is a customised product which has to be delivered to a consumer. We can thus view it as a business case. On the one hand we have a set of new technologies and on the other we have a set of consumer needs. These needs are latent because the consumers are not aware of what the technologies can deliver. They may be content with what they have or they may not be content but they see no way out of the situation. Take the simple case of obtaining a land record. Till a GIS based solution was found consumers accepted the reality of having to wait for days after submitting a requisition to the local land registry office. Today with a GIS database and a printer the same task is achieved over the counter. These types of business opportunities can be visualised through a combination of the felt needs and the available technologies. Innovative strategies to convert these opportunities into viable businesses can then be developed.

The strategies need to take into account that the services will require a range of technologies to be called up. We can therefore state that the technological resources are global. This would call for collaboration between technology service providers who may be companies, NGOs or different departments of the government or a mix of all three. At the same time we need understand at how these technologies could be draw upon them to arrive at a decision which is unique to his needs.

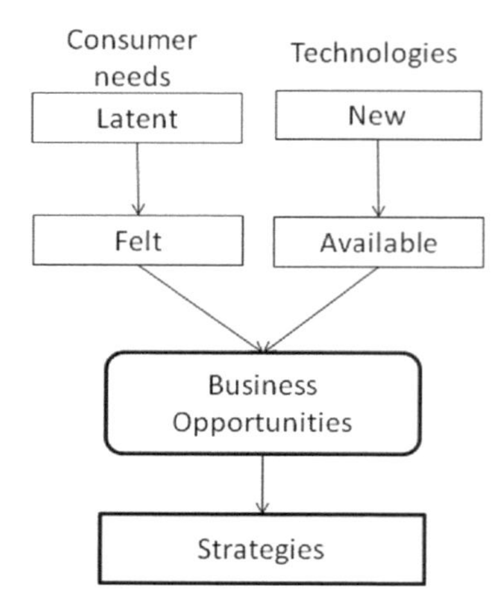

Fig. 1:
Conceptual model

Since we are discussing geospatial systems it is clear that the basic data format has to be digital and therefore usable on computing platforms. The services have to be configured to meet the local conditions and requirements. The delivery of services has to be to the potential customer and therefore there is a need for connectivity and finally we need to be able to emulate the traditional support groups that we find in every society. These groups provide a social support system whenever a new system has to be absorbed. They debate and discuss and are very important in the final acceptance of the technology. Figure 2 summarises the development strategy as discussed above.

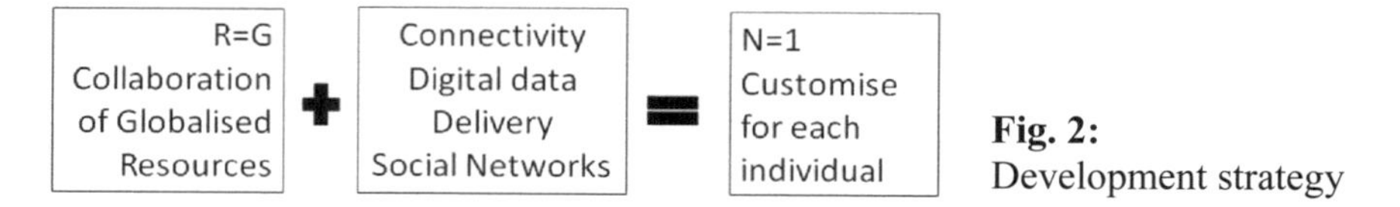

Fig. 2:
Development strategy

In this discussion we assume that the individual is capable of accessing such a system. Such an assumption is valid because it has been observed that the adaptability of the individual to new technologies is dependent on the utility of that technology to his work. The successful proliferation of mobile phones in rural areas is one such case in point.

3 Collaboration of Globalised Resources

The globalised resources will consist of two groups. One will be the set of data which is nationally available. Geospatial data falls in this group. As these data sets are aggregated at a national level they may be considered to be the top-down group. The other sets of data are the local data collected in the course of projects and surveys. These data sets are localised and most importantly, they represent the needs and aspirations of the local people. This group of data is the bottom-up group. There is a need to study both these data sources and to evolve a way of integrating them such that they can be successfully utilised to develop services which can be customised for individual applications.

3.1 The top-down resources

Geospatial resources are essentially the top-down resources. Remotely sensed data, thematic maps and attribute tables are all geo-referenced to a base map and are made available to users. Such resources are usually organised as a Spatial Data Infrastructure, SDI. Figure 3 is one such organisation of data planned for an Indian National SDI. Data from a range of data providers are implicit in this model. While the major data providers will be government departments it also includes data from other nongovernmental organisations. The architecture of such a system is not apparent in this model.

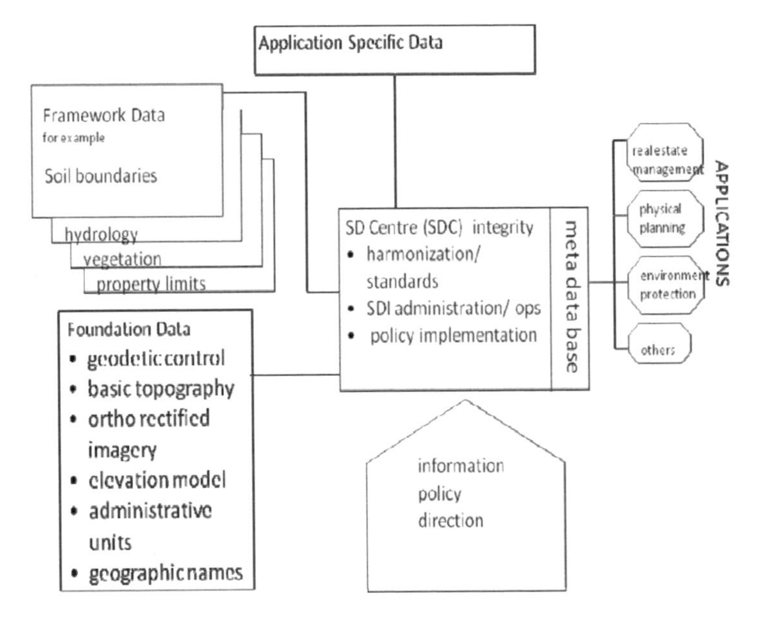

Fig. 3: Structure of NSDI in India

3.2 The bottom-up resources

As mentioned earlier local projects begin with a participatory appraisal. One of the most important tools used is to develop a local map with the help of the participants using a technique called Resource Mapping (IAPAD). Resource Mapping is a method for collating and plotting information on the occurrence, distribution, access and use of resources within the economic and cultural domain of a specific community by the community. Such maps may not be cartographically accurate but they contain a wealth of information in terms of local knowledge which is not available in the top-down surveys. The difficulty with such maps is that they cannot be integrated with other information because of the lack of cartographic quality, hence a wealth of information available in the top-down resources in not directly usable.

3.3 Integrating the two resources

The way to integrate the two data resources such that the best of both data sets can be used effectively is to follow a two-step approach. In the first step the local resources map is prepared. In the second step it is adjusted to fit into the cartographic framework using reference points. The technological difficulty is that most cartographic frameworks stop at 1:50,000 scale while resource maps will be typically in the range of 1:4000 scale. The solution here is to use high resolution remotely sensed data which is already registered to the framework, and adjust the resources map to the image using visible and identifiable features. An intermediate step is to register the official village cadastral map to the image. The resources map, which usually covers a village or a group of villages, can now be registered to the village map. Figure 4 shows a part of a cadastral map registered to a high resolution satellite image, in this case a 23 metre multispectral image sharpened with a 5 metre panchromatic image.

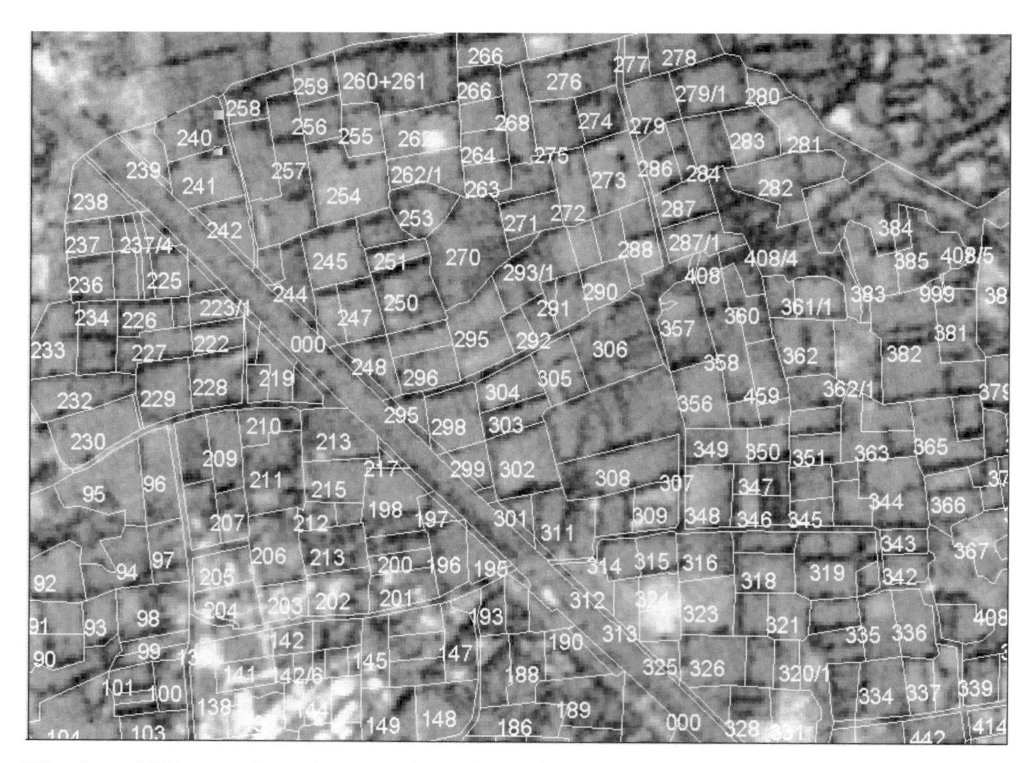

Fig. 4: Village cadastral map registered to a high resolution satellite image

Both high resolution data and the technology to register maps to such data already exist. Once this is achieved the wealth of data available from the NSDI can now be used for local planning and support. What remains is the development of services which can be customised by the end user. This is the key. An excellent example of such a service is the *e-Choupal* an IT service for farmers run by a leading agro-industry company in India (ITC). A local farmer acts as a nodal point. He aggregates the needs of the local farmers and sources the required good and services directly from the company. In turn the company directly sources agricultural produce from the farmers. The system removes middlemen and also provides advisory services during the crop growth period. The system does not use geospatial data but the model can easily be adapted for geospatial services using the integrated approach.

4 Delivering Globalised Resources

Figure 2 lists out four components required to deliver these resources to the consumer. They are connectivity, digital data, delivery and social networks.

4.1 Connectivity

The Indian government has provided connectivity through fibre optics such that a 2 Mbps capacity is available at each district town. Many states have connected villages using VSAT. The *e-Choupal* system uses telephone lines and VSATs. However, these systems require a mediator like the *Sanchalak* in *e-Choupal*. The individual farmer would need information on a 'here and now' basis. The mobile telephone network provides this level of connectivity. The mobile network covers 85% of India and there are 100 million rural connections and growing. Handset manufacturers are coming out with models suited to Indian user needs. Most handsets now have built in GPS receivers and are GPRS enabled. 3G services are about to be launched in India.

4.2 Digital data

The need for high resolution satellite imagery and digital cadastral maps has already been discussed. In addition it is necessary to include some more digital datasets which are now available through new technologies. The first is a sensor network for weather data acquisition. Such a network can provide localised weather information and can be used to create local weather advisories. The *Bhuvan* system implemented by the Indian Space Research Organisation makes available such data on a near real-time basis. Another data set is the local information which can be fed into systems like *Bhuvan* by local person who can be trained as neo-geographers. Apart from these data like market prices, market arrivals, market demand projections and other economic data should also be available in a digital formal to enable their use in models

4.3 Delivery

The technology to bring together all the data sources has to be given adequate attention. Given the context of the services it is envisaged that the system has to be organised as a

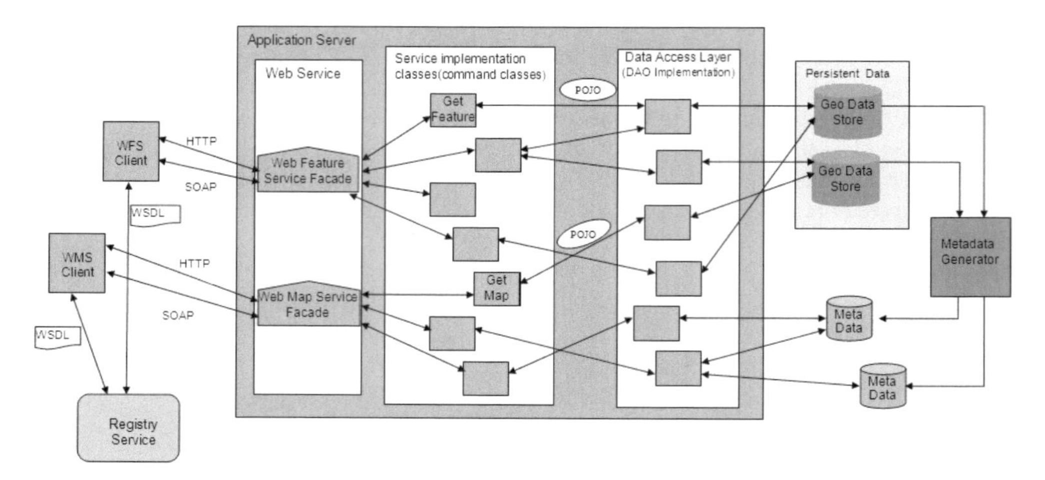

Fig. 5: GRID architecture for NSDI

GRID. This will enable the building of a virtual organization of NSDI data provider agencies. It will also enable an adequate coverage of data and identify missing data. GRID architecture will provide a non trivial quality of service and provide effective integration of existing services. Use of OGC compliant products and the configuration of these products and data sources can also be easily achieved in a GRID environment. Figure 5 shows a typical architecture which has been proposed for the NSDI by BHATT& DASGUPTA (2008). This is based on a Service Oriented Architecture which provides sufficient independence for individual systems and at the same time provides the required connectivity between systems. Application developers will need to use such architecture to develop interactive services.

4.4 Social networks

The proposed system could be used to create social networks for rural audiences. The network could be used to collect, collate and store traditional local knowledge and to bring out the rich store of knowledge in rural agricultural research institutes. Such networks could also provide extension service through two way communications between farmers and the rural agricultural research institutes and a database of innovations and best practices could be built up.

5 Customising for the individual

The applications which have to be run on these systems to provide the information required are many. At present there are simple applications which require the individual to access a system through a mediator. At the next level we need applications that can be accessed by an individual. These have to be interactive and should be able to provide decision alternatives and also enable the individual to navigate through the alternatives and make an intelligent choice which minimises his risk. A typical scenario could be that of a marginal farmer seeking advice on the best crop to grow in a given agro-climatic setting considering his own economic status and future expectations. Such applications do not exist and this is a challenging task for NGOs and others who can build up such interactive models using the diverse databases. Additional requirements are the need to develop interfaces in local languages using local terminology.

Acknowledgement

This concept paper is based on a seminal talk by the late C K Prahlad, Paul & Ruth McCracken Distinguished Professor, Stephen M. Ross School of Business, University Of Michigan at the Map World Forum 2009 at Hyderabad.

References

BHATT, H. C. & DASGUPTA, A. R. (2008), Design considerations for realising a web enabled Spatial Data Infrastructure. Journal of Geomatics, (1) 3, pp. 113-124.

IAPAD (2010), Resource Mapping, Participatory Avenues, the Gateway to Community Mapping, PGIS & PPGIS. – http://www.iapad.org/resource_mapping.htm (accessed 25-04-2010).

ITC Agricultural Business Division (2010), About eChoupal. – http://www.itcibd.com/e-choupal1.asp (accessed 25-04-2010).

PRAHLAD, C. K. (2009), Potential of Geospatial technology to transform business operations. – http://www.mapworldforum.org/2009/conference/speakers_abstract.htm (accessed 25-04-2010).

Detecting Characteristic Scales of Slope Gradient

Clemens EISANK and Lucian DRĂGUŢ

The GI_Forum Program Committee accepted this paper as reviewed full paper.

Abstract

Very high resolution (VHR) DEMs such as obtained from LiDAR (Light Detection And Ranging) often present too much detail for various applications. Finding the right spatial scale for analysis is a challenging task, especially in landscape ecology. It still seems to be unsettled, whether scales in digital representations of the land surface are explicitly detectable. We applied the statistical method of local variance (LV) to explore the data-inherent scale structure. Scale levels of slope gradient were produced by 1) degrading the initial LiDAR DEM at 1 m resolution through resampling to successively coarser resolutions (further referred to as scale levels) and then calculating slopes from the degraded DEMs, and 2) performing multi-resolution segmentation on slope gradient in Object-Based Image Analysis (OBIA). LV was calculated as mean standard deviation. Values of LV were plotted against scales to derive scale signatures. Obtained graphs exhibited peaks and stepwise changes. These thresholds can be interpreted as characteristic spatial scales. Results demonstrated the potential of LV for identifying characteristic scales of continuous terrain data within multi-scale analysis of slope gradient. For the object- and the cell-based approach we could identify 4 respectively 5 characteristic scales. We found out that scale detection in OBIA-based LV graphs is easier to perform than in the one obtained from resampling, due to differences in underlying aggregation techniques.

1 Introduction

In this paper we address the important question of finding characteristic spatial scale(s) for analysis. From a landscape ecology perspective, scale refers to both grain and extent. Grain is defined as "the finest level of spatial resolution possible with a given data set, e.g. pixel size for raster data" (TURNER et al. 1989). Extent indicates the size of the study area. Characteristic scale may be defined as level, where the grain size of the digital representation allows for meaningful mapping of real-world units with similar properties. Thus, the term "meaningful" is user-dependent, as it relates to individual modelling objectives. In landscape ecology and related fields such as remote sensing and geomorphometry, scale issues have been widely discussed among scientists, since phenomena and processes on the earth's surface operate at specific spatial scales (EVANS 2003). Modelling outcomes are restricted to the scale of investigation, which in many cases is not chosen appropriately (DRĂGUŢ et al. 2009a, DRĂGUŢ et al. 2009b).

The problem of selecting characteristic scales mainly arises from ever finer spatial resolutions of digital elevation models (DEMs) and derived parameters (e.g. for Austria whole provinces have already been covered by LiDAR data at 1 m resolution). Land-surface models are increasingly used as input for ecological modelling, but in many cases they hold too much detail (i.e. noise) for given applications. If the original data set is too detailed, the grain size has to be degraded to coarser ones. Recently, LE COZ et al. (2009) have demonstrated the need for aggregating even the relatively low resolution SRTM to address the scale specificity of large basins. Hence, comprehensive methods for revealing changes in the structure of continuous terrain data as a function of grain size, and that allow fast detection of characteristic scale levels for a given application, are required.

There are several authors who promoted the use of statistical methods to obtain scale signatures, graphs that show the behaviour of statistical indicators across a defined range of scale (WOOD 2009). These signatures exhibit some thresholds that indicate more appropriate scales for analysis, i.e. where groups of real-world units with similar spatial, temporal, and statistical properties occur (WU & LI 2006). HILL (1973) was one of the first who recognizes the suitability of variance measures for scale detection in landscape ecology. CARLILE et al. (1989) suggested the use of spatial variance and, in addition, correlation estimates to determine the inherent scale of ecological processes. Based on this, appropriate levels of resolution for measuring plant distribution could be identified. Later, CULLINAN et al. (1997) compared the methods developed by the afore-mentioned authors. They used satellite images to investigate scales of vegetation patterns. Therefore, they calculated statistics for ever larger window sizes and plotted the values against scales. All obtained graphs showed thresholds, whereas peaks in the variance and correlation curves, and troughs in graphs from Hill's method corresponded with characteristic scales of vegetation pattern.

The statistical approaches for detecting characteristic scales presented in the previous section are mostly applied on digital images. Of course, images from remote sensing are continuous digital representations of the earth's surface. However, they exhibit more pronounced boundaries and transitions between spatial units than land-surface models such as DEMs and slope gradient that show much smoother characteristics (HENGL & EVANS 2009). As has been proved, the identification of characteristic scales works well on satellite images, since there are more differences in spectral and spatial properties of image units resulting in higher variations of statistical estimates across scales. Though, little efforts have been undertaken to transfer these approaches to smoother land-surface models. Still, it seems to be unsettled if spatial scales in digital representations of the land-surface are explicitly detectable or if scale is just a "window of perception" (MARCEAU & HAY 1999).

The presented research was based on WOODCOCK & STRAHLER (1987), who measured local variance (LV) as a function of resolution to detect characteristic spatial scales in digital images. We went one step further by measuring LV on continuous land-surface models rather than digital images. Slope gradient is one prominent example for a continuous land-surface model that influences ecological patterns (e.g. vegetation) and processes (e.g. erosion).The main objectives of this study were 1) to investigate the potential of LV to detect characteristic scales in a multi-scale analysis of slope gradient, and 2) to compare between cell- and object-based scaling methods. Following the work of DRĂGUŢ & BLASCHKE (2006) we applied multi-resolution segmentation in OBIA. Additionally, we simulated scale levels using the method of resampling. Finally, we plot local variance graphs as examples for scale signatures to identify scale thresholds that might correspond with characteristic spatial scale levels.

2 Material and Methods

2.1 Study Area and Input Data

The study area "Schlossalm" is located in the Gasteinertal, an alpine valley in the southern part of the province of Salzburg, Austria (Fig. 1). It is 3 × 3 km in size and characterized by mountainous terrain features such as glacial cirques, ridges, gullies and steep slopes. The elevation ranges from 1,635 to 2,578 m a.s.l. The federal government of Salzburg provided us with a VHR digital elevation model, namely a LiDAR DEM at 1 m resolution acquired during a flight campaign in 2006.

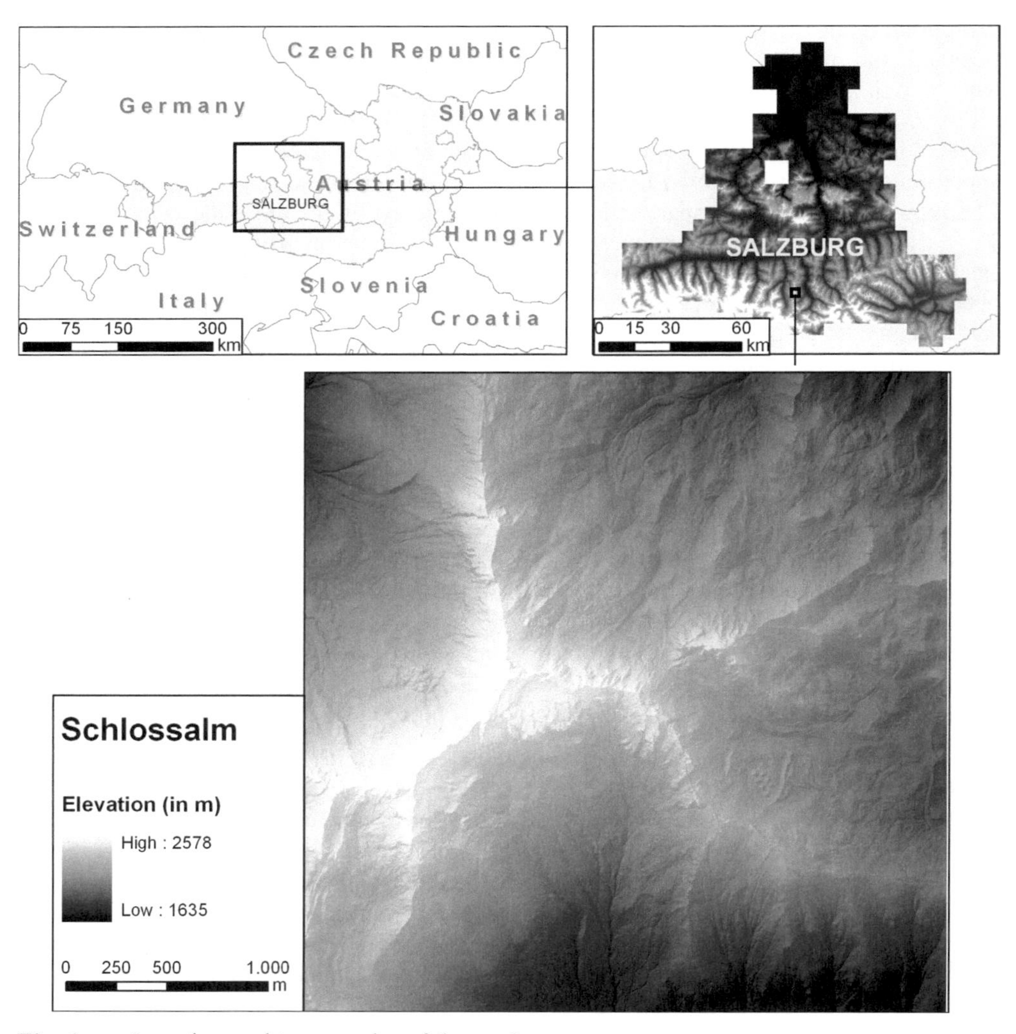

Fig. 1: Location and topography of the study area

2.2 Local Variance

Local variance is a statistical indicator that measures the spatial variation of values in a scene. It is calculated as the mean value of standard deviation in a defined neighbourhood over a scene. The higher the LV value, the higher the variation. One example for a scale signature is a graph showing how LV changes across scales. Again, we consider scale a function of grain size (mean object size and moving window size respectively) that specifies the neighbourhood size for calculating LV. More details about the method can be found in WOODCOCK & STRAHLER (1987).

2.3 Scaling

Scaling was performed in a cell- and an object-based environment (Fig. 2).

2.3.1 Cell-based

Cell-based: In previous studies traditional pixel-based techniques have been applied to simulate multiple scale levels of the same input data to evaluate the effects of scale and to find at least one characteristic scale level for analysis (BIAN & WALSH 1993, WOODCOCK & STRAHLER 1987). We used the technique of resampling. Resampling changes the proportion of a raster data set by transforming the input cell size to a user-specified cell size without altering the extent. For the transformation process several algorithms are implemented in standard GIS software. We decided on bilinear interpolation, which determines the new value of a cell based on a weighted distance average of the four nearest input cell centres. The bilinear option is useful for continuous data and causes some smoothing of the data.

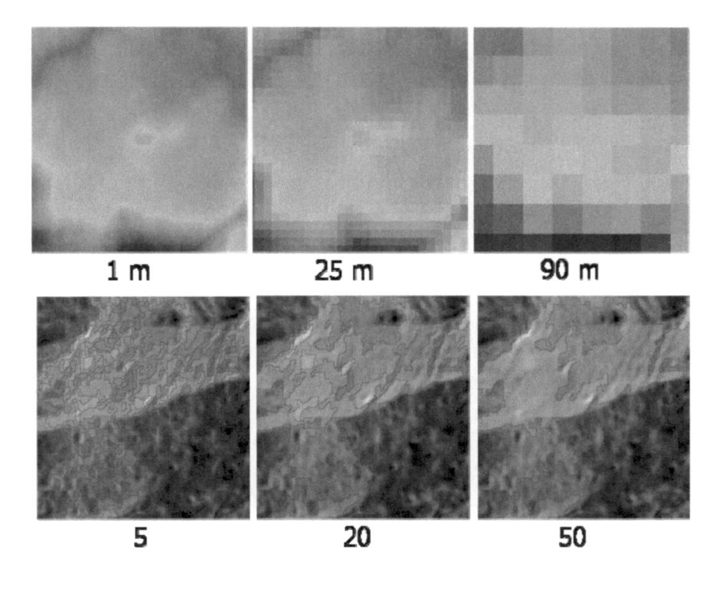

Fig. 2:
Spatial scale levels from resampling (top) and multi-resolution segmentation (OBIA, bottom). Values indicate resolution in m (top) and scale parameter, as used in OBIA (bottom).

2.3.2 Object-based

Object-based: Since several years, the object-based approach offers a powerful framework to overcome some of the cell-based limitations in multi-scale analysis of complex systems. The basic processing units in OBIA are segments, so called image objects. Through image segmentation the input layer is subdivided into regions of minimum heterogeneity based on several user-defined parameters. Thus, input layers are segmented into more 'realistic' irregular-sized objects rather than in regular-sized pixels. Heterogeneity refers to both spectral and shape properties and threshold for each must be set. The value assigned to an image object is the mean of the aggregated pixel values. The most crucial factor influencing the segmentation result is the scale parameter. Its value defines the threshold of the maximum increase in heterogeneity when two objects are merged (BENZ et al. 2004). A well-designed software package we used for OBIA is provided by Definiens AG (http://www.definiens.com). Originally, OBIA was introduced for the use with remote sensing and aerial images (BAATZ & SCHÄPE 2000, BLASCHKE & HAY 2001, LANG & LANGANKE 2006). We applied multi-resolution segmentation to generate multiple scale levels of slope gradient.

3 Implementation

In order to derive scale signatures for further scale detection, we generated a wide range of scales from the initial VHR data.

In the first step, cell-based scaling was performed applying the technique of resampling. Within multiple resampling operations the initial high resolution LiDAR DEM at 1 m was constantly generalized by an increment of 2. The process was stopped at a cell size of 49. This threshold is in line with WOODCOCK & STRAHLER (1987) who suggested at least 60 pixels at each side of a raster to get significant results for LV. For each of the 25 transformed DEMs slope gradient was derived using the 3 × 3 moving window algorithm implemented in a standard GIS. Then, LV was measured as the mean layer value of standard deviations of cells, whereas standard deviation was calculated from slope values within a 3 × 3 neighbourhood. The size of the neighbourhood for calculating slope gradient and LV values ranged from 3 m (3 × 1 m) to 147 m (3 × 49 m). Consequently, the area of the neighbourhood increased from 9 to 21,609 m^2.

In the second step, the slope gradient layer, as derived from the initial LiDAR DEM at 1 m, served as input for scaling in OBIA. Multiple scale levels were produced by increasing the scale parameter from 1 up to 170 within a multi-resolution segmentation process using an increment of 1 (DRĂGUŢ et al. 2010). We selected 170 as the upper threshold for comparison reasons, because at this level the mean object size is similar to the size of the maximum moving window in cell-based scaling. For each level LV was measured as the mean value of standard deviations of objects, whereas standard deviation was derived from slope gradient values within each object.

For both methods values of LV were plotted against scale levels. In order to make the resulting graphs comparable, we introduced a common denominator for scale: the mean area size of reference units for LV derivation, i.e. the size of the moving window and the mean size of objects, respectively. For assessing the LV dynamics from a scale level to the

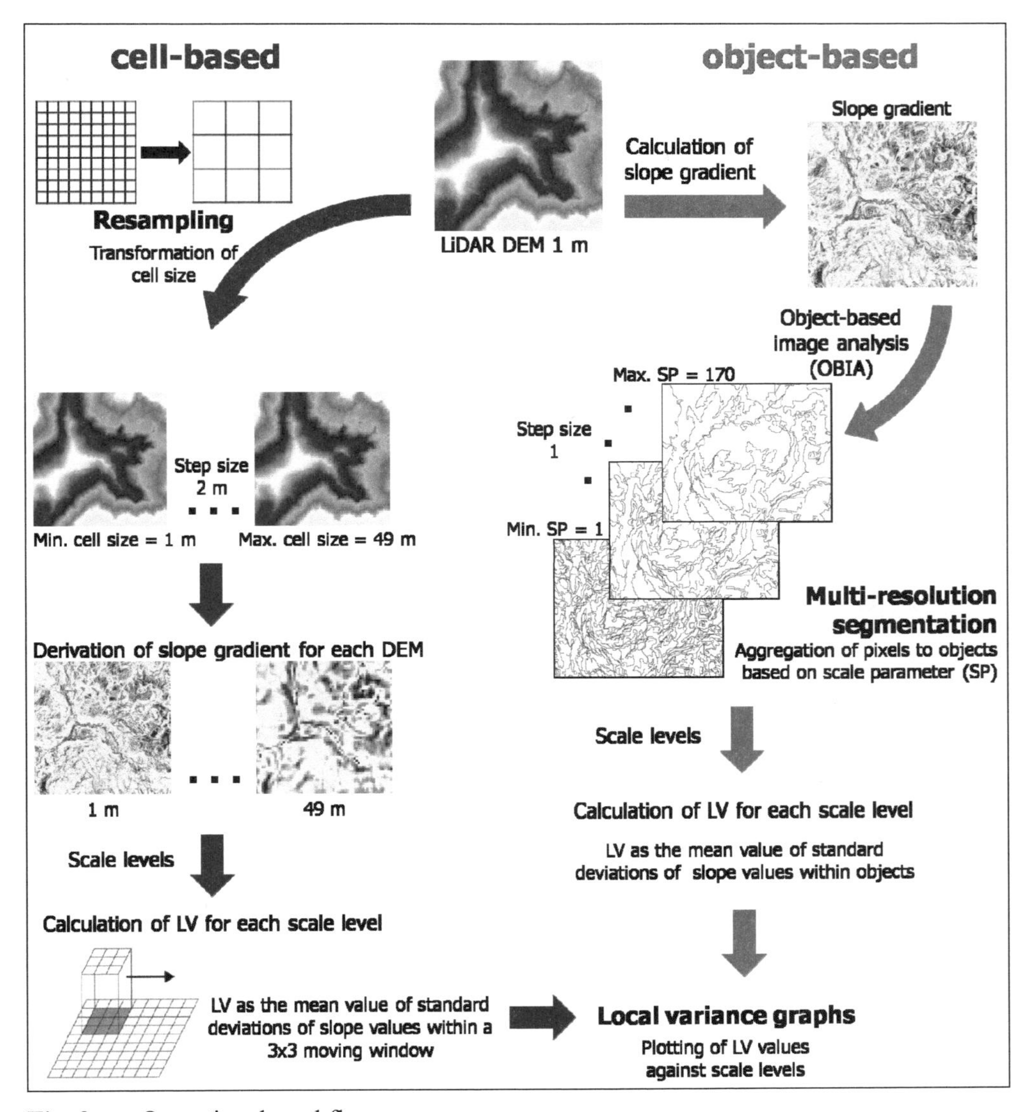

Fig. 3: Operational workflow

next higher level, we introduced a measure called rate of change of LV (ROC-LV; DRĂGUŢ et al. 2010). ROC-LV was plotted against scale levels as well. Figure 3 presents the operational workflow of our study.

4 Results

In general, LV graphs of slope gradient layers obtained from cell- and object-based scaling displayed ascendant trends with more generalized scale levels (Fig. 4). For both methods the rise in LV values from one level to another was much higher at finer scales and

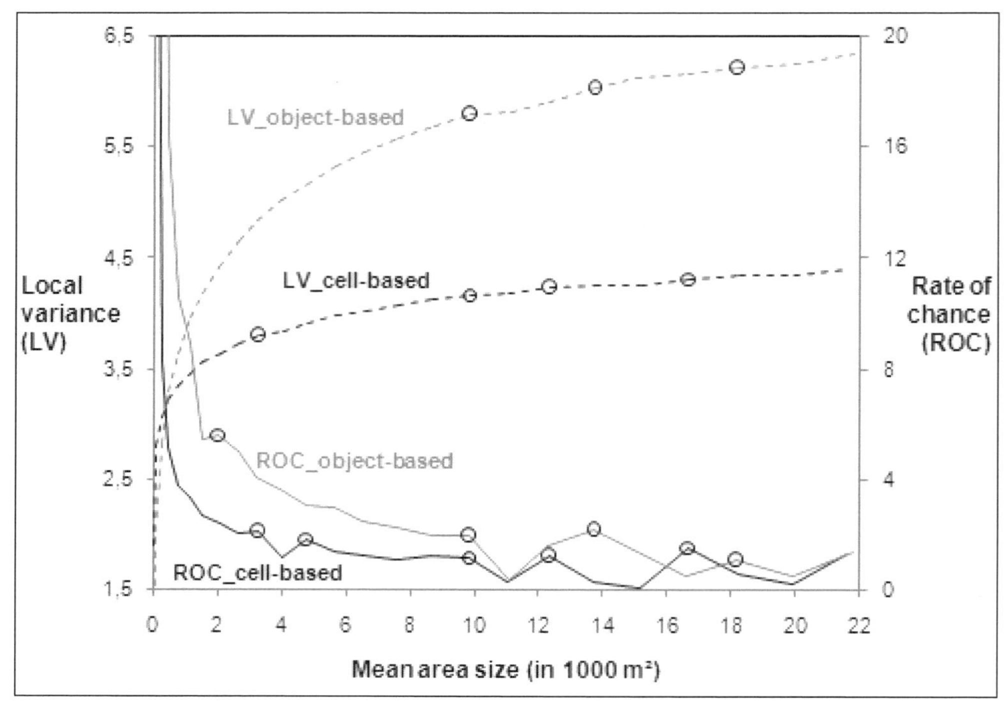

Fig. 4: LV graphs from cell- and object-based scaling and correspondent ROC-LV curves. Black circles mark observed scale thresholds

significantly lower for coarser ones. Except for finest scale levels, where LV values are nearly the same, OBIA levels of slope gradient showed higher LV than cell-based levels and this margin increased with scales. The maximum LV values at the coarsest scale (mean area size = 22,000 m²) were 6.3 and 4.3 respectively. This is one remarkable difference between the two scaling approaches.

Despite the fact that slope gradient is smoother than digital images, LV graphs exhibited some steps in the spatial scale continuum, even though they were not well pronounced and a few were not yet observable. However, in ROC-LV curves these steps were better indicated as peaks or plateaus. The measure ROC-LV was introduced to calculate the difference in LV values from one level to the next higher level, thus showing the variation across scales. In ROC-LV graphs, we were able to identify 5 thresholds for the cell-based approach at window sizes of 3.5, 4.5, 10, 12.5 and 16.5 thousand m² corresponding to resolutions of 19, 23, 33, 37 and 43 m. The resulting ROC-LV obtained from multi-resolution segmentation in OBIA indicated 4 thresholds at mean object size of 2, 10, 14 and 18 thousand m² corresponding to scale parameters of 46, 80, 108 and 151. Most of the detected peaks in ROC were represented in LV curves as well. Only two thresholds – one in the object- and one in the cell-based LV graph – could not be identified.

As we have assumed, in most cases thresholds in cell- and object-based graphs did not occur at the same scale. Only at mean area size of 10 thousand m² detected thresholds were identical, though they were not very pronounced at this level.

5 Discussion

In this experimental research we tested the potential and usability of LV for revealing the inherent scale structure of continuous terrain models for further detection of characteristic scales. The thematic focus was on ecological terrain-related analysis. In landscape ecology, breaks in statistical properties of land-surface parameters such as slope gradient across scales might reveal levels of organization in the structure of data as a consequence of the occurrence of similar sized spatial objects (DRĂGUŢ et al. 2010). We simulated scale levels using both a cell-based method (resampling) and an object-based approach (multi-resolution segmentation).

Detection of characteristic scales has become necessary due to increased integration of VHR DEMs into ecological modelling. Digital elevation data often hold too much detail for the application of interest. Hence, there is the need for more generalized scale levels, i.e. coarser resolutions in cell-based approaches and larger objects in multi-resolution segmentation. In our study we proved that scale signatures such as the suggested LV graphs indicate scale thresholds. These breaks mark characteristic spatial scales, where groups of real-world objects are more appropriately imaged than for other scale levels. Therefore, these levels could be more appropriate for analysis.

Both methods – resampling and segmentation – try to emulate real-world units by aggregating cells. While resampling operates locally on the basis of quadratic windows that do not account for spatial anisotropy of real-world units in the scaling process (SCHMIDT & ANDREW 2005), segmentation produces more realistic objects based on minimum heterogeneity. Due to the smoothing of DEMs through resampling, variation of slope gradients decreases with coarser scales, while this variation is preserved in objects, regardless their size. This difference between object-based and cell-based method eventually leads to significant differences of LV values for the coarsest scales (Fig. 4). Visual evaluation of segmentation results for the detected scale levels confirms good matching of slope values and their aggregation into objects (Fig. 5).

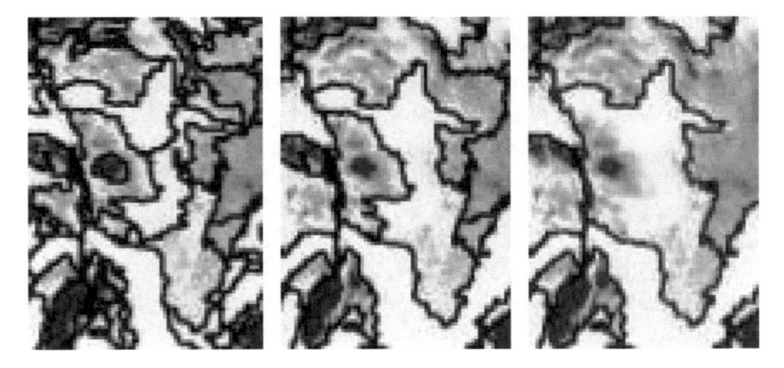

Fig. 5: Characteristic scale levels of slope gradient at scale parameters of 46, 108 and 151 (left to right) as derived from the OBIA-based LV graph.

6 Conclusion and Outlook

As suggested by LI (2008) we applied LV method on land-surface models and demonstrated the usability of LV for the detection of characteristic scales within a multi-scale analysis of slope gradient. Due to differences in the underlying aggregation techniques, scale detection works better in LV graphs obtained from OBIA scales than LV graphs obtained from resampling-based levels.

Our future work will apply the same methodology to other areas. We have already defined three additional test sites. Each of them shows different terrain characteristics ranging from flat to mountain. We will aim for comparing detected scales across terrain types. For both scaling approaches, we will increase the scale ranges, since we assume that we have not captured all the characteristic scales with the chosen range.

Acknowledgements

Work has been carried out within the frame of the research project SCALA (Scales and Hierarchies in Landform Classification, Einzelprojekt, No. P20777) funded by the Austrian Science Fund (FWF).

References

BAATZ, M. & SCHÄPE, A. (2000), Multiresolution Segmentation – an optimization approach for high quality multi-scale image segmentation. In STROBL et al. (Eds.), Angewandte Geographische Informationsverarbeitung. Wichmann, Heidelberg, pp. 12-23.

BENZ, U. C., HOFMANN, P., WILLHAUCK, G., LINGENFELDER, I. & HEYNEN, M. (2004), Multi-resolution, object-oriented fuzzy analysis of remote sensing data for GIS-ready information. ISPRS Journal of Photogrammetry and Remote Sensing, 58(3-4), pp. 239-258.

BIAN, L. & WALSH, S. J. (1993), Scale Dependencies of Vegetation and Topography in a Mountainous Environment of Montana. The Professional Geographer, 45 (1), pp. 1-11.

BLASCHKE, T. & HAY, G. (2001), Object-oriented image analysis and scale-space: theory and methods for modeling and evaluating multiscale landscape structure. International Archives of Photogrammetry and Remote Sensing, 34 (4), pp. 22-29.

CARLILE, D. W., SKALSKI, J. R., BATKER, J. E., THOMAS, J. M. & CULLINAN, V. I. (1989), Determination of ecological scale. Landscape Ecology, 2 (4), pp. 203-213.

CULLINAN, V. I., SIMMONS, M. A. & THOMAS, J. M. (1997): A Bayesian test of hierarchy theory: scaling up variability in plant cover from field to remotely sensed data. Landscape Ecology, 12 (5), pp. 273-285.

DRĂGUŢ, L. & BLASCHKE, T. (2006), Automated classification of landform elements using object-based image analysis. Geomorphology, 81 (3-4), pp. 330-344.

DRĂGUŢ, L., EISANK, C., STRASSER, T. & BLASCHKE, T. (2009a), A comparison of methods to incorporate scale in Geomorphometry. In PURVES et al. (Eds.), Proceedings Geomorphometry2009. University of Zurich, Zurich, pp. 133-139.

DRĂGUŢ, L., SCHAUPPENLEHNER, T., MUHAR, A., STROBL, J. & BLASCHKE, T. (2009b), Optimization of scale and parametrization for terrain segmentation: An application to soil-landscape modeling. Computers & Geosciences, 35 (9), pp. 1875-1883.

DRĂGUŢ, L., TIEDE, D. & LEVICK, S. (2010), ESP: a tool for estimating the scale parameter in multiresolution image segmentation of remotely sensed data. International Journal of Geographical Information Science (in press).

EVANS, I. S. (2003), Scale-specific landforms and aspects of the land surface. In EVANS et al. (Eds.), Concepts and Modelling in Geomorphology: International Perspectives. Terrapub, Tokyo, pp. 61-84.

HENGL, T. & EVANS, I. S. (2009), Mathematical and digital models of the land surface. In HENGL et al. (Eds.), Geomorphometry – Concepts, Software, Applications. Developments in Soil Science, 33. Elsevier, Amsterdam, pp. 31-63.

HILL, M. (1973), The intensity of spatial pattern in plant communities. The Journal of Ecology, 61, pp. 225-235.

LANG, S. & LANGANKE, T. (2006), Object-based mapping and object-relationship modeling for land use classes and habitats. Photogrammetrie – Fernerkundung – Geoinformation, 1, pp. 5-18.

LE COZ, M., DELCLAUX, F., GENTHON, P. & FAVREAU, G. (2009), Assessment of digital elevation model (DEM) aggregation methods for hydrological modeling: Lake Chad basin, Africa. Computers & Geosciences, 35 (8), pp. 1661-1670.

LI, Z. (2008), Multi-scale digital terrain modelling and analysis. In ZHOU et al. (Eds.), Advances in Digital Terrain Analysis. Springer, Berlin, Heidelberg, pp. 59-83.

MARCEAU, D. & HAY, G. (1999), Remote sensing contributions to the scale issue. Canadian Journal of Remote Sensing, 25 (4), pp. 357-366.

SCHMIDT, J. & ANDREW, R. (2005), Multi-scale landform characterization. Area, 37 (3), pp. 341-350.

TURNER, M., DALE, V. & GARDNER, R. (1989), Predicting across scales: Theory development and testing. Landscape Ecology, 3 (3), pp. 245-252.

WOOD, J. (2009), Geomorphometry in LandSerf. In HENGL & REUTER (Eds.), Geomorphometry – Concepts, Software, Applications. Developments in Soil Science, 33. Elsevier, Amsterdam, pp. 333-349.

WOODCOCK, C. E. & STRAHLER, A. H. (1987), The factor of scale in remote-sensing. Remote Sensing of Environment, 21 (3), pp. 311-332.

WU, J. & LI, H. (2006), Concepts of scale and scaling. In WU et al. (Eds.), Scaling and uncertainty analysis in ecology: methods and applications. Springer, Dordrechts, Netherlands, pp. 3-15.

A method for the Inverse Reconstruction of Environmental Data Applicable at the Chemical Weather Portal

Victor EPITROPOU, Kostas KARATZAS and Anastasios BASSOUKOS

The GI_Forum Program Committee accepted this paper as reviewed full paper.

Abstract

It is common practice to disseminate Chemical Weather (air quality and meteorology) forecasts to the general public, via the internet, in the form of pre-processed images which differ in format, quality and presentation, without other forms of access to the original data. As the number of on-line available Chemical Weather (CW) forecasts is increasing, there are many geographical areas that are covered by different models, and their data cannot be combined, compared, or used in any synergetic way by the end user, due to the aforementioned heterogeneity. This paper describes a series of methods for extracting and reconstructing data from heterogeneous air quality forecast images coming from different data providers, to allow for their unified harvesting, processing, transformation, storage and presentation in the Chemical Weather portal.

1 Introduction

In the field of web-based CW forecasts, air quality information is often presented to the end user in the form of pre-processed images representing pollutant concentrations over a geographically bounded region, typically in terms of maximum or average air pollution concentration values for the time scale of reference, which is usually the hour or day (KARATZAS 2005, SAN JOSÉ et al. 2008). A typical implementation can be seen in the case of the majority of data providers involved in the CW Portal, developed in the frame of COST Action ES0602 (www.chemicalweather.eu, KUKKONEN et al. 2009). These providers present their air quality forecasts almost exclusively in the form of preprocessed images with a color index scale indicating the concentration of pollutants.

Furthermore, providers of CW forecasting arbitrarily choose the resolution of their images, the color scale and color depth employed for visualizing pollution loadings, the covered region and geographical map projection for the spatial representation of the information, etc. In addition, one should take into account that each provider may apply different CW models, with different time scales, to take into account. In the end, the user is presented with a processed image synthetically showing the chemical weather forecast as produced by the used model, often superimposed with geographical markers and other spatial information. The actual mode of presentation may vary from simple web images to more elaborated AJAX, Java or Adobe Flash viewers (KARATZAS & KUKKONEN 2009).

While this approach is sufficient for the casual web site visitor/web surfer and requires minimal client and standard server side infrastructure, it has the drawback of the data being presented in a wide range of highly heterogeneous forms and interpretations: in practice, no two providers' images look alike, and there is no direct way to intercompare or combine their results. If multiple such images from different providers are presented and superimposed simultaneously on a map, the end result will be a heterogeneous collage of differently colored images, which represents a hindrance to any meaningful unified presentation or interpretation attempt. Also, some of the images are permanently altered with visible watermarks, compression artifacts, blurring, noise, symbols, text, lines etc. that would make a unified presentation even more visually unappealing.

Furthermore, the used bitmap image formats impose a finite quantization and rasterization step on the data. The variable scale, pixel/geocoordinates ratio and map projection used means that the data from different providers cannot be easily presented in a unified form, which is one of the goals of the CW portal (http://www.chemicalweather.eu/Domains). Having access to raw forecast data in a standard vector geoformat would get around many of these limitations. In practice only a few of the data providers will make available some means of access to their raw forecast data, and they are actually not even expected to do so. Although a number of initiatives have been undertaken in order to make model data (output) available for other services and users the issue of harmonized, seamless access to chemical weather forecasts remains open.

To overcome these limitations, we have developed a specialized configuration-driven image parsing system. Its job is to convert the bitmap data from different data providers into an internally unified, index-based, hybrid, point/raster-oriented representation of forecast data which can be made homogeneous when displayed, taking different scale, projections, value ranges etc. into account. This will allow for the development of a new service in the CW portal system that will allow users to combine and to compare forecasting information coming from various providers, in a harmonized way. This may be done without requiring access to the raw data, but only on the basis of information being publicly available on the Internet, thus respecting data access policies and practices

The system combines elements of screen scraping, image processing, data reconstruction and map transformations in order to produce uniform, indexed data using a unified format and geographical projection. This process is referred to as "inverse data reconstruction" in this paper, and its final goal is providing users of the Chemical Weather portal with a unified overview of data harvested daily from the various providers of model results (forecasts), in the context of the Chemical Weather portal's parsing system.

2 Screen Scraping

The first step in the data reconstruction phase is to crop the original image to a region of interest and parse it into a 2D data array directly mapped to the original image's pixels (Fig. 1, right). Image elements not carrying map/forecast data are discarded by cropping, and only the portion that actually carries information is retained, hereby called the "cropped region".

The cropped region is then directly parsed into a 2D index data array, based on a chroma key constructed from the color-based legend and the pixels of the original image. This chroma key is an RGB color/value-index pairing map, under which the colors of the original images' pixels are mapped to an integer index (0, 1, 2, 3 etc.) based on their actual RGB value and the relative order of the legend's colors. Colors are also associated with minimum/maximum value ranges of the air pollutant concentration levels presented. Special values are provided for cases such as invalid data (like data gaps due to the removed watermarks and geomarkers) and transparent regions that are easily recognizable, as they have no matching with the color palette used for the graph. The final result of the above mentioned mapping procedure is called "recovered data" (Fig. 2), which are better suited for further processing and comparison with other available data because now each pixel is mapped to a precise integer index with an associated range of real values

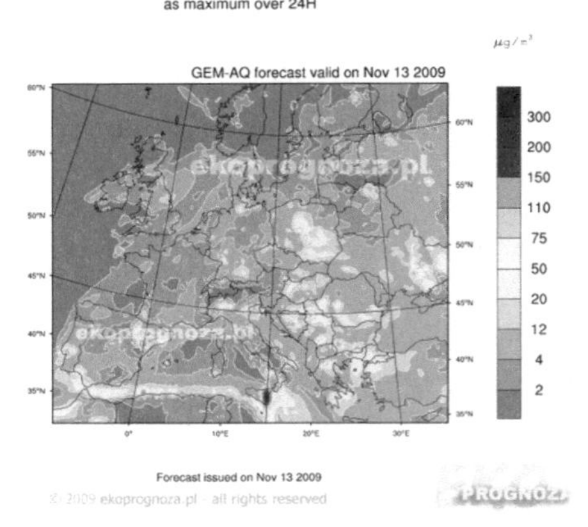

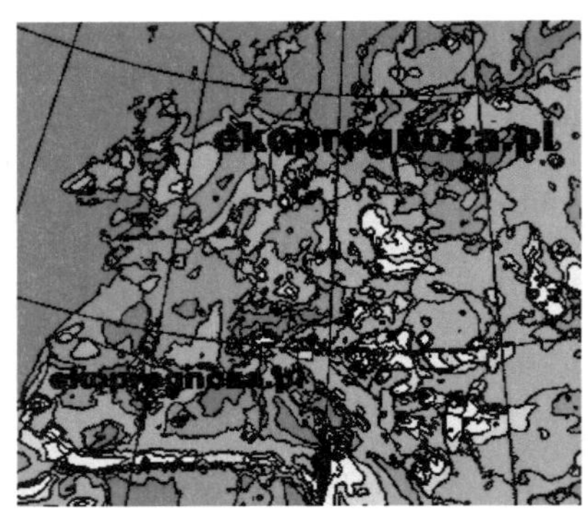

Fig. 1: A sample image taken from the publicly available EKO Prognoza chemical weather forecasting system (ekoprognoza.pl, right picture). The typical elements of air quality forecasts, such as map region, color scale, watermarks etc. are easily discernible. Recovered data after cropping and parsing with a RGB minimum Euclidian color distance criterion are presented to the left (original images in color, reproduced in B&W).

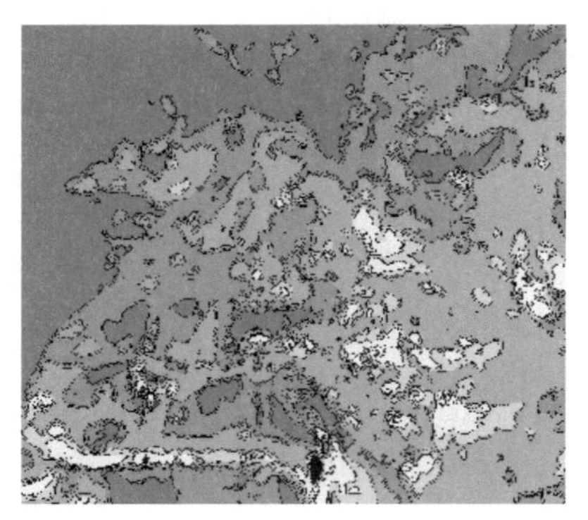

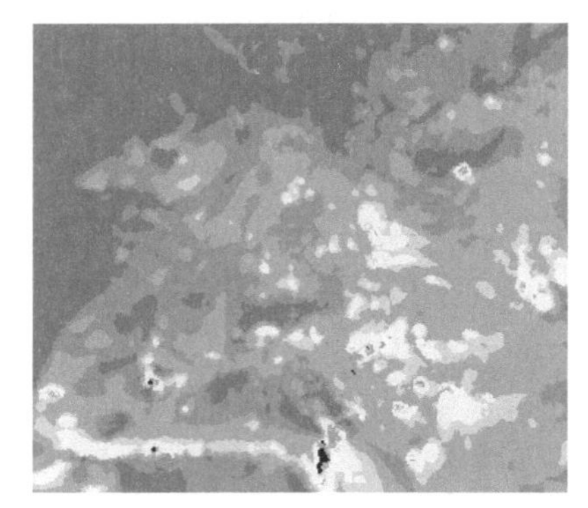

Fig. 2: Left: Result obtained after parsing figure 1, using a minimal hue error criterion. Right: Compared to the Euclidean RGB distance criterion, semi-transparent watermarks and geomarkings are eliminated almost effortlessly, resulting in less unclassifiable pixel and thus making the subsequent data interpolation less error prone (original images in color, reproduced in B&W).

It should be noted that our image parsing system allows for distinction between actually invalid data (which must be filled in with valid values) and regions which should be treated as transparent (which are never supposed to carry valid values, e.g. areas outside a forecast's intended coverage but which appear on the map's bitmap nonetheless). Transparent areas are treated similarly to valid data regions in that they are associated to a range of colors on the original image, they can be parsed, can contain data gaps themselves, and may be expanded and filled-in into these gaps. The information about where to crop, where each color on the legend lies, to which index it should correspond, what map projection the original image is in, etc. are all stored in external XML configuration files, which must be tailor-made for each data provider's image format, but are flexible enough to be easily reconfigured for different schemas.

3 Reconstructing Missing Values and Data Gaps

A problem that often arises when parsing the original images of the data providers, in order to create the indexed recovered data, is that the original images usually carry unwanted elements such as legends, text, geomarkings and watermarks, as well as regions that are not part of the forecast area. These data are similar to random impulse noise in manifestation, for they stand out from the data they corrupt, have a random location (like salt and pepper or streak noise), and their removal is required in order to proceed with further image processing. In order to display the recovered data in a complete, continuous and homogeneous form, the resulting data gaps must be somehow filled-in.

3.1 Color classification

A first approach to minimizing the amount of data classed as "invalid" is to use a color classification criterion that is appropriate for the image at hand: for example, some images use strictly the colors present in their legends, while others may display intermediate hues or exhibit color artifacts due to lossy compression or post-processing. The parsing system allows for the following adjustments for what regards color parsing:

- Palette interpolation: the original legend's color-range pairs are linearly interpolated by a constant factor, allowing for parsing more colors and values from the original image, and thus leaving less unclassified "data gaps".
- Color parsing tolerance: the maximum Euclidian distance (or a similarly suitable metric) between two similarly classified pixels' color components can be fine-tuned in order to allow for stricter or laxer classification.
- Choice of appropriate color space: while the default RGB color space works well for images with few color artifacts, other color spaces may be more effective for certain particular cases (cf. GONZALEZ 2007) such as partially transparent watermarks.

3.2 Data interpolation

The second part of the data reconstruction process consists in filling/reconstructing data gaps left over by the color classification step, which are inevitable for certain classes of images. The procedure shares some similarities with the field of image inpainting/interpolation, which aims at reconstructing missing or corrupt parts in still images and videos. Inpainting was performed manually, while algorithmic interpolation can range from simple linear interpolation to complex techniques combining Dirichlet boundary conditions, Gaussian mixtures, isophote directions, neural networks, etc. (cf. SHEN 2003, PETERSON 2002, PHANI DEEPTI et al. 2008)

Due to the specific needs of the Chemical Weather portal, a rapid and automated, unsupervised method is required for data interpolation. Also, since most data gaps are expected to be small (regions of 3×3 pixels or less) and the detail level of most forecast images is usually low, simple region expansion methods were given development priority.

3.3 Data interpolation in other settings

It's useful to consider how data reconstruction and interpolation of missing values is implemented in similar domains such as sea surface temperature (cf. PISONI et al. 2008), navigational maps, as well as remote sensing imagery (cf. MAXWELL et al. 2006). In particular, the method proposed in MAXWELL et al. 2006 was based on the assumptions that:

- There was a historical archive of older, complete images using the same projection, resolution, perspective etc. from which to extrapolate the missing data for future ones.
- The nature of the application (remote sensing data on macroscopic, long term ground features) justified the assumptions of short-medium term constancy in the data.
- The RS images themselves were pre-segmented and divided into clear-cut regions (crop fields, woodland etc.) which could be assumed to have a homogeneous visual signature and benefit from a medium-term immutability (the segments' boundaries themselves could be considered totally immutable). In addition, the regions themselves could be quite large.

Other data interpolation methods applied to datasets rather than pure imagery, like those mentioned in KAPLAN et al. (2002), PISONI et al. (2008) and STÖCKLI (2007), also rely on the assumption that data interpolation can be supported by looking up to archived, complete datasets.

3.4 The case of Chemical Weather portal

In the case of the images used in the CW portal, none of the above assumptions are valid:

- There's no historical archive to speak of. Even if some of the involved data providers do maintain one, said archive is often in the form of processed images that also need to be parsed and cleaned with the same method that led to the missing data problems.
- Unlike geographical features, environmental pollution data like air quality is ephemeral. Assumptions of constancy or periodicity can only be made within comparatively narrow time intervals (days, or even just hours).
- Data gaps appear systematically at the same positions in all available images for a given provider, unless the presentation format changes dramatically. This means that in the general case there's no complete dataset within one data provider's domain, therefore any supporting data must necessarily come from third-party data providers.

- Even if a certain degree of correlation between pollution levels and certain regions can be made (e.g. urban centers) such regions are often very limited compared to the map as a whole (usually with an area of less than 100 square pixels), unlike the ETM Landsat case where the regions were relatively extended and were assumed to exhibit property constancy. This means that reconstructed data over sufficiently large regions would be largely arbitrary, as pollution can be carried and travel long distances by the wind etc.

3.4.1 Working assumptions

Therefore, within the Chemical Weather portal, the interpolation of missing data is instead shaped around the following assumptions:

- Most data gaps appear in a systematic way on the same spots on past and future pollution forecasts, therefore there will never be any complete reference dataset (except for ensembling and cross-validation with other providers' data).
- It's reasonable to assume that in the case of thin discontinuities (e.g. 1 pixel) neighboring values can be used as a basis for filling in, with a small margin for error, something which was confirmed experimentally. This is especially true on low-concentration regions and especially in images with low color counts.
- The color-value range association is very variable, non-linear and typically coarse: in some cases a significant portion of a scale's possible values may be mapped to a single color. This might make ensembling and cross-validation of data between providers problematic, but on the other hand allows for continuity assumptions near data gaps.
- The gap-filling procedure should be holistic: any updates should be limited to single pixels/gaps (combined with an iterative method) after all of the data has been examined independently and any and all changes should be written en-bloc, in order to avoid "data dragging" phenomena, where recursive computations generate vicious feedback circles which bias successive ones. If insufficient data is available to make a reliable decision about a gap, this will be reconsidered on a future iteration, where the expansion of other data will possibly lead to a more solid decision-making.
- Due to the expected daily volume of images to process ,the image parsing and data interpolation filling methods used should be reasonably fast (possibly parallelizable), direct and work unsupervised once the configuration files are created.

3.5 Map projection correction

Another aspect of parsing heterogeneous chemical weather data from images is that they correspond to different geographic regions, and that the data might not be in a convenient common projection (like the equirectangular projection). The parsing system developed allows for transforming the data on demand (by preserving the original raster resolution) or by permanently altering and resampling the internal data to a desired resolution, using a variety of input and output map projections. The rasterized chemical weather data, derived from images, is then effectively treated as geocoordinate-based texel data, as demonstrated in Fig. 3. This allows for combining CW forecasts from various providers, and constructing synthesis maps, thus harmonizing the way that information is being accessed and presented.

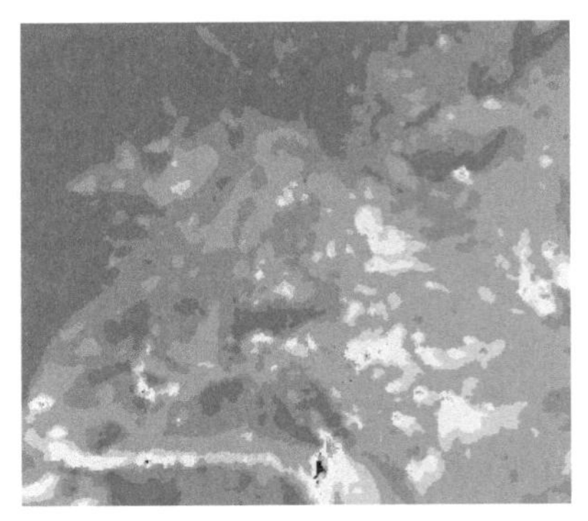

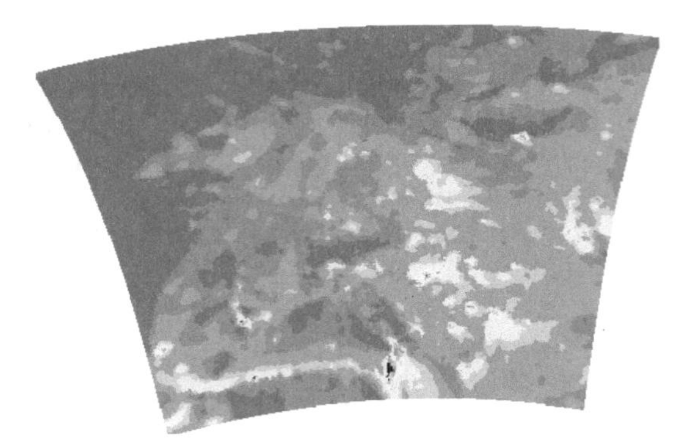

Fig. 3: Image of the EKO Prognoza system, with a conformal conical transformation (left), and its transformation into an equirectangular projection system, right (original images in color, reproduced in B&W).

4 The used data interpolation algorithm

The data interpolation algorithm used in our prototype system is designed to operate on indexed data that can be projected on a 2D grid (images, dense arrays etc.) and where 8-connectivity/movement can be defined on every element except the border elements, which have decreased connectivity and are treated accordingly.

The initial data elements are grouped in two classes: valid and invalid. The algorithm operates by iteratively reducing the set of invalid data until it is empty. This is achieved by replacing data elements classed as invalid with valid values, based on a variety of decision rules, whose common trait is that all neighbors of an invalid data element within a certain distance are considered in a voting or decision system in order to decide with which new value to replace it. The distance and metric used in our prototype system is the Tchebychev distance of 1, which is equivalent to the Moore neighborhood on a 2D grid.

Since the algorithm is iterative and multi-pass, it is expected by design that not all gaps will be filled in one pass. Avoiding doing all of the processing in a single pass reduces the "data dragging" phenomenon and "flat looks" mentioned in MAXWELL et al. (2006), and precludes that a particular boundary, e.g. the left one, will dominate the majority of a scan line, by allowing the other boundaries to expand too, in all 8 directions.

The maximum number of iterations required to interpolate an invalid data element is indicated with k and called "obscurity rank". The computational complexity of the algorithm in terms of iterations (and therefore the maximum obscurity rank k) is proportional to the maximum size of present data gaps: if all gaps have at least one neighboring pixel, then they will be filled within a single iteration (plus one for final checking), else their filling will be delayed.

The number of required iterations k in the case of images with at least one data gap is given by 2 plus maximum existing Tchebychev distance $\max(D_{Tchebychev})$ within the data gaps (formula 1), or just by $k=1$ if there are no data gaps.

$$k = 2 + \max(D_{Tchebychev}) \quad (1)$$

The voting strategies employed in the currently used versions of the algorithm are all guaranteed to terminate if there is at least one valid data element. Even if no updates are required (zero-sized gaps), at least one iteration scanning through all elements for data gaps will be required. The algorithm's coarse grained complexity is proportional to the number of data points in a given image of size $m \times n$ and the obscurity rank k just for checking for invalid data elements. The complexity of actually interpolating missing data is dependent on the number of invalid data points e, where $e < m \times n$. Assuming that checking for a missing data point has a fixed maximum cost of c, then an upper bound in complexity is given by $O(ce+mnk) < O((1+c)(mnk))$, while additional storage requirements are $O(mn)$ words, since processed data is entirely buffered at each iteration before changes are finally written. It is therefore evident that regardless of the decision rules applied, the proposed algorithms are of linear complexity, thus always capable of providing with final results, in term of the computing time being required.

5 Effectiveness of Data Interpolation

In this section, a series of experimental results obtained from applying the aforementioned algorithms is presented. The sample data consisted of reference "clean" images with pattern distributions and shapes similar to those encountered in CW forecasting and consisted of a pre-cleaned real forecast image as well as a low and a high detail fractal image, modified with opaque noise and opaque or partially transparent watermarks. All of the images used had a raster resolution of 344 per 298 pixels, for a total of 102512 pixels. The watermarks and noise were always kept fixed with only color, opacity and/or hue alterations applied to them. As a metric of the effectiveness of the various data interpolation methods, the number of erroneously classified pixels versus the total number of pixels in the reconstructed data was used, expressed as a percentage.(e_{dirty} and e_{clean}, respectively). The equivalent PSNR metrics are also included, $PSNR_{dirty}$ and $PSNR_{clean}$. Lower percentages and higher PSNR values indicate better performance.

The minimal hue criterion can be expressed as follows:

$$A_{color} = B_{color} \Leftrightarrow \left| HSV(A)_{Hue} - HSV(B)_{Hue} \right| < \varepsilon \tag{2}$$

where the colors of two pixels A and B are considered to be the same color if their hue value, taken from an RGB to HSV conversion function, is smaller than a predetermined tolerance level ε. The minimum RGB Euclidean distance between two pixels is defined in a similar manner:

$$A_{color} = B_{color} \Leftrightarrow \sqrt{(r_A - r_B)^2 + (g_A - g_B)^2 + (b_A - b_B)^2} < \varepsilon \tag{3}$$

Where $r_{A,} g_{A,} b_{A,}$ are the RGB components of pixel A, and $r_{B,} g_{B,} b_B$ are the RGB components of pixel B. The allowed ranges of values are $[0,2\pi)$ and 0-255 for RGB components, for which the tolerances ε_{Hue} and ε_{RGB} are calibrated, respectively. During the parsing procedure, pixel A is taken from the image data to parse, and pixel B is taken from a point in the color legend. The search does not stop when the first A-B pair satisfying condition (2) or (3) is found, but only when the best such pair is found (minimum possible distance).

From the results in Table 1, it is evident that the minimal hue criterion is able to deal very effectively with certain kinds of transparent watermarks which do not affect hue directly (grayscale, whites, blacks etc.), even allowing 100% data recovery in certain cases (infinite PSNR) However its performance declines sharply with increases in opaqueness, as well as with hue shifts (in the case of cyan colorized watermarks). By comparison, using the RGB Euclidean criterion (Table 2), results almost always in a systematically larger number of unclassified pixels and is largely independent of transparency for white hues, but a near constant performance.

As a test of the data interpolation limits, we applied the above methodology on the image of a colorized fractal using a limited range of valid colors, and adding some opaque noise such as lines and text watermarks. This choice was inspired from the fact that a few CW forecasts exhibit fractal-like patterns, although not as extended. In that case, with a constant 7.461% corruption percentage, recovered data exhibited an error percentage comprised between 3.6 and 3.9%, as can be seen from Table 3.

Table 1: Results for pre-cleaned forecast image, using the minimal Hue error criterion and majority rule as a gap-filling method

Corruption introduced	ε_{Hue}	e_{dirty} (%)	e_{clean}(%)	$PSNR_{dirty}$(dB)	$PSNR_{clean}$(dB)n
White watermarks, 25% opacity	0.001	0.620	0.129	22.05858	28.90201
White watermarks, 25% opacity	0.002	0.000	0.000	∞	∞
White watermarks, 50% opacity	0.001	0.426	0.076	23.68833	31.1868
White watermarks, 50% opacity	0.002	0.002	0.000	47.08285	∞
Cyan watermarks, 25% opacity	0.002	2.866	0.610	15.38757	21.84052
White watermarks, 100% opacity	0,001	2.322	1.863	16.32738	17.29741
White watermarks, 100% opacity	0,002	2.115	1.835	16.73055	17.36386

Table 2: Results for pre-cleaned forecast image, using min RGB Euclidean criterion and majority rule as a gap-filling method

Corruption introduced	ε_{Hue}	e_{dirty} (%)	e_{clean}(%)	$PSNR_{dirty}$(dB)	$PSNR_{clean}$(dB)n
White watermarks, 25% opacity	15.00	2.911	0.604	15.34498	22.18963
White watermarks, 25% opacity	10.00	2.798	0.576	15.51692	22.39578
White watermarks, 50% opacity	15.00	2.826	0.587	15.47368	22.31362
White watermarks, 50% opacity	10.00	2.872	0.598	15.40356	22.23299
Cyan watermarks, 25% opacity	15.00	2.645	0.531	15.76115	22.74905
White watermarks, 100% opacity	15.00	2.911	0.605	15.34498	22.18245
White watermarks, 100% opacity	10.00	2.946	0.616	15.34498	22.18963

Table 3: Results for a high-detail fractal image, using the minimum RGB Euclidean criterion with and majority rule as a gap-filling method

Corruption introduced	ε_{Hue}	e_{dirty} **(%)**	e_{clean}**(%)**	$PSNR_{dirty}$**(dB)**	$PSNR_{clean}$**(dB)n**
Majority rule (most common)	Exact	7.461	3.839	11.25743	14.15782
Quantized median	Exact	7.461	3.837	11.25743	14.16008
Minority rule (most rare)	Exact	7.461	3.804	11.25743	14.19759
Random neighbor (means)	Exact	7.461	3.601	11.25743	14.43577
First found, clockwise priority	Exact	7.461	3.702	11.25743	14.31564

6 Conclusion and Outlook

The prototype image parsing system exhibited good performance in parsing images originating from the full spectrum of the CW data providers, ranging from crisply colored images in equirectangular geographical projection, all the way up to fuzzy colored images in more complex projections. The data interpolation mechanisms proved effective in recovering missing data from the color parsing procedure, and even reduced the error rates in highly detailed test images with a complexity exceeding the typical one found in provider images, despite their relative algorithmic simplicity.

The combination of the above functionality will provide a mean of getting around the current highly heterogeneous state of CW forecasts, and allow a unified presentation of the available data, as well as allow the creation of new services.

The inclusion of more flexible color space criteria during parsing, as well as artificially intelligent data interpolation algorithms for the cleanup phase are currently being investigated. Future expansions to the system include ensembling (combining the data from multiple providers) and implementing the system as a service with a public interface, as well as promoting its application in other domains.

Acknowledgements

The authors greatly acknowledge COST action ES0602 (Chemical Weather COST action), as well as the Finnish Meteorological Institute for supporting their work on the CW portal.

References

COST action ES0602 (2007-2011), Towards a European Network on Chemical Weather Forecasting and Information Systems (ENCWF). – http://www.chemicalweather.eu/.

GONZALEZ, R. C. (2007), Digital Image Processing. Prentice Hall, Richard Eugene Woods.

KAPLAN, A., KUSHNIR, Y. & CANE, M. A. (2002), Reduced Space Optimal Interpolation of Historical Marine Sea Level Pressure. Journal of Climate, (13) 16, pp. 2987-3002.

KARATZAS, K. (2005), Internet-based management of Environmental simulation tasks. In FARAGO I., GEORGIEV K. & HAVASI A. (Eds.), Advances in Air Pollution Modeling for Environmental Security (NATO Reference EST.ARW980503, 406 p.) Springer, pp. 253-262.

KARATZAS, K. & KUKKONEN, J. (Eds.) (2009), COST Action ES0602: Quality of life information services towards a sustainable society for the atmospheric environment. Sofia Publishers, Thessaloniki, Greece (118 p), ISBN 978-960-6706-20-2. – http://www.chemicalweather.eu/material/COST.Action.ES0602.Quality.of.life.info.services.Karatzas.Kukkonen.eds.2009.pdf.

KUKKONEN, J., KLEIN, T., KARATZAS, K., TORSETH, K., FAHRE VIK, A., SAN JOSÉ, R., BALK, T. & SOFIEV, M. (2009), COST ES0602: Towards a European network on chemical weather forecasting and information systems, Advances in Science and Research Journal, 1, pp. 1-7. – www.adv-sci-res.net/1/1/2009/.

MAXWELL, S. K., SCHMIDT G. L. & STOREY, J. C. (2006), A multi-scale segmentation approach to filling gaps in Landsat ETM+ SL-off images. International Journal of Remote Sensing, 28 (23), pp. 5339-5356.

PETERSON, I. (2002), Filling in Blanks. Science News, 161 (19), pp. 299-300.

PISONI, E., PASTOR, F. & VOLTA, M. (2008) Artificial Neural Networks to reconstruct incomplete satellite data: application to the Mediterranean Sea Surface Temperature. Nonlinear Processes in Geophysics, (15) 1, pp. 61-70.

PHANI DEEPTI, G., BORKER, M. V. & SIVASWAM, J. (2008), Impulse Noise Removal from Color Images with Hopfield Neural Network and Improved Vector Median Filter. Proc. of the IEEE Sixth Indian Conference on Computer Vision, Graphics and Image Pro–cessing (ICVGIP 2008), pp. 17-24, 2008.

SAN JOSÉ, R., BAKLANOV, A. SOKHI, R. S., KARATZAS, K. & PÉREZ, J. L. (2008), Computational Air Quality Modeling. In Developments in Integrated Environmental Assessment, 3, Environmental Modeling, Software and Decision Support. Ed. by JAKEMAN, A. J., VOINOV, A. A., RIZZOLI, A. E. & CHEN, S. H.

SHEN, J. (2003), Inpainting and the fundamental problem of image processing. SIAMNews, 36 (5).

STÖCKLI, R. (2007), LBA-MIP driver data gap filling algorithms. Technical report, Department of Atmospheric Science, Colorado State University, 23 May 2007.

VestiGO! – More Than an Adaptable Location-Based Mobile Game

Christoph ERLACHER, Karl-Heinrich ANDERS and Simon GRÖCHENIG

The GI_Forum Program Committee accepted this paper as reviewed full paper.

Abstract

This paper illustrates the user centred design and development of the mobile Location-Based Game "VestiGO!" that represents a treasure-hunting game. Pupils, students, teachers, professors and research assistants are involved in the game development process. The target group of the game are pupils as well as adults who like to play games. Users of this game have the opportunity to define their own game location and game content that is facilitated by a web based game editor. Furthermore, this game is playable for different mobile platforms and provides a component framework consisting of modules, which are expansible and reusable for other mobile Location-Based Games and Location-Based Services too.

1 Introduction

Throughout the last years mobile Location-Based Games gained importance for the sectors economy and research. Reasons are the increasing demand of the end users for mobile applications as well as the technological progress for the client development (e.g. Smartphones and PocketPCs).

Therefore, the project "Applications on the Move" was initiated and financially supported by the Sparkling Science research programme. Sparkling Science is a funding source supported by the Austrian Federal Ministry of Science and Research to combine research and educational institutions through joint projects. The aim of the programme is to create a strong linkage between research and education with the target to offer a specific source for young academics. The Carinthia University of Applied Sciences, department of "Geoinformation" and the "Höhere Technische Bundes-Lehr- und Versuchsanstalt Villach" (HTL), department "EDV & Organisation" (EDVO) are developing together a framework for Location Based Games. Furthermore, a series of Location Based Games have been designed, implemented and analysed by the pupils from the HTL and other higher vocational schools within two years that includes business plans, use cases, the design and implementation as well as tests and analysis. Through the support of Location Based Games interest should arise in handling location based data and the students will get an overview about the topic Geoinformation Systems (GIS).

"VestiGO!", an adaptable treasure-hunting game that combines education with fun, represents such a mobile Location-Based Game that is designed and developed by pupils, students and researchers of the department "Geoinformation". In the initial phase of the game development the target groups were represented by pupils only, but through iterative game evaluation and analysis processes the target group was extended to tourists and people who are interested in geocaching or mobile games. This game is not limited to any location or obstacles in the real world. Furthermore, players of the game obtain the possibility to create their own context information. *A system is called adaptable if it provides the user with tools that make it possible to change the systems characteristics* (OPPERMANN 1994, 456).

2 Theoretical Background and Scope

This chapter illustrates relevant literature on Location-Based Service, Location-Based Gaming including some examples as well as design- and developing guidelines, which represents an important source for implementation of the mobile Location-Based Game "VestiGO!".

2.1 Mobile Location Based Services

*The term **L**ocation-**B**ased **S**ervices (LBS) is a recent concept that denotes applications integrating geographic location (i.e., spatial coordinates) with the general notion of services (*SCHILLER et al. 2004, 1*)*. This definition incorporates a wide variety of application fields such as car navigation systems, location allocation planning for companies (PREM 2009), routing planers for transportation reasons or tour planning for tourists (SORNIG 2008). This paper focuses on mobile location-based applications that represent a subset of the general LBS term. These applications *can be defined as services that integrate a mobile device's location or position with other information so as to provide added value to a user* (SCHILLER et al. 2004, 10).

LBSs are formed by some domains and technologies (see figure 1) such as **G**eographical **I**nformation **S**ystems (GIS) and spatial databases, **I**nformation **C**ommunication **T**echnologies (ICTs & New ICTs) like mobile telecommunication systems, mobile devices like Smartphones or PocketPCs, systems to determine the position as well as the global network "Internet".

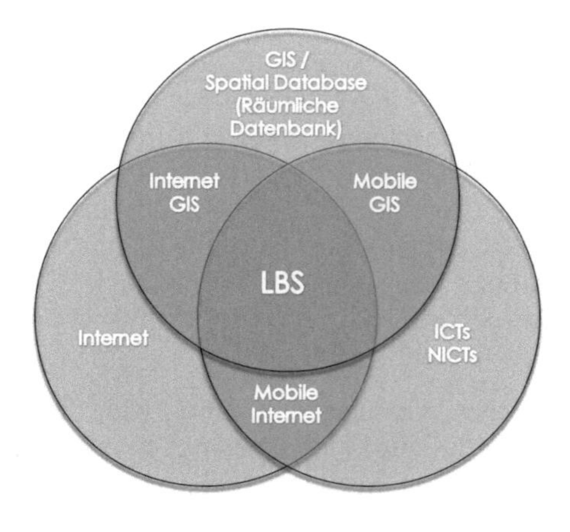

Fig. 1:
Fields & Technologies that form a LBS (Source: BRIMICOMB 2002)

Technologies to determine the position of a mobile device, a user and a group of users respectively represent the basis of a general LBS communication model (positioning layer). Such technologies are the **G**lobal **P**ositioning **S**ystem (GPS), the cellular technology, **W**ireless **L**ocal **A**rea **N**etwork (WLAN) or Bluetooth. Detailed explanations to these technologies can be found in BULLERDIEK (2006) and SORNIG (2008). This paper focuses on the GPS technology to determine the position of mobile devices. The development of a spaceborne positioning system started in the early 1970s and was directed by the U. S. Department of Defence (DOD) (HOFMANN et al. 2001). This system consists of the following segments (HOFMANN et al. 2001, 12):

- The space segment incorporates a set of satellites that broadcast signals and *have nearly circular orbits with an altitude of about 20,200 km above the earth.*
- The control segment manages and operates the positions of the satellites respectively.
- The user segment includes different kind of receivers (GPS-Mouse or internal GPS-Antenna for Smartphones or PocketPCs).

The accuracy of determining the position (x, y and z) depends on the number of satellites (at least four are required) and their position to each other, surface of the earth and obstacles such as high buildings in urban areas, weather conditions as well as on the GPS receiver. More details on GPS in general and the accuracy of determining the position can be found in HOFMANN et al. (2001).

As mentioned before, the positioning layer represents the basis of a general LBS communication model. Furthermore, a LBS communication model consists of an application layer that refers to the client application (e.g. Smartphone) and a middleware layer that can be necessary if complex algorithms or operations have to be accomplished too. The GIS component in a general LBS communication model translates the raw network information, which comes from the positioning layer, into geographical coordinate values (longitude and latitude) and processes computations or spatial analysis that can be transmitted either to the middleware layer or directly to the application layer (SCHILLER et al. 2004).

2.2 Mobile Location Based Games

Mobile LBGs refer to conventional video-games or traditional games like treasure-hunting (Geocaching), paper chasing or playing cops and robbers that incorporates the current real world position of the users. Games *that are just delivered on mobile platforms* (PSP, handhelds or Gameboys) are mobile games and not mobile LBGs (PAELKE et al. 2008, 310).

LBGs include the same segments, respectively layers like a general LBS communication model. Additionally, the components User Interface (UI), which represents and shows the context of a game, and content, which refers to the game play, are integral part of a LBG model (PAELKE et al. 2008). For both LBGs and LBSs the graphical representation (e.g. digital maps) is a central element that covers the presentation of the location, the spatial game content, map-based authoring, the GIS-based game content and the map-based position and self location techniques (PAELKE et al. 2008). PAELKE et al. (2008) found out the key subjects for the implementation of a mobile LBG as follows:

- The user interaction illustrates the content presentation and interaction that combine traditional maps (e.g. raster- or vector images) with augmented reality game styles.
- The content authoring and acquisition covers all actives in respect to the design time, which incorporates acquisition of real-world environments, the integration of virtual media and content elements and existing GIS mapping tools.
- The content representation and management refers to the run time handling of mobile LBGs like spatial queries.
- The positioning and communication segment incorporates position technologies (BULLERDIEK 2006, SORNIG 2008, PAELKE et al. 2008) to determine the position of a player. These positions, which can be sights, restaurants, cinemas, the current position of a player or any place of interest, can be seen as Points of Interest (POI). A detail explanation of such POIs can be found in SCHILLER et al. (2004).

2.3 Examples of LBGs

A huge bandwidth of developed mobile location based- and pervasive games are available for different mobile platforms (MS Windows Mobile, Android, J2ME or Mac OS mobile). Some of them are commercial games and some of them are completely free available without any payment. Furthermore, such mobile location based games differ in complexity such as multi- or single player mode, integrated client-server communication or offline/online versions.

A famous example of a mobile LBG is “Can You See Me Now?” that represents a hide and seek application. Players of this game are chased through a virtual model of a city by ‘runners’ (TANDAVANITJ et al. 2006) equipped with GPS- and WiFi technologies. Another example is “Human Pacman” that refers to the content of the arcade game “Pacman” published in the 1980s that incorporates mobile computing, WLAN, ubiquitous computing and motion tracking technologies (CHEOK et al. 2004). The characters of the traditional game (pacmen and ghosts) are represented by human players in the real world. This game was played on several locations including the Central Park of New York City. A list of games including instructions and descriptions can be found in BULLERDIEK (2006) and PAELKE et al. (2008).

2.4 Designing and Developing a Mobile Location Based Game

Game development for game pads, personal computers, mobile games or mobile location based games refer to similar steps. For a typical game design process these steps incorporate the development of the core idea, writing of a game concept, producing the artwork, programming the game engine, production of the game content, testing the game, balancing and bug fixing (MASUCH et al. 2005). Furthermore, criteria *such as players and their roles, objectives, procedures, rules and underlying game mechanics conflicts, obstacles and opponents and an outcome* are critical parts for the development of a game (MASUCH et al. 2005, 70). Another aspect refers to game developers that can be either professionals or educated persons such as students. Students, for instance, do not have such experiences and resources as professionals, but they are highly motivated.

Game development for location based mobile games implies the position of a client device (e.g. Smartphones or PocketPCs) in contrast to mobile games. Mobile LBGs utilize the player's context that can be derived from a GPS sensor (PAELKE et al. 2008). The fundamental aspects of a mobile LBG are the size and duration of the game, the infra-structure available in the game location and the role that place has within the game play (REID 2008). Additionally, mobile LBGs incorporate further components, e.g. positioning (see chapter 2.1), connectivity (e.g. GPRS, UMTS), UI, spatial interaction, distributed infrastructure and custom game-engines (PAELKE et al. 2008). The UI design for mobile- and for desktop applications differs primarily in respect to the size, limited resolution and range of available colours of the graphical display for cellular phones, Smartphones or PocketPCs. The resolution for mobile handhelds ranges from 100*80 pixels for mobile phones and 480*800 pixels for Smartphones (e.g. HTC HD2). The information of the position that can be derived for example from a GPS sensor represents the input for spatial interaction. The proximity, the direction or the path to POIs can be seen as spatial interaction (PAELKE et al. 2008). Spatial interaction tools like a proximity sensor are very helpful in respect to spatial- and location awareness respectively. Last but not least the possibility of single- and multi player games influences the development, because of the implementation of web services to allow the communication between components (PAELKE et al. 2008).

3 From the first Vision to the mobile LBG "VestiGO!"

3.1 The Original Idea

The original idea of this game named "*Goldrush*" referred to a virtual paper chasing mobile LBG for pupils, in order to combine fun with education. The idea of the game was elaborated by pupils, teachers and scientists of the Carinthia University of Applied Sciences, Department of Geoinformation. This game was designed as treasure hunt to solve puzzles and tasks. The puzzles represented hidden references to POIs and the tasks incorporated historical questions or questions to a specific topic (e.g. natural trails). Equipped with mobile phones several groups of pupils (two to four players per group) tried to find all given POIs, in order to obtain many credits. For each group a set of navigation tools was provided (e.g. a digital map – see figure 3 left and a digital compass – see figure 4 centre) to identify the POIs in less time, but the usage of these support helps decreased the credits for founded POIs. At the end of the game, which was represented through the last POI, the pupils retrieved the treasure.

3.2 The Current Version

The goal of entertainment and pleasure is an important factor for a successful game or mobile LBG (PAELKE et al. 2008). Therefore, an iterative refinement of the game content, the visual representation of the content (e.g. digital map), and the game location as well as mobile environment are necessary to meet the player's needs. The development of the mobile LBG "VestiGO!", which is the new game name, incorporates a user-centred design process and prototype implementation to collaborate between developers, designers and end users (players) (BULLERDIEK 2006, PAELKE et al. 2008). The first series of prototypes were designed for the city of Villach and for the Villach Technology Park in Carinthia. However, there were interested users who wanted to play the game "VestiGO!" in other regions than mentioned above. Therefore, the concept model of the mobile LBG was redesigned to play "VestiGO!" almost everywhere. Furthermore, the new concept model incorporates authoring tools that facilitates the generation of new game versions of the mobile LBG "VestiGO!". PAELKE et al. (2008) highlights some authoring tools for mobile LBG and describes the "Caerus" tool that allows users to add further digital maps and POIs. A similar approach was used for the "VestiGO!" game creation. In order to enable a free available authoring tool for the general public, a web GIS application (Editor) was first integrated into the new conceptual model and later implemented. This procedure requires an interface between the web GIS- and client application. Such interfaces are commonly realized with the XML standard for configuration reasons or for a service oriented architecture to exchange information between the middleware and the mobile client via web services (SANCHEZ-NIELSEN et al. 2006). The interface definition between the mobile client and the web GIS application incorporates the folder structure including the digital map and the pictures of the POIs as well as the XML configuration, which stores the content of the game play. The implementation of these measurements will allow players of the mobile LGB to generate their own games for almost each location.

3.3 The Program Flow

The client application of the mobile LBG "VestiGO!" is modularly designed, which implies the advantage of fast source code adaptation and adding further components to the system. Figure 2 displays the program flow of the game that illustrates the conceptual model too. This concept framework is independent in respect to the any mobile platform (e.g. Windows Mobile, Android or Mac OS X). Each process in the flow diagram (see figure 2) represents a module that consists of several sub-modules and sub-processes. First, the client application verifies the accurateness of the folder structures including the files and the content of the game, which is documented within the XML configuration. Afterwards, the content of the game is interpreted, prepared and stored in the **R**andom **A**ccess **M**emory (RAM) for a fast interaction when playing the game and in a database for regarding the system stability. The next process incorporates the GPS device detection and connection, in order to receive the satellite information. The processes "Position Locator Activator" and "Tracker Activator" try to start the threads for determining the current position (Latitude and Longitude of the WGS84 – World Geodetic System – coordinate system) of the client and track recording. After starting the threads the user could start the POI search. For the first POI the user gets a message (e.g. in form of a spoonerism) that incorporates the task for the first POI search. Then, they have the possibility to use supportive tools such as a digital map, which shows the current position of the player and the destination (POI). The usage of the supportive gadgets is possible through the entire game, but decreases the credits for founded POIs. The POI search uses trigger elements that describe a spatial event such as geofencing, which can be *a circular geofence including centre coordinates, a radius and the actual trigger like entering, leaving or residing inside that geofence for a certain amount of time* (MARTINS et al. 2009, Article No.: 38 p. 3). In this case the trigger is activated once the user enters the buffer zone (circle) around the POI and a question appears on the mobile screen. The player has to answer the question, in order to retrieve the next task. At the end, when all POIs are found, the module "Game Statistic Viewer" is presented to the player, which incorporates the time for each POI, the correct answers and the total amount of credits.

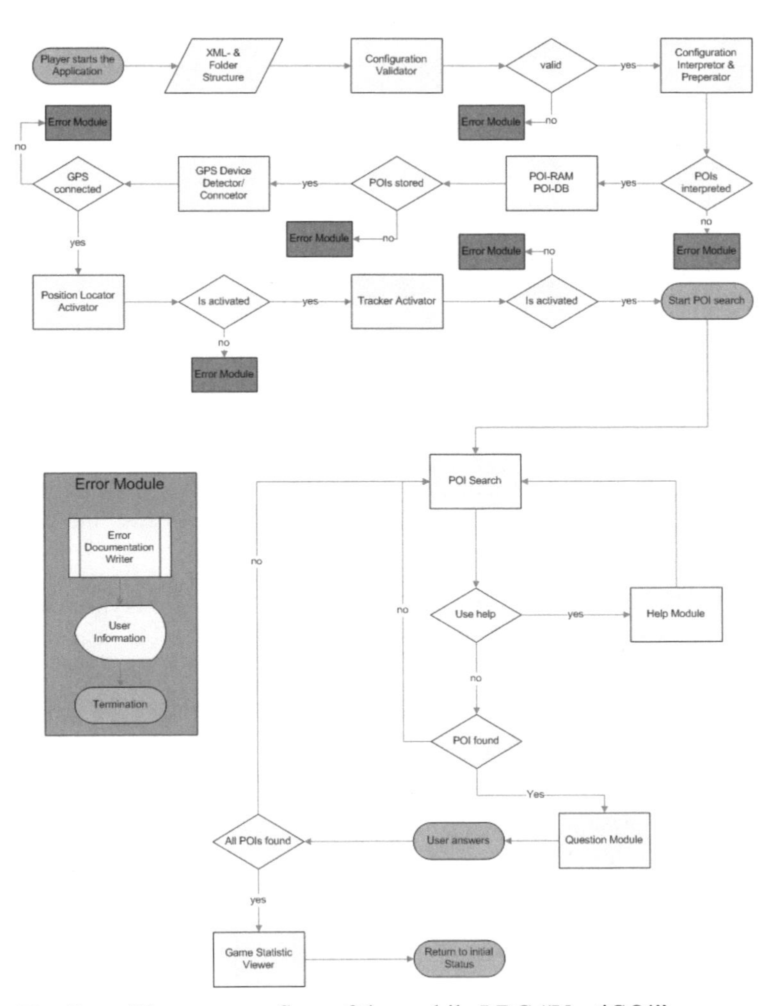

Fig. 2: The program flow of the mobile LBG "VestiGO!"

4 Feedback and Upcoming Versions

The game will be developed for different mobile platforms through the development of a series of prototypes (e.g. Windows Mobile, Android or iPhone OS). The first two prototypes for the city of Villach and the Villach Technology Park are already implemented, tested and evaluated. GPS connection, user interaction during the game, user comments and suggestions for improvements were documented. Problems concerning the GPS connection during the game could occur along narrow streets and high buildings. Therefore, users have the possibility to define the destination radius to the POIs for each game in order to deal with the GPS imprecision. Pupils, students and professors of the University of Applied Sciences were involved to evaluate and test "VestiGO!". The outcome of this evaluation was very positive as pupils, students and professors were highly enthusiastic by the game. Especially games with a historical- and cultural content (e. g. historic city centre of Villach,

Fig. 3: The left image displays a support module of the mobile LBG "VestiGO!" that represents a digital map of the game location. The right image shows the web based game editor.

see figure 4 right) as well as natural trails raised enormous interest. These prototypes were developed for the MS Windows Mobile 5 platform and tested with the HTC PocketPC "P3300 Premium" (see figure 3 left that represents a digital map).

The second series of prototypes incorporates a mobile LBG "VestiGO!" version, which can be used for almost each location. Furthermore, this series provides an authoring tool, packed within a web GIS application that allows players to generate their own game context for any area (see figure 3 right). The new prototypes are ported at least on the mobile platforms "Windows Mobile 6.5" (e.g. Smartphone HTC HD2) and "Android 2.0" (e.g. Smartphone HTC Hero). The aim is to provide the first prototype for the Windows mobile platform by the end of March 2009 and the first prototype for the Android platform by the end of June 2009. Another valuable output of this project is the component framework (see figure 2) that consists of modules for verification-, configuration- and interpretation tools, GPS detection and connection, position determination, tracking and POI search. These modules are reusable and extendable, which can be used for other LBSs and LBGs.

5 Conclusion and Outlook

This paper shows the development of the mobile LBG "VestiGO!" that can be played in almost each location and provides an authoring tool for creating user specific game content. Furthermore, component framework of this game is portable for any mobile platform and provides useful extendable modules for other LBGs as well as LBSs in the fields of tourism or sport events. The project "Applications one the Move" is a two year funded project. Now, after one year we can state that we are within our proposed time schedule. The basic modules of our framework described in chapter 3 are implemented for the Windows Mobile platform and used in our application VestiGO!. Currently we are adopting the modules to the Android platform. But still we have some modules on our roadmap to be finished by the end of this year (at least for one mobile platform). These missing modules are described briefly in the following paragraphs.

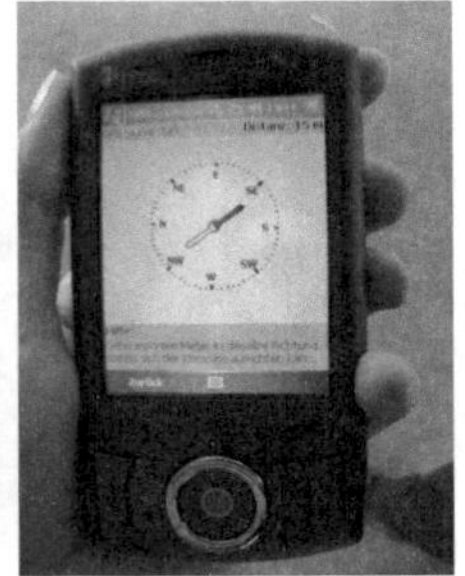

Fig. 4: The left image represents a mountain chain in Carinthia and virtual objects representing the peaks of the mountains "Goldeck" (large virtual object that is closer) and "Oisternig" (small virtual object that is far away) through a digital camera. The image placed in the centre displays the compass help functionality and the right image shows a historical map of Villach that is used as help too.

It is scheduled that the third series of building prototypes will start at the end of this year. This series will provide an offline and online version of the mobile LBG "VestiGO!" as single- and multi player mode. In the online version it should be possible to trace the POI search with the help of a web GIS application, which can be useful for sport events. Furthermore, an online version of this game provides the possibility to use different digital map representations (e.g. Google Maps or MS Bing Maps) via a browser. For a multi-user mode a refinement of the conceptual design has to be done. PAELKE et al. (2008) illustrates example architecture of a custom mobile LBG, which uses standard components and incorporates a client server architecture that communicates over the HyperText Transfer Protocol (HTTP). Additionally, a further support tool can be integrated within the component framework that could view a virtual path from the current position of a player to a specific POI through the digital camera of a Smartphone. Figure 4 displays a mobile Android application that adds virtual objects representing the peaks of two mountains to the digital camera perspective.

Furthermore, the integration of an advanced information module for POIs can be implemented as hypertext, speech output, images or movies. Another additional help module can be a spatial decision support tool that computes the nearest POI, the shortest path to the POI or the best POI in order to the distances and credits. SORNIG (2008) describes a mobile spatial decision support tool, which indicates the best order of POIs in respect to the criteria such as culture, events, parks or weather conditions. All modules mentioned before can be integrated into the current component framework of the mobile LBG "VestiGO!" and are reusable for other mobile LBGs or LBSs.

Acknowledgment – Many thanks to the Austrian Federal Ministry of Science and Research for the financial support through the Sparkling Science research program.

References

BENFORD, S. A., CRABTREE, M., FLINTHAM, A., DROZD, R., ANASTASI, M., PAXTON, N., TANDAVANITJ, M., ADAMS, M. & ROW-FARR, J. (2006), Can you see me now? ACM Transactions on Computer-Human Interaction, 13 (1), pp. 100-133.

BRIMICOMBE, A. J. (2002), GIS – Where are the frontiers now? GIS 2002, Bahrain, pp. 33-45.

BULLERDIEK, S. (2006), Design und Evaluation von Pervasive Games. Unpublished Diploma Thesis, Institut für Multimediale und Interaktive Systeme, Universität zu Lübeck, Lübeck.

CHEOK, A. C., GOH, K. H., LIU, W., FARBIZ, F., FONG, S. W., TEO, S. L., LI, Y. & YANG, X. (2004), Human Pacman: a mobile, wide-area entertainment system based on physical, social, and ubiquitous computing. Personal and Ubiquitous Computing, 8 (2), pp. 71-81.

HOFMANN-WELLENDORF, B., LICHTENEGGER, H. & COLLINS J. (2001), GPS Theory and Practice. Fitfth edition. Springer-Verlag, Vienna.

MARTENS, J. & BARETH, U. (2009), A declarative approach to a user-centric markup language for location-based services. International Conference On Mobile Technology, Applications, And Systems, Proceedings of the 6th International Conference on Mobile Technology, Application & Systems, Nice, France.

MASUCH, M. & RUEGER, M. (2005), Challenges in collaborative game design developing learning environments for creating games. In: Third International Conference on Creating, Connecting and Collaborating through Computing, IEEE Computer Society Washington, DC, USA, 67- 74

OPPERMAN, R. (1994), Adaptively supported adaptability. International Journal of Human Computer Studies, 40 (3), pp. 455-472.

PAELKE, V., OPPERMANN, L. & REIMANN, C. (2008), in MENG, L., ZIPF, A. & WINTER, S. (Eds.), Map-based Mobile Services Design, Interaction and Usability. Springer-Verlag, Berlin/Heidelberg, pp. 310-333.

PREM, R. (2009), Identifikation optimaler Standorte für Sonnentor-Franchisepartner in der Region Wien mit Hilfe des Analytischen Hierarchieprozesses (AHP). Unpublished Diploma Thesis, Institut für Marketing und Innovation, University of Natural Resources and Applied Life Sciences, Vienna.

REID, J. (2008), Design for coincidence: incorporating real-world artefacts in location-based games, in: Proceedings of ACM International Conference on Digital Interactive Media in Entertainment and Arts. DIMEA 2008, Athens, Greece.

SANCHEZ-NIELSEN, E., MARTIN-RUIZ, S. & RODRIGUEZ-PEDRIANES, J. (2006), An Open and Dynamical Service Oriented Architecture for Supporting Mobile Services. International Conference On Web Engineering, Proceedings of the 6th International Conference on Web Engineering, Palo Alto, California, USA, 263, pp. 121-128.

SCHILLER, J. & VOISARD, A. (2004), Location-Based Services. Elsevier, San Francisco.

SORNIG, J. (2008), Development of a distributed Service Framework for Location-based Decision Support. Unpublished Diploma Thesis, School of Geoinformation, Carinthia University of Applied Sciences, Villach.

Challenges in Planetary Mapping – Application of Data Models and Geo-Information-Systems for Planetary Mapping

Stephan VAN GASSELT and Andrea NASS

The GI_Forum Program Committee accepted this paper as reviewed full paper.

Abstract

Mapping of solid surfaces in the context of geoscientific exploration programs nowadays requires the use of high-performance GIS environments to cope with the amount of data, different geographic and projected reference systems, their relationships and the needs for data analysis, data organization and exchangeability. While for the Earth projects are mostly localized and targeted on specific questions and needs, geoscientific mapping of planetary surfaces often focuses on much larger, partly even on global scales. The large amount of new data requires integrated systems to be able to handle different sensor data (physical data or their representation) as well as a variety of body references. There is a growing need for exchangeability of mapping efforts and their integration in a consistent and homogeneous way. We here report on the design of a geodatabase model as part of larger-scale system composed of applications and database elements supporting a hybrid approach for conducting geological as well as geomorphologic mapping of planetary surfaces and aiding the process in maintaining and organizing data and map products.

1 Background and Aim

Planetary geologic mapping has evolved tremendously since first planetary fly-by missions and orbiting spacecraft provided detailed data of planetary surfaces in the 1960s, beginning with a variety of mission concepts including fly-bys, impactor and robotic lander/rover missions to the Moon and culminating in the US-American manned Apollo program and the sample returns from the Soviet Luna mission in the 1960s and early 1970s. Few years after that, large-scale mission programs for mapping terrestrial planets started, delivering a wealth of data of new worlds. The data basis has been growing considerably since the late 1990s with several new missions to Mars, to the Moon, and to Mercury that remained unvisited for over 30 years, and, especially, to the outer icy satellites orbiting Saturn and Jupiter.

In the context of such programs, mapping efforts, in particular geologic mapping projects, conducted by the United States Geological Survey (USGS) and under financial support by NASA have provided numerous thematic maps of the terrestrial planets. Most of these products were created in the 1970s and 1980s with several updates at the end of the millennium and complete re-mapping of, e.g., Mars, as a response to high-resolution data that became available recently (TANAKA et al. 2005).

Despite such large-scale attempts and achievements in planetary mapping there are no firm guidelines or official requirement catalogs targeted on mapping conduct so that the research community is working on such issues on a best-knowledge basis with some outlines summarized by the USGS (NASA/USGS 2009). Substantial guidelines are, however, difficult to assemble and require community-driven efforts and many years of round-table discussions. There is a need, however, as ongoing efforts in, e.g., Europe show as documented in active EU projects (European Planetary Network, EU FP-7) and new proposals. In the context of ongoing missions and in response to the wealth of new data, commercial and open-source GI systems are used nowadays to account for the growing requirements. User-definable properties of GIS-based projects allow expanding the usage of such systems to other planetary objects (RANA & SHARMA 2006).

The aim of this contribution is to present a GIS-integrated and exchangeable geodatabase model which assists planetary mappers to

- inquire about the recent status and coverage of geologic and geomorphologic mapping for any planetary object (beyond Earth) with meta-information on maps, reference systems and map-projection data,
- perform data-coverage queries on observations by all major imaging instruments as represented by sensor-footprint data and driven by a back-end data model that contains homogenized metadata information and attributes for all imaging detectors and imaging spectrometers as well as derived products such as digital terrain models,
- perform geologic and geomorphologic mapping (for any terrestrial object) using predefined (but customizable and "targetable") relationships and domain constraints as well as symbologies (NASS 2010) defined by genetic context,
- perform data analyses using integrated tools and workflows; some of which are known from terrestrial data analyses, some of which are specific to planetary sciences (e.g., crater-size frequency distributions for age determination).
- organize, i.e., add and modify, data layers and data relationships, import and export mapping data using XML metadata interchange (XMI) formats (OMG 2007),
- visualize mapping efforts and results of data analysis.

The geodatabase model was setup within ESRI's ArcGIS environment and is controlled by standard relationships as well as topology and specific domain constraints (ESRI 2010).

2 Requirements

Performing geologic and/or geomorphologic mapping on planetary surfaces within a GIS environment requires some tasks to be performed before the actual mapping process can be conducted. Such tasks deal with (a) setting up a working environment by querying and defining raster data (as represented by sensor footprints) from a variety of planetary missions to be processed and included, (b) importing auxiliary and ancillary data (e.g., organizational information, instrument kernel data, details on specific observations, additional map data), (c) defining projection parameters for map layers (and raster/vector datasets/feature classes), (d) ingesting processed data, and (e) defining different sorts of vector shape geometries for mapping in terms of feature type, representation and attributes in a consistent way.

As each mapper uses different techniques to perform such tasks, subsequent work dealing with combining such heterogeneous map data for further studies and exploitation is time-consuming and often limited due to data and interpretation ambiguities. In larger-scale projects several people work on different planets and areas, so that integration and homogenization of map data becomes important to allow continuation of work within a project's frame. Proper maintenance of such data is paramount, if mapping projects do not only deal with delineating map units, but if additional attributes and pieces of information on, e.g., chronostratigraphic relationships, needs to be related to map units.

Data are stored within a geodatabase context which allows not only to keep and maintain attribute data and meta-information but also to query and relate these pieces of information within a spatial context for locating and managing data. Such geodatabase models are commonly developed and applied in terrestrial applications and form the backbone for any geo-information-system (e.g., ZEILER 1999, ARCTUR & ZEILER 2004).

In order to overcome problems concerning map homogenization, [a] a mapper can either stick to a certain definition, i.e., internal guidelines, that have to be implemented individually, or [b] the mapper makes use of pre-defined templates and hard-wired definitions allowing to import mapping representation and styles as well as a backbone database model to start working by making use of the provided infrastructure. The latter approach is frequently used for larger-scale mapping and analysis purposes in terrestrial environments and it makes full use of the data infrastructure modern GIS provide.

An additional need to provide such an environment is the wealth and variety of different-resolution and different-type data for the terrestrial planets in the course of recent missions which has led to reconsidering mapping approaches in terms of map-unit definitions and data to be utilized and integrated into a mapping project. In order to be applicable in a wide context and to allow, e.g., geologic as well as geomorphologic mapping in a combined form, any system must cope with such requirements and must be expandable by the user. Many maps frequently on display deal with a hybrid approach of mapping geological as well as geomorphologic units in order to depict not only lithologies and time-stratigraphic units and their relationships but also to highlight geomorphologic characteristics related to a unit's shape and its general appearance. Such needs are challenging for the definition of a cartographic map product and employed symbol sets and are also a challenge for the software system in which such requirements need to be addressed, handled and managed in a consistent way. Consistency is therefore a driving requirement in order to be able to allow comparability of map products provided by different mappers and generated at [a] different scales and [b] at different locations of a planetary surface with [c] different mapping foci. A datamodel infrastructure must therefore support the mapper in his/her task to conduct mapping and to be able to expand a system, but it also needs to support the administrator in homogenizing/combining different mapping efforts.

In order to support the mapper in his work, the database model should, e.g., contain as much information about prior mapping efforts and data coverage, i.e., extent and meta-information, in order to query, relate and efficiently analyze data. It should furthermore provide a high-level of standardized representations and mapping symbologies as defined by, e.g., FGDC (FGDC 2000) or any other local authority/organization in order to facilitate the mapping process and avoid inconsistencies. Such sections should be based upon pre-selecting genetic terrain types and selecting appropriate unit representations. It should furthermore support the selection of stratigraphic schemes and definitions by providing planet-specific and domain-driven attributes. It should provide a high-level of topological constraints in order to avoid errors/inconsistencies which become visible during a homogenization process. Such basic requirements are database-driven while other requirements and add-on tools need to be defined on the overall datamodel level and which deal with handling and maintaining metadata information. On the last basis-requirement level, the mapper should not be confronted with the database model upfront but the mapper should always be able to expand the functionality to specific needs without introducing integrity problems.

The overall design of the geodatabase model is driven by requirements summarized as use-case diagram (figure 1). In this model, mapper and researcher community interact with data processing and database administration level through data and information exchange. Both levels make use of a common query component to update their specific requirements and tasks. In practice, all three actor levels are commonly combined within a small group so that the seamless and easy exchange of a data model handling such tasks is paramount to successful mapping efforts.

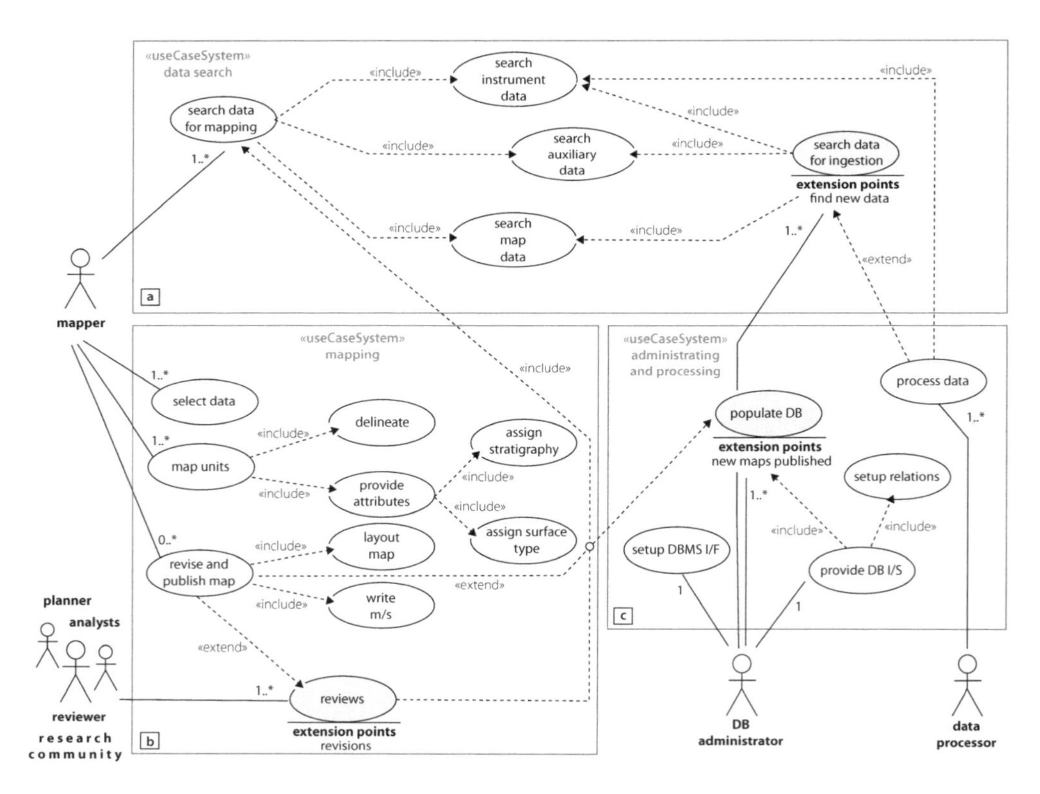

Fig. 1: UML use-case diagram depicting requirements and needs for planetary mapping at different actors' levels. Data processor and DB/DBMS manager (right) interact with mappers and mapper/science community (left). Data queries are common tasks conducted by both actor sides.

3 Conceptual Design and Integration Aspects

The conceptual geodatabase design involves the design of the main object and data layers and consists of objects, object types, their relationships and additionally the formulation of integrity conditions on a level which is in principle independent of the exact implementation and its environment. The conceptual design has been assembled using ESRI's ArcGIS File Geodatabase (FGDB) environment but it can, in principle, be exported to any other DBMS capable of handling spatial data.

The FGDB concept is helpful and allows users to exchange their data model and contents via GIS application easily so that there is no initial effort to setup and configure a DBMS fur multi-user access. On the other hand, FGDBs are known to lead to inconsistencies quickly when the system is instable and data recovery becomes a major problem. Furthermore, the centralized storage and access of many gigabytes or even terabytes of raster datasets cannot be handled efficiently within the FGDC concept so that external DBMS are a much better choice. Finally, when it comes to querying spatial data, the user of a FGDC-based data model is limited to the menu-driven approach of selecting data by locations or attributes. A query for image-data coverage across multiple feature classes that intersect a mapping area leads to many mouse-clicks, temporary table exports and re-querying to ex-

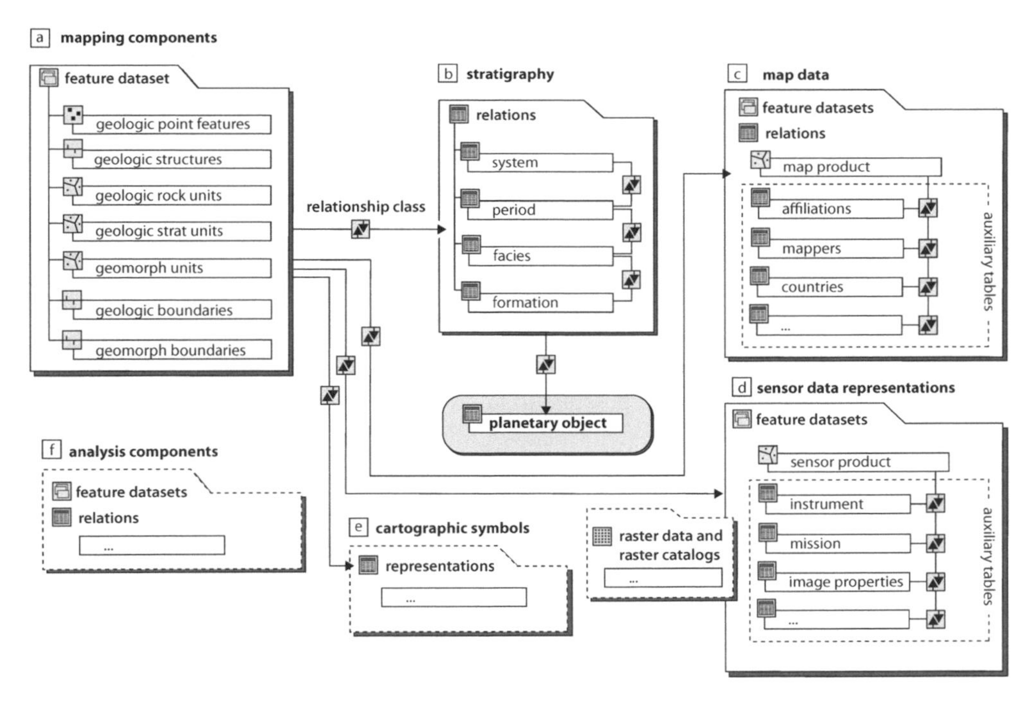

Fig. 2: Simplified ER diagram showing relations and major components of a planetary data model for mapping. Individual components labeled [a-d] are discussed in the main text. Additional elements, such as analysis components [f] can be included via additional feature classes and datasets. Cartographic elements dealing with symbolization are dealt with in detail by NASS 2010.

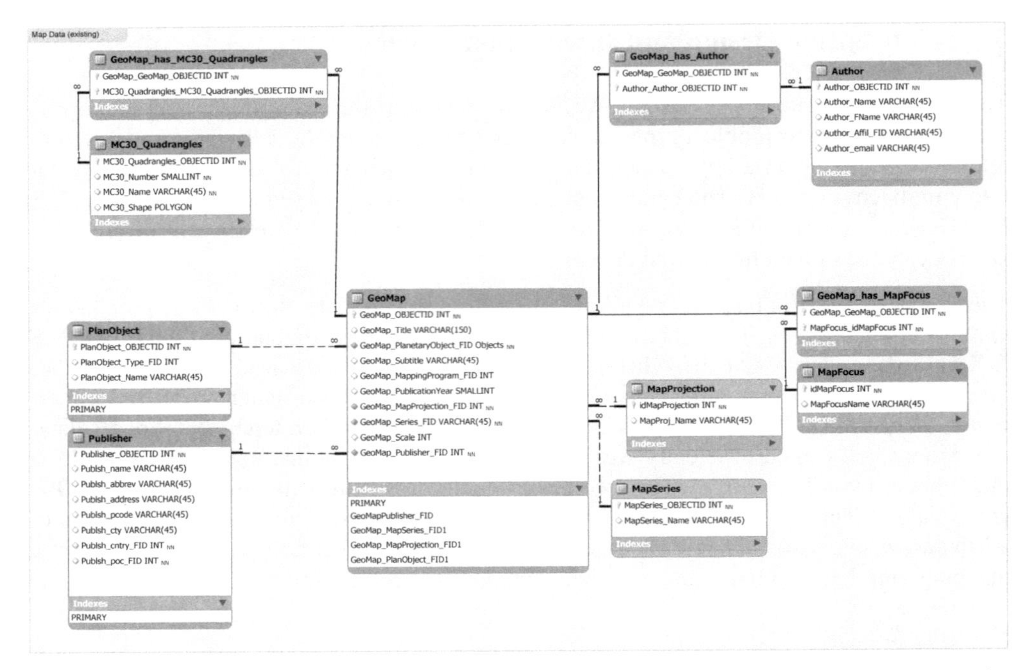

Fig. 3: An example ER-diagram showing relations concerning the query-process for existing map data and data products. It forms a small part of the data-search component depicted in figure 1.

tract (project) the required attributes. This could easily be overcome by making use of the spatial extensions of any external DBMS by simply formulating (and storing) complex SQL queries.

However, no matter the exact engine specification that needs to be employed, the data model remains essentially the same. For the setup of the data model, proprietary GIS functionalities were avoided whenever possible to allow for later integration into different GIS environments.

The overall layout consists of four main elements or entity groups represented by feature classes and relations concerning

1. available map products consisting of thematic maps and cartographic representations as depicted in figure 2c,
2. sensor-data footprint representations including all meta information and homogenized relationships (figure 2d),
3. stratigraphic definitions and data domains for each planet (information on facies, formation, systems, and groups, figure 2b),
4. mapping units and their representation by specific sets of symbols that are selected from a pre-defined domain-driven genetic context and which are controlled by a set of topologic constraints (figure 2a).

Relationship classes frame the main mapping units feature datasets and relations which deals with the main mapping process. Domains and subtypes as well as a set of relationship and topologic constraints define their interaction in order to limit inconsistencies.

Each of the ER components outlined in figure 2 are implemented via standard feature and relationship classes (outline example in figure 3). As independence of reference frames are a driving requirement when dealing with mapping of planetary objects, the usage of feature datasets is limited to sensor and map data that is stored within the same reference frame. Offsets and rotations as well as higher-level distortions due to poor knowledge and determination of early navigation-data introduced errors that can only be handled using tie-point matching and new georeferencing.

4 Summary and Outlook

The proposed geodatabase model is a part of a mapping environment under development in the context of providing tools and definitions for mapping, cartographic representations and data exploitation of planetary data.

The database model forms an integral part and is designed for portability with respect to geoscientific mapping tasks in general and can in principle be applied to every GIS project dealing with solid-surface planets. It will be accompanied by definitions and representations on the cartographic level (NASS 2010) as well as tools and utilities for providing easy accessible workflows focusing on querying, organizing, maintaining, and integrating planetary data and meta-information. At the current stage, hybrid geologic and geomorphologic mapping can be conducted and data can be exchanged between similar FGDC-driven projects.

The data model's layout is modularized with individual components dealing with symbol representations (geology and geomorphology), metadata accessibility and modification, definition of stratigraphic entities and their relationships as well as attribute domains.

Extensions for typical planetary mapping and analysis tasks and integration of additional utilities for easy accessible querying, as well as offline-data processing in connection within the ISIS environment (USGS ISIS 2010) and in combination with the GDAL (GDAL 2010) environment are currently tested.

Major components that need to be addressed deal with the accessibility of symbolizations for (a) stratigraphic units as well as (b) lithologies and geormorphologic context in a proper, i.e., topologically consistent way.

References

ARCTUR, D. & ZEILER, M. (2004), Designing Geodatabases, ESRI Press, Redlands, CA.
ESRI (2010), Data Models – ESRI Support. – http://support.esri.com/index.cfm?fa=downloads.datamodels.gateway (28.01.2010).
FGDC (2000), Federal Geographic Data Committee, Digital Cartographic Standard for Geologic Map Symbolization, FGDC-STD-013-2006. – http://www.fgdc.gov/standards/projects/FGDC-standards-projects/geo-symbol (28.01.2010).
GDAL (2010), GDAL – Geospatial Abstraction Language. – http://www.gdal.org/ (28.01.2010).
NASA/USGS (2009), NASA/USGS Planetary Geologic Mapping Program. – http://astrogeology.usgs.gov/Projects/PlanetaryMapping/ (20.01.2010).
NASS, A. (2010), Planetare Kartographie: Entwicklung einer modularen GIS Umgebung für die Planetengeologie. In AGIT 2010 proceedings, Wichmann, Heidelberg.
OMG (2007), MOF 2.0 / XMI Mapping Specification, v2.1.1. – http://www.omg.org/spec/XMI/2.1/PDF (26.01.2010).
RANA, S. & SHARMA, J. (Eds.) (2006), Frontiers of Geographic Information Technology. Springer, 329 p.
TANAKA, K. L., SKINNER, J. A. & HARE, T. M. (2005), Geologic Map of the Northern Plains of Mars. US Geological Survey, Scientific Investigations Map 2888, USGS.
USGS ISIS (2010), USGS Isis: Planetary Image Processing Software. – http://isis.astrogeology.usgs.gov/ (26.01.2010).
ZEILER, M. (1999), Modeling our World, ESRI Press, Redlands, CA.

Mixed Geographically Weighted Regression for Hedonic House Price Modelling in Austria

Marco HELBICH and Wolfgang BRUNAUER

1 Background

According to hedonic price theory (ROSEN 1974), real estate is valued for its utility-bearing characteristics. Because a property is fixed in space, a household implicitly chooses many different goods and services by selecting a specific object. From the methodological point of view, this can be explained with the hedonic price function f, describing the functional relationship between the real estate price P and object characteristics $X_{O1},...,X_{On}$ as well as neighbourhood characteristics $X_{N1},...,X_{Nm}$. Traditional approaches use a log-linear model structure (with the price and some of the continuous covariates logarithmically transformed), which reduces heteroscedasticity and nonlinearity. Nevertheless, locally varying equilibria or "submarkets" can be expected. If not accounted for, this leads to biased results and falsely induced spatial autocorrelation. Therefore, the literature provides a variety of local and global models (e.g. ANSELIN 1988, LESAGE & PACE 2009). One cutting-edge methodology that explicitly models heterogeneity is the geographically weighted regression (GWR, FOTHERINGHAM et al. 2002). Numerous applications (e.g. YU et al. 2007) show the usefulness of this technique, as discussed later on. The main purpose of this research is therefore to define a hedonic pricing model that explains transaction prices for family dwellings in Austria accurately, taking into account structural and locational differences as well as spatial heterogeneity in intercept and slope parameters.

2 Study Site and Data

The data set consists of 3,892 locations of family dwellings situated in Austria for the purchase period of 1998 to 2009 and is provided by the Bank Austria. For GWR analysis, a random sample of 35% (1,393 objects) of the population is used in order to make computation feasible. Additional to the transaction prices, the data comprise 24 structural attributes (e.g. condition of the house, quality of heating system, floor space), as well as characteristics of the surroundings (e.g. proportion of academics, purchase power index).

3 Methodology

The results of global models, in particular the feasible generalized least square and the simultaneous autoregressive error model, indicate serious problems concerning heteroscedasticity and autocorrelation, which advocates the use of GWR. Efficiency can be gained using a mixed GWR model (MGWR, FOTHERINGHAM et. al 2002), a semi-local approach where coefficients with small variation over space are kept constant over Austria. The decision whether an effect is global or local can be carried out on the basis of LEUNG's et al.

(2000) statistic. A model with adaptive Gaussian kernel functions is used, which accounts for irregular densities of observations over space. MGWR models can be written as follows:

$$y_i = \sum_{j=1}^{k} a_j x_{ij} + \sum_{l=1}^{m} b_l(u_i, v_l) x_{il} + \varepsilon_i \tag{1}$$

where in this case y_i is the logarithmically transformed sales price of observation i, $i \in 1,\ldots,n$, a_j the global coefficients of covariates x_{ij}, $j \in 1,\ldots,k$ and $b_l(u_i, v_l)$ the local coefficients of covariates x_{il}, where the pair (u_i, v_l) are the coordinates of observation i. In matrix notation, this model is written as

$$\mathbf{y} = \mathbf{X}_1\mathbf{a} + (\mathbf{B} \circ \mathbf{X}_2)\mathbf{1} + \boldsymbol{\varepsilon} \tag{2}$$

where $\mathbf{y}$ is the $n \times 1$ vector of responses, $\mathbf{X}$ is an $n \times k$ matrix of covariates with the respective $k \times 1$ vector of global coefficients $\mathbf{a}$, and $\boldsymbol{\varepsilon}$ is the usual *iid* vector of error terms. $\mathbf{B}$ is an $n \times m$ matrix whose i-th row is given by $\mathbf{b}(i) = (\mathbf{X}_2^{\mathrm{T}}\mathbf{W}(i)\mathbf{X}_2)^{-1}\mathbf{X}_2^{\mathrm{T}}\mathbf{W}(i)\mathbf{y}$, where $\mathbf{W}(i)$ is the diagonal spatial weighting matrix at point i. $\mathbf{B} \circ \mathbf{X}_2$ is the Hadamard product of the matrices

$$\mathbf{B} = \begin{bmatrix} b_0(u_1,v_1) & b_1(u_1,v_1) & \cdots & b_m(u_1,v_1) \\ b_0(u_2,v_2) & b_1(u_2,v_2) & \cdots & b_m(u_2,v_2) \\ \vdots & \vdots & \ddots & \vdots \\ b_0(u_n,v_n) & b_1(u_n,v_n) & \cdots & b_m(u_n,v_n) \end{bmatrix}, \tag{3}$$

and $\mathbf{X}_2$, the $n \times m$ matrix of explanatory covariates with spatially varying coefficients. Here, $\mathbf{B}$ is multiplied entry-wise with the corresponding elements of $\mathbf{X}_2$, i.e. $(\mathbf{B} \circ \mathbf{X}_2)_{ij} = \mathbf{B}_{ij} \times \mathbf{X}_{2ij}$. $\mathbf{1}$ is an $m \times 1$ vector of ones. We write $\boldsymbol{\Gamma} = (\mathbf{B} \circ \mathbf{X}_2)\mathbf{1}$ and define $\mathbf{H}_{\mathbf{X}_2}$ as the partial hat matrix that projects the partial residuals of the response variable given the global part of the model onto $\boldsymbol{\Gamma}$, resulting in $\hat{\boldsymbol{\Gamma}}$, i.e.

$$\hat{\boldsymbol{\Gamma}} = \mathbf{H}_{\mathbf{X}_2}(\mathbf{y} - \mathbf{X}_1\mathbf{a}). \tag{4}$$

The Frisch-Waugh-Lovell theorem states that pre-multiplying the equation with the orthogonal complement of this hat matrix leads to the same result for the parameters $\mathbf{a}$ as estimated in an equation with all covariates:

$$(\mathbf{I} - \mathbf{H}_{\mathbf{X}_2})\mathbf{y} = (\mathbf{I} - \mathbf{H}_{\mathbf{X}_2})\mathbf{X}_1\mathbf{a} + (\mathbf{I} - \mathbf{H}_{\mathbf{X}_2})\boldsymbol{\varepsilon}. \tag{5}$$

Using the estimated global parameters $\hat{\mathbf{a}}$ in turn, one can subtract $\mathbf{X}_1\hat{\mathbf{a}}$ from both sides of equation (2) and estimate a basic GWR model. Therefore, in contrast to the GWR without any fixed coefficients, MGWR is estimated in following steps: (a) Regress each vector $\mathbf{x}_j$, $j \in 1,\ldots,k$ on $\mathbf{X}_2$ using GWR, obtaining the residuals $\tilde{\mathbf{x}}_j$ that form the columns of $\tilde{\mathbf{X}}_1$. (b) Regress $\mathbf{y}$ on $\mathbf{X}_2$ using GWR, obtaining the residuals $\tilde{\mathbf{y}}$. (c) Regress $\tilde{\mathbf{y}}$ on $\tilde{\mathbf{X}}_1$, which yields the correct coefficients $\hat{\mathbf{a}}$ for the non-varying part of the model. (d) Calculate the residuals $\tilde{\tilde{\mathbf{y}}} = \mathbf{y} - \mathbf{X}_1\hat{\mathbf{a}}$ and regress them on $\mathbf{X}_2$. This yields the correct local coefficients. An in-depth discussion can be found in FOTHERINGHAM et al. (2002). Common techniques to transfer the pointwise parameter estimations on non-observed locations are interpolation (e.g. YU et al. 2007) or model estimation on a regular grid (FOTHERINGHAM et al. 2002). For our application, ordinary kriging was employed. All calculations are accomplished in the R environment for statistical computing.

4 Results

After some model selection procedures (minimizing AIC), the final model consists of seven global predictors and nine significant non-stationary variables. Table 1 gives an overview concerning the parameter estimations and confirms the pre-assumed relationships.

Table 1: Parameter estimations

		Global parameters			**Local parameters**		
		Estim.	**Std. err.**	**t-val.**	**1. QT**	**Med.**	**3. QT**
Non-stationary	Purchase power 2009 (M)***	0.005	0.001	4.497	0.000	0.003	0.006
	Share of academics 2001 (M)***	0.012	0.003	3.897	0.010	0.016	0.020
	Age index (M)***	-0.045	0.005	-8.635	-0.040	-0.030	-0.025
	Ln populat. density 2009***	0.066	0.009	7.459	0.038	0.075	0.083
	Condition house 3 (D)*	-0.044	0.019	-2.236	-0.098	-0.085	-0.004
	Attic 1 (D)**	-0.061	0.019	-3.137	-0.063	-0.039	-0.032
	Ln of plot space***	0.092	0.020	4.527	0.040	0.112	0.151
	Ln of total floor area***	0.466	0.029	15.906	0.375	0.423	0.518
	Age of building***	-0.006	0.001	-11.205	-0.006	-0.005	-0.005
Stationary	Intercept	-0.007	0.008	-0.828			
	Unemployment rate 2009 (M)	-1.232	1.735	0.478			
	Share academ. deviation from municipality mean 2001 (C)***	0.008	0,002	3.939			
	Quality heating syst. 3 (D)***	-0.132	0.038	-3.491			
	Quality bathroom 3 (D)	-0.017	0.035	-0.484			
	Existence cellar (D)***	0.117	0.022	5.368			
	Quality garage 3 (D)***	-0,094	0,020	-4.765			
	Terrace (D)*	0.044	0.019	2.257			

M = Municipal. level, C = Census tract level, D = Dummy, QT = Quantile, Signif.: '***' 0.001, '**' 0.01, '*' 0.05

For instance, the global predictor "quality of the garage" has a negative effect on the dwelling's price. A garage in bad condition reduces the price approximately about 9%. The same can be said for the quality of the heating system, whereas bad conditions yield a 13% reduction of price. Deeper insights can be gained from Figure 1, where exemplarily four kriged parameters are mapped. It can be clearly seen that there is significant spatial heterogeneity in the predictors. Thus, the relationship between these variables and the house price depends on the geographical location, which cannot be explored with a global stationary model. For example, the index of purchase power has a negative effect on the price in and around the metropolitan areas of Vienna and Salzburg, elsewhere the effect is positive.

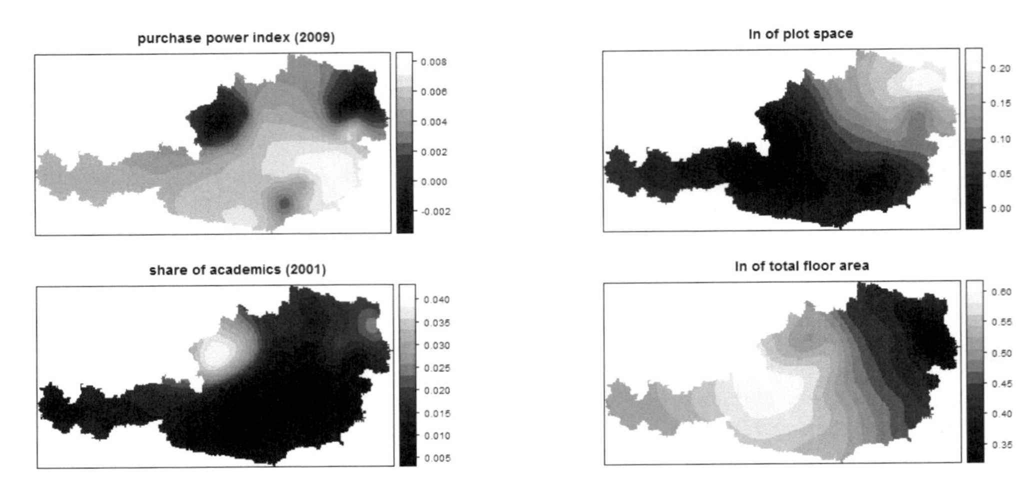

Fig. 1: Spatial distribution of four non-stationary parameters as kriged surfaces

The model fit, indicated by local R^2, varies between 0.25 and 0.50, whereas the lowest values are located in Vienna and its surroundings, as well as the south of Austria. The highest fit is achieved in northern areas. Finally, it is certainly worth noting that MGWR is a useful method to explore heterogeneity, although with limited applicability for large datasets. Hence, the application of computationally more efficient algorithms seems promising, particularly the tensor product smooths approach (WOOD 2006), where a penalized regression approach is adopted in which low-rank, scale-invariant tensor product smooths are constructed. The smooths can be written as components of (generalized) additive mixed as well as of standard (generalized) linear mixed models, allowing them to take advantage of the efficient and stable computational methods that have been developed for such models.

References

ANSELIN, L. (1988), Spatial Econometrics. Methods and Models. Dordrecht, Kluwer.

FOTHERINGHAM, S., BRUNDSON, C. & CHARLTON, M. (2002), Geographically Weighted Regression: The Analysis of Spatially Varying Relationships. Wiley, Chichester.

LESAGE, J. & PACE, K. (2009), Introduction to Spatial Econometrics. Boca Raton, CRC.

LEUNG, Y., MEI, C.-L. & ZHANG, W.-X. (2000), Statistical Tests for Spatial Nonstationarity Based on the Geographically Weighted Regression Model. Environment and Planning A, 32, pp. 9-32.

ROSEN, S. (1974), Hedonic Prices and Implicit Markets: Product Differentiation in Pure Competition. Journal of Political Economy, 82, pp. 34-55.

WOOD, S. (2006), Low-Rank Scale-invariant Tensor Product Smooths for Generalized Additive Mixed Models. Biometrics, 62, pp. 1025-1036.

YU, D., WEI, Y. D. & WU, C. (2007), Modeling Spatial Dimensions of Housing Prices in Milwaukee, WI. Environment and Planning B: Planning and Design, 34, pp. 1085-1102.

Spatial Association of Geotagged Photos with Scenic Locations

Hartwig H. HOCHMAIR

The GI_Forum Program Committee accepted this paper as reviewed full paper.

Abstract

Web 2.0 provides a continuously growing source of volunteered geospatial information through numerous interactive Web applications. It complements geographic information that can be extracted from traditional sources, such as Earth imagery. This paper analyses for a sample of scenic routes whether footprints of geotagged photos posted on Web 2.0 are more frequently found along scenic routes than along their corresponding fastest routes. If the footprint frequency is higher for scenic routes, locations and street segments within close distance to footprints can be assumed to contribute to route scenery, and therefore be utilized as an additional data source within software based trip planners for the computation of scenic routes.

1 Introduction

The development of Web 2.0 allows the Web community to interact with each other, provide information to central sites, and thus become a significant source of information. More and more Web 2.0 platforms use geographic information as a key factor in their information, which lead to the term Geospatial Web (SCHARL & TOCHTERMANN 2007). Especially the development of the Global Positioning System (GPS) and its integration into cell phones, photo cameras, and other mobile devices has made it easy even for non professionals to measure latitude and longitude and to share geotagged information with the Web community. Such collection of spatial information which is built in a collective effort from voluntary Web 2.0 users has been coined Volunteered Geographic Information (VGI) (GOODCHILD 2007). Well known examples of Web 2.0 applications are Wikimapia[1], where users provide descriptions of places of interest to them, or OpenStreetMap[2] (OSM) which builds a public-domain street map of the whole world. Other applications allow to map favorite locations through geotagged photos, e.g., flickr[3], or Panoramio[4], or to map favorite routes, e.g., RouteYou[5].

This paper focuses on volunteered geotagged photos and examines whether the location of photo footprints is generally associated with scenic portions of a street network. Knowledge

---

[1] http://wikimapia.org
[2] http://www.openstreetmap.org
[3] http://www.flickr.com
[4] http://www.panoramio.com
[5] http://www.routeyou.com

about scenic segments is of importance for computer assisted trip planning since route scenery and attractiveness have been identified as prominent route selection criteria besides travel time, simplicity, and safety (HOCHMAIR 2004, HOCHMAIR 2008). A series of earlier empirical studies identified parameters for landscape scenery (APPLETON 1975, STEINITZ 1990, BISHOP & HULSE 1994, BISHOP 2003) which can be used in a GIS to predict the scenery for given locations using viewshed analysis. 3D models allow for a more detailed approach to automated analysis of views, including human depth-variation preferences (BISHOP et al. 2000), or modeling the impact of nearby objects, such as transmission towers (GROSS 1991). However, these approaches require detailed spatial information, such as existing vegetation at fore and middle ground, and complex algorithms for information extraction, e.g. to identify folded landscapes (STEINITZ 1990). As opposed to this, geotagged images are readily available and can be downloaded from Web 2.0 sources using APIs.

Although it is intuitive that volunteered geotagged photos are preferably taken along scenic street segments, this assumption has not been empirically tested so far. Thus the following working hypothesis is formulated: Geotagged photos on Web 2.0 applications are found more frequently along scenic routes than along fastest routes. This hypothesis will be tested for Panoramio photos. Using the concept of proof by contradiction, a null-hypothesis can be formulated which states that the number of panoramio photos found along scenic and fastest routes is equal, or that it is even higher for fastest routes. To describe the number of photos along a route, we use a measure called linear footprint density. It is obtained by creating a buffer around the route, followed by counting the number of photo footprints within that buffer, and by dividing this number by the route length. The linear footprint density captures therefore the number of photos per route length unit.

2 Potential Web 2.0 Data Sources Related to Route Scenery

Testing the hypothesis requires two data sets: a set of scenic and corresponding fastest routes, and a set of volunteered geotagged photos associated with these routes.

2.1 Point Data

Two types of geotagged point data provided through Web 2.0 sources can be distinguished:

1) Point data that are added in a collaborative effort to map the whole world, where points represent primarily the location of physical objects, such as parks, buildings, view points, or sea ports. Examples are the categories shop or hotel in OSM and Wikimapia.

2) Point data that are based on personal experiences, such as mapping one's favorite locations, visited places, or footprints of photos taken. These points do not necessarily map a physical object. Examples are stopovers in travel diaries, such as in mapvivo[6], or footprints of photos in flickr or Panoramio.

With respect to scenery, the first group of points has the advantage of being assigned to a more or less pre-defined set of categories when uploaded by the Web user. This

[6] http://mapvivo.com

categorization allows filtering feature classes when downloading scenic points from these portals and building the database for scenic route planning. The disadvantage with this type of point data, and in general with all types of data that aim to provide a map like depiction of the world, is that subjective relevance, i.e., the perceived scenery associated with such features, is not reflected in the map. This is because even attractive features and locations, such as a nice view point, are only mapped once. Whereas OSM and Wikimapia in general host this kind of spatial data, Wikimapia provides a point feature class "interesting place" which partially relates to scenery. It could express some subjective preference of Web users for a place since more than one feature can be pinned onto a single location. However, points in this feature class are scarce. For example, only three features are found in the Port of Miami area (Fig. 1a). Visualizing OSM point data that could be considered related to scenery gives a similar picture. Fig. 1b shows OSM point data from categories "tourism", "historic", and "leisure", which includes parks, museums, and theaters, among others. The problem of point scarcity with reference to identification of scenic route segments remains.

Footprints from geotagged flickr and Panoramio photos are used as examples for the second type of point data. Both flickr and Panoramio allow uploading geotagged pictures as well as manually georeferencing of photos on a basemap. Whereas the flickr website provides also access to images that are not geo-referenced, Panoramio is more of a geolocation-oriented website where all published images are geo-referenced.

Flickr allows to share geotagged photographs taken both outdoors and indoors. It is apparent that only the outdoor photos can be potentially utilized to determine route scenery. Flickr users can assign uploaded photos to one or several groups to more clearly specify their content. Searching groups for photos provides more accurate and complete search results than utilizing photo tags, because many users do not annotate their photos, or they use the same tags for all their photos uploaded. To assess the density of available flickr photos for a given geographic area, the flickr API was used to download photos from 27 scenery related groups, including "scenicwater", "scenery", "scenic-outdoors", scenicareas", "landscape", "worldlandscape", "mountain_water", "heights", or "world". Flickr photos are stored with an accuracy value between 1 and 16, where world level is 1, city is 11, and level 16 is considered to be accurate at the street level. The map in Fig. 2 visualizes flickr footprints that reveal somewhat reliable accuracies, i.e., levels between 14-16.

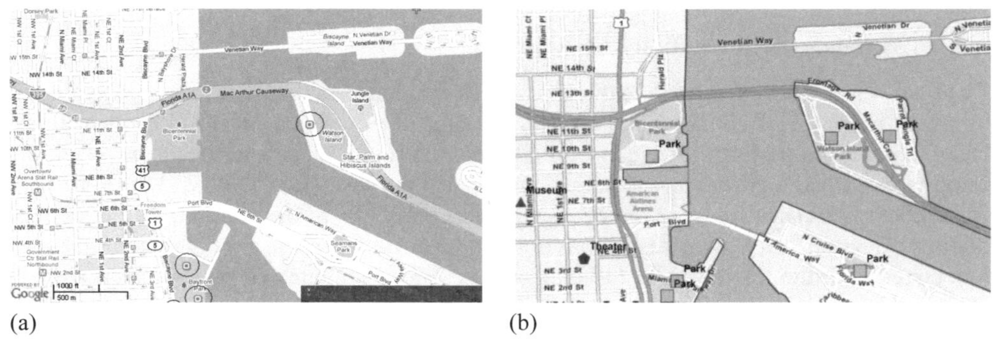

(a) (b)

Fig. 1: Points from category "interesting place" on Wikimapia (circled) (a), and points from categories tourism, leisure, and history in OSM (b)

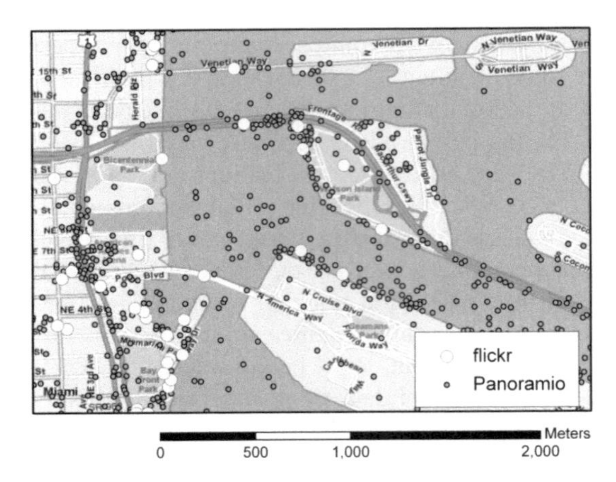

Fig. 2:
Points from Panoramio and 27 flickr groups in the Port of Miami area

In Panoramio, photos are organized exclusively through tags, not through groups. All Panoramio photos are showing outdoor scenes which makes them potentially associated with scenery. No accuracy information is provided at this point, therefore all footprints are used for this study. Fig. 2 shows footprints for Panoramio (small dots) and flickr (large dots) in the Port of Miami area. When compared to Fig. 1 it can be seen that the point density for both data sources is higher than for OSM or Wikimapia data. Panoramio's point density is higher than the point density for the 27 flickr groups (even when mapping all accuracy levels 1-16 in flickr). Because of this we expect more discernible results for Panoramio photos. Therefore the focus of point analysis in this study will be on Panoramio photos although a similar study could be repeated for flickr photos.

2.2 Scenic Routes

Various Web 2.0 sites allow registered users to upload waypoints of their favorite routes. For this study 32 user suggested motorcycle routes for Florida, Germany and Austria were downloaded from the EveryTrail[7] portal. Although the downloaded routes are not explicitly labeled scenic, the descriptions for some of the routes and the fact that they generally avoid highways suggest that they pass through a nice landscape or urban environment, respectively. In addition, nine favored on-street bicycle routes were downloaded for Florida from the GPSies[8] portal. These bicycle routes were relatively short, so that none of the corresponding fastest route counterparts were running on a highway segment. All together a total of 41 routes were used for the analysis.

Planning a route between two locations involves the optimization of various criteria, such as travel time, travel comfort, and scenery, among others. Since in most cases available route alternatives perform differently on a given criterion, the traveler needs to weigh the importance of each criterion, depending on the trip purpose (Bovy & Stern 1990). If a route chosen deviates from the shortest or fastest route, this means that the benefits of the alternative routes, such as increased travel safety or better scenery, are higher than the cost associated with the longer travel distance or time. Since volunteered photos taken along a chosen route primarily reflect landscape, we can assume that a linear footprint density

[7] http://www.everytrail.com/
[8] http://www.gpsies.com

along alternative routes can be attributed to route scenery, thus indicating more scenic routes. We use the fastest routes, computed from Google maps directions[9], as reference routes where route origin and destination are adapted from downloaded scenic routes. For the comparison of geotagged photos, only non-overlapping portions between scenic routes and their fastest routes were considered. Table 1 summarizes the lengths of the 41 downloaded scenic and computed fastest routes after removal of route overlaps. It further shows how much longer (in %) scenic routes are in comparison to their corresponding fastest routes.

Table 1: Characteristics of used route set. Route lengths are given in km

Mode	Location	Number	Mean Distance			Median Distance		
			Scenic	*Fast*	*Diff.[%]*	*Scenic*	*Fast*	*Diff.[%]*
Motorbike	*Florida*	16	205.7	165.0	24.7	159.2	110.7	43.8
	Germany	12	137.0	97.6	40.4	94.7	69.8	35.7
	Austria	4	184.8	150.6	22.7	195.9	152.3	28.6
	All routes	32	177.3	137.9	28.6	145.6	99.7	46.0
Bicycle	*Florida*	9	24.4	20.5	19.0	18.2	16.3	11.7

3 Panoramio Footprints along Shortest and Alternative Routes

This section introduces two methods for counting Panoramio footprints within route buffers, i.e., the base method and the filter method.

3.1 Base Method

Since footprints of geotagged photos do not exactly overlay with route geometries, a buffer distance around routes needs to be defined. The buffer distance accounts for the fact that georeferencing methods for photos, i.e., GPS or using a basemap, have some inaccuracies. It also expresses the fact that a scene captured by the photographer may be similarly attractive some distance away from the location where it was taken. The latter distance will depend on the local surroundings, such as building heights, canopy, or topography, but in general we assume that the scenery changes slower in less densely built areas than in high density housing areas. To simplify terminology in the remainder of the paper we refer to the first type of areas as "rural environments", and to the second type as "urban environments", although, strictly said, lower housing and street densities can also occur within urban boundaries, e.g. around parks or lakes. We use street density as a proxy variable for housing density. To assess the street density along the analyzed routes, the 82 scenic and fastest routes were overlaid with a 500m × 500m square grid. Then the street density was computed within each grid as total street length in the grid divided by 25,000. Visual inspection of these density values along with satellite images suggests that a density threshold of 6,200 is appropriate to separate higher density from lower density housing areas. For route segments above the 6,200 threshold a route buffer of 100m and for route segments below this threshold, a 300m buffer was assumed to be appropriate. In addition to

[9] http://maps.google.com

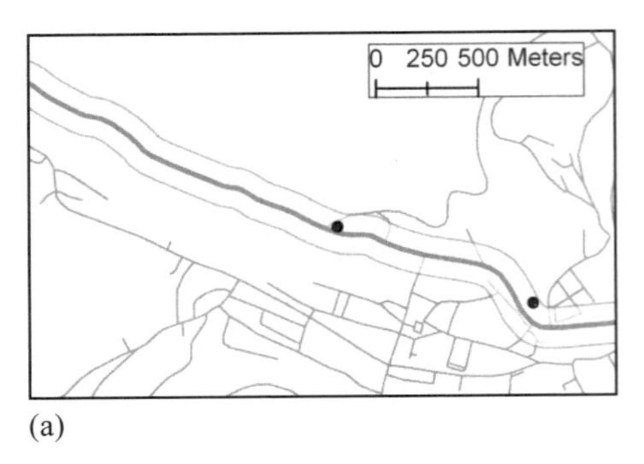

(a)

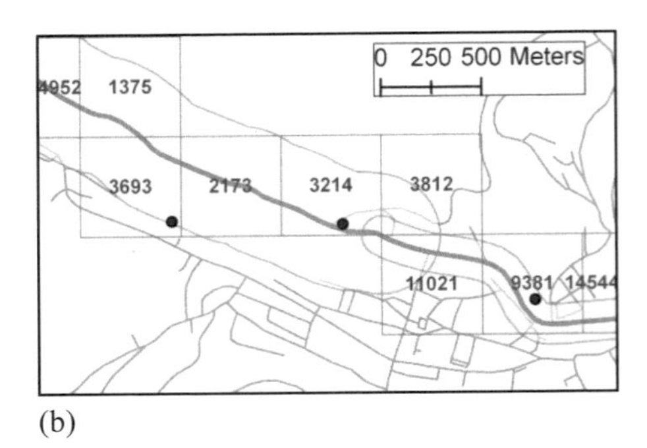

(b)

Fig. 3: Base method applied on 100m buffer (a) and combined 100m/300m buffer (b)

this, the working hypothesis was also tested with a constant 100m buffer. Fig. 3 visualizes these concepts. The thick line denotes the route, the shaded area is the route buffer, and dots show the footprints of Panoramio photos. In Fig. 3a a constant buffer size is used. In Fig. 3b the numbers in the grid cells indicate the street density, based on which the buffer distance varies between 100m and 300m.

The point counts per route km, i.e., the linear point density, were compared between buffers around the scenic and fastest routes for all 41 route pairs described in Table 1. Since linear point densities were not normally distributed, a non-parametric test was used. For the 100m buffer (Fig. 3a) results show that the linear point density associated with scenic routes (Mdn=0.553) is significantly higher than for fastest routes (Mdn=0.391) (Wilcoxon Signed Rank, N=41, Z=3.842, p=<0.001, 2-tailed). Results also reveal that for the combined 100m/300m buffer approach (compare Fig. 3b) the linear point density associated with scenic routes (Mdn=1.058) is significantly higher than for fastest routes (Mdn=0.738) (Wilcoxon Signed Rank, N=41, Z=3.738, p=<0.001, 2-tailed).

3.2 Filter Method

Whereas the previous results support the hypothesis in this paper, the association of Panoramio photos with scenic routes compared to fastest routes can still be improved. Casual comparison of Panoramio footprints with the photo content and photo tags leads to some observations relating to scenery. It appears that some users upload photos showing their home or work place which is unrelated to scenery. Such pictures show, for example, a house owner's garden. Other pictures appear to show some random content, such as gas stations or fast food restaurants, which were probably taken to test the photo camera or GPS device. On more frequented roads, particularly highways, more pictures of such random content appear to be taken as well, which includes photos of other cars, traffic signs, or the sky. To more precisely separate scenic photos from irrelevant photos, a simple rule can be applied. Since information provided by two users is more reliable than information from just a single user, it is suggested to retain only a footprint if it is within a given distance from the footprint of a photo provided by a different Web user. The buffer distances from Fig. 3 can be used as distance thresholds.

Fig. 4 gives an example of the proposed filter method. Labels next to footprints show the username of the person who uploaded a photo. A buffer distance is assigned to each footprint, depending on the street density (Fig. 4a). 100m buffers are in black, and 300m

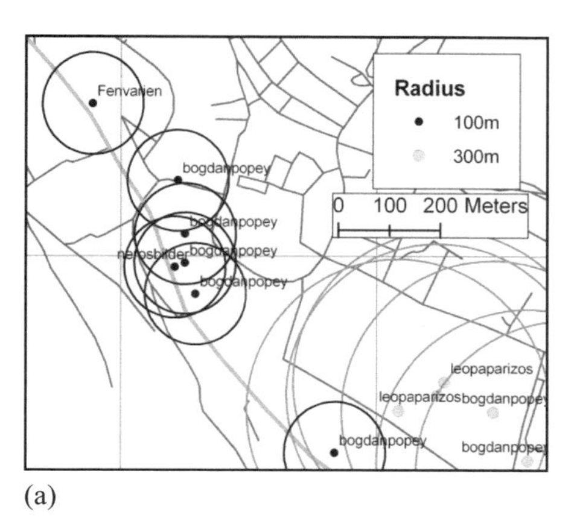

(a)

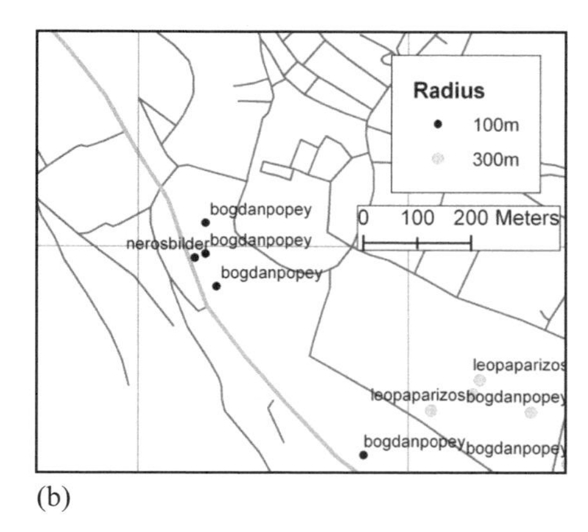

(b)

Fig. 4: Distance thresholds for footprints (a) and result of filter process (b)

buffers in gray. The two point clusters in the middle left and the lower right area contain photos from at least two different users. That is, each footprint is within at least one circle associated with a second photo footprint that was uploaded from a different user. As opposed to this, the two footprints on the upper left are not contained by any other circle. Therefore they are removed within the filter process, whereas all other footprints are retained (Fig. 4b).

The hypothesis of this paper was also tested on footprints modified by the filter method using a constant 100m buffer radius and the combined 100m/300m buffer radius. For the 100m buffer the linear point density associated with scenic routes (Mdn=0.220) was found to be significantly higher than for fastest routes (Mdn=0.084) (Wilcoxon Signed Rank, N=41, Z= 4.235, p=<0.001, 2-tailed). Also for the combined 100m/300m buffer approach the linear point density associated with scenic routes (Mdn=0.547) was significantly higher than for fastest routes (Mdn=0.258) (Wilcoxon Signed Rank, N=41, Z= 4.412, p=<0.001, 2-tailed).

3.3 Comparison of Methods

Further it was tested whether observed differences in linear point density between scenic and fastest route increase if the filter method is used instead of the base method. For a route pair consisting of a scenic and a fastest route, the normalized difference measure m can be computed as $m=(d_s-d_f)/(d_s+d_f)$, where d stands for the linear point density of the scenic (d_s) and fastest route (d_f), respectively. If $d_s=d_f=0$, then $m=0$ by definition. A higher m thus means a higher difference in linear point density between scenic and fastest route. The m values of all 41 route pairs were compared for the base and the filter method using a constant 100m buffer and a combined 100m/300m buffer. With the constant 100m buffer m was significantly higher for the filter method (Mdn=0.504) than for the base method (Mdn=0.201) (Wilcoxon Signed Rank, N=41, Z=3.428, p=0.001, 2-tailed). Also for the combined 100m/300m buffer approach m was significantly higher for the filter method (Mdn=0.329) than for the base method (Mdn=0.182) (Wilcoxon Signed Rank, N=41, Z=3.092, p=0.002, 2-tailed). This indicates that the filter method provides a better distinction between scenic and fastest routes based on Panoramio photo footprints.

Table 2: Results for tests of hypothesis

Buffer distance	Method	d (Mdn)		Z	p	m (Mdn)	Z	p
		Scenic	Fast					
100m	Base	0.553	0.391	3.842	*<.001*	0.201	3.428	*.001*
100m	Filter	0.220	0.084	4.235	*<.001*	0.504		
100m/300m	Base	1.058	0.738	3.738	*<.001*	0.182	3.092	*.002*
100m/300m	Filter	0.547	0.258	4.412	*<.001*	0.329		

Table 2 summarizes the results of the various empirical tests. Values in the two *d* columns show that for both buffer types and point selection methods the linear point density is higher for scenic than for fastest routes, which supports the hypothesis of this paper. Values in the *m* column show that the relative difference between the number of points associated with scenic and fastest routes is higher for the filter method than for the base method. This means that the distinction between scenic and fastest routes based on Panoramio footprints is more accurate when applying the filter method.

Fig. 5 illustrates the patterns numerically described in Table 2. The two highlighted routes in Fig. 5a and b show part of one of the 41 route pairs used for hypothesis testing. The route labeled "Scenic" runs along a scenic river and partially within a national park in Austria (Nationalpark Gesäuse). The route labeled "Fast" runs along a highway (Pyhrnautobahn). Dots along routes indicate the location of Panoramio footprints when using the combined 100m/300m buffer distance. Fig. 5a uses the base method whereas Fig. 5b shows the footprints that are retained after the filter method.

Reflecting the value pattern in the *d* columns in Tab. 2, the linear point density of footprints is higher for the scenic than for the fast route both in Fig. 5a and b. With the base method (Fig. 5a) this difference is less obvious than for the filter method. The latter removes most of the isolated footprints along the highway while preserving many footprints along the scenic route. Thus the difference between scenic and fast route regarding footprints is more evident with the filter method, which is reflected by the higher *m* values associated with the filter method in Tab. 2.

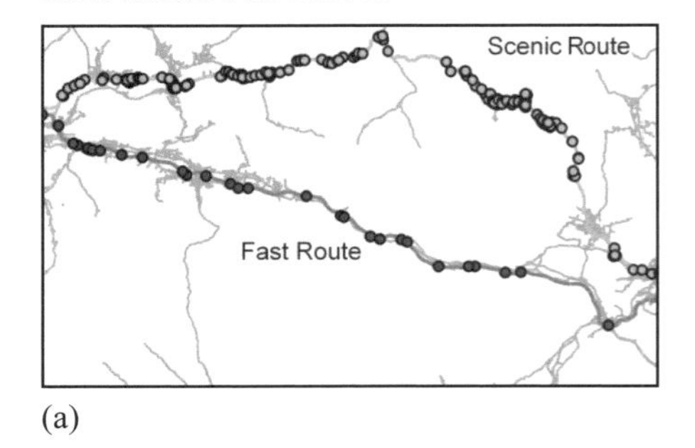

(a)

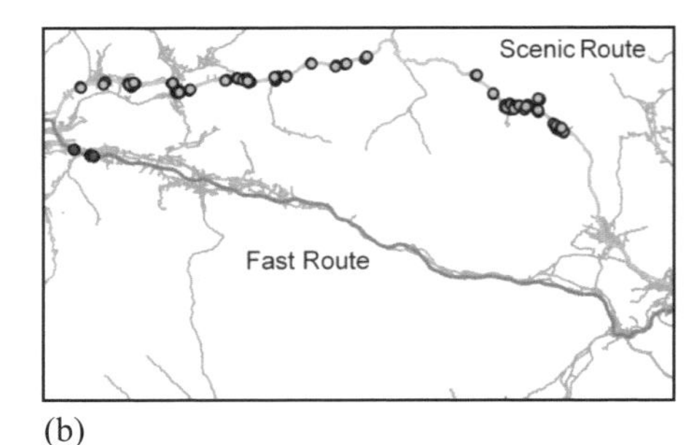

(b)

Fig. 5: Panoramio footprints within 100m/300m before (a) and after (b) the filter method

4 Discussion

The results of this empirical study reveal that geotagged Panoramio photos show a higher spatial association with user posted routes on Web 2.0 when compared to fastest routes. This is indicative of street segments near Panoramio photo footprints exhibiting some sort of scenery. In areas with an abundance of geotagged photos, such as in Fig. 5, the footprints could be readily utilized as a data source to plan a scenic route when applying a corresponding algorithm. This approach can be applied in all geographic regions where a sufficient number of photo footprints is available. A more centralized approach is provided through various organizations and initiatives, such as the National Scenic Byways Program in the United States[10], which, after a thorough collaborative screening process, assign a scenic label to selected routes in their administrative boundaries. In the latter approach the coverage is somewhat coarser, generally limited to selected linear routes, and neglecting scenic local streets beside the labeled routes. In geographic regions with scarce geotagged photos, the photo footprints will not be so helpful in planning a scenic route. Such regions may, for example, be relatively homogeneous without any remarkable scenery. Or the area is secluded and therefore less frequently visited. In either case, some general proxy criteria, such as avoidance of highways or industrial zones, or preference for nearness to water bodies (lakes, rivers) may be used for the automated computation of scenic routes instead (HOCHMAIR & FU 2009).

Various topics for future work can be identified. One task is to analyze how footprints of geotagged photographs overlap with scenic landscapes as defined by criteria identified in literature (APPLETON 1975, STEINITZ 1990, BISHOP & HULSE 1994, BISHOP 2003). A significant overlap would facilitate an automated identification of scenic street segments with a more parsimonious set of landscape/topographic/landcover variables when used in combination with photo footprints. The filter method introduced improves matching results with scenic routes significantly. Nevertheless, further refinement of this strategy will help to remove undesired footprints not associated with scenery, such as clusters of photo footprints around railway stations. Another topic to be addressed is the accuracy of geospatial data contributed by individual Web users (GOODCHILD 2008), more specifically, the distance between the location where a photograph was taken and its mapped position. This will help to determine whether, for example, a 100m buffer distance is meaningful to count footprints along a route.

Whereas this research found that geotagged images are an additional potential data source to identify street segments of relative scenery in a given area, future work also needs to address the question of which algorithm for the computation of scenic routes is appropriate given the location of geotagged images alone, or in combination of different data, respectively. Two general approaches are commonly used. Given scenic points along the street network an algorithm that maximizes the number of such points along a route while minimizing the path length could be applied, e.g., using integer programming methods or genetic algorithms (CURRENT et al. 1985). Alternatively, if scenery is an attribute of route segments, modified edge weights within a single-criterion shortest path algorithm could be utilized (HOCHMAIR & NAVRATIL 2008).

[10] http://www.byways.org

References

APPLETON, J. (1975), The Experience of Landscape. London, Wiley.

BISHOP, I. D. (2003), Assessment of visual qualities, impacts, and behaviours, in the landscape, by using measures of visibility. Environment and Planning B, Planning and Design, 30 (5), pp. 677-688.

BISHOP, I. D. & HULSE, D. W. (1994), Prediction of scenic beauty using mapped data and geographic information systems. Landscape and Urban Planning, 30 (1-2), 59-70.

BISHOP, I. D., WHERRETT, J. R. & MILLER, D. R. (2000), Using image depth variables as predictors of visual quality. Environment and Planning B, Planning and Design, 27 (6), pp. 865-875.

BOVY, P. H. L. & STERN, E. (1990), Route choice: Wayfinding in transport networks. Dordrecht: Kluwer Academic.

CURRENT, J. R., REVELLE, C. S. & COHON, J. L. (1985), The Maximum Covering/Shortest Path Problem: A Multiobjective Network Design and Routing Formulation. European Journal of Operational Research, 21, pp. 189-199.

GOODCHILD, M. F. (2007), Citizens as Voluntary Sensors: Spatial Data Infrastructure in the World of Web 2.0 (Editorial). International Journal of Spatial Data Infrastructures Research (IJSDIR), 2, pp. 24-32.

GOODCHILD, M. F. (2008), Spatial accuracy 2.0. In J.-X. ZHANG & M. F. GOODCHILD (Eds.), Spatial Uncertainty, Proceedings of the Eighth International Symposium on Spatial Accuracy Assessment in Natural Resources and Environmental Sciences (1, pp. 1-7). Liverpool: World Academic Union.

GROSS, M. (1991), The analysis of visibility – environmental interactions between computer graphics, physics, and physiology. Computers and Graphics, 15, pp. 407-415.

HOCHMAIR, H. H. (2004), Towards a classification of route selection criteria for route planning tools. In P. FISHER (Ed.), Developments in Spatial Data Handling, Berlin, Springer, pp. 481-492.

HOCHMAIR, H. H. (2008), Grouping of Optimized Pedestrian Routes for Multi-Modal Route Planning: A Comparison of Two Cities. In L. BERNARD, H. PUNDT & A. FRIIS-CHRISTENSEN (Eds.), The European Information Society – Taking Geoinformation Science One Step Further (Lecture Notes in Geoinformation and Cartography. Berlin, Springer, pp. 339-358.

HOCHMAIR, H. H. & FU, J. (2009), Web Based Bicycle Trip Planning for Broward County, Florida. Proceedings of ESRI User Conference, San Diego, CA, July 13–17, 2009 (CD-ROM).

HOCHMAIR, H. H. & NAVRATIL, G. (2008), Computation of Scenic Routes in Street Networks. In CAR, A., GRIESEBNER, G. & STROBL, J. (Eds.), Geospatial Crossroads@GI_Forum '08: Proceedings of the Geoinformatics Forum Salzburg. Heidelberg, Wichmann, pp. 124-133.

SCHARL, A. & TOCHTERMANN, K. (2007), The Geospatial Web: How Geobrowsers, Social Software and the Web 2.0 Are Shaping the Network Society. Berlin, Springer.

STEINITZ, C. (1990), Toward a sustainable landscape with high visual preference and high ecological integrity: the loop road in Acadia National Park, U.S.A. Landscape and Urban Planning, 19 (3), pp. 213-250.

Using Geospatial Data for Assessing Thermal Stress in Cities

Sebastian HOECHSTETTER, Tobias KRÜGER, Valeri GOLDBERG,
Cornelia KURBJUHN, Jörg HENNERSDORF and Iris LEHMANN

The GI_Forum Program Committee accepted this paper as reviewed full paper.

Abstract

The assessment of the bioclimatic situation in cities is fundamental for sustainable urban planning. Due to the consequences of climate change, the frequency of extreme heat events is likely to increase during the summer months, even in the temperate zones. Therefore, standardised procedures for monitoring the thermal stress affecting human health are needed. In this study, we present a multi-scale approach for identifying urban heat islands, modelling bioclimatic parameters and assessing the cooling effect of urban green spaces. We demonstrate how different sources of widely available geospatial data can be combined with urban climate models in order to obtain a sound knowledge basis for planning mitigation measures over a range of spatial scales.

1 Introduction

Climate change is clearly one of the major ecological and socio-economic challenges of our time. The connections between climate change and sustainable development of cities and regions are well-studied, and strategies for limiting the adverse impacts of climate change are needed (BEG et al. 2002). Therefore, governmental bodies on different organisational levels are increasingly concerned about these effects and recognise the need of considering changing climatic conditions in adopting sustainable approaches to the use of natural resources and to spatial planning (DEFRA 2007).

Especially urban environments have been identified as being particularly vulnerable to climatic changes. In general, cities are warmer compared to the surrounding landscape and may exhibit the so called urban heat island effect (OKE 1987). Especially during the night the reduced cooling is significant in urban areas. Heat island intensity of cities and, consequently, urban warming mainly depends on the density of urban structures and the sealing of urban areas.

This urban heat island effect is likely to intensify climatic stress for humans. The increase of cardiovascular mortality and respiratory illnesses due to heatwaves in cities is the consequence of lowered evaporative cooling and intensified heat storage resulting from increased impervious cover and complex surfaces of the cityscape (PATZ et al. 2005). During the European heatwave of 2003, these interrelations could be studied in a particularly drastic way. KOSATSKY (2005) provides a review of reports on the estimated excess deaths during this event in western Europe, with the numbers of people dying due to direct and indirect heat effects being in the range of several tens of thousands. He also finds evidence for the fact that the mortality impact was greatest on elderly people.

Hence, spatial planning in urban regions needs strategies for identifying vulnerable settlement areas regarding thermal stress. This information is needed in order to define the urban districts where the mitigation of those adverse effects—for example by creating green spaces or by improving ventilation—must be granted a high priority. Apart from meteorological data, climatic simulation models used for this task usually also require data on settlement structures, land use and the vegetation inventory of green spaces.

In this paper, we outline which kinds of these geospatial data serve as a suitable input for some of the commonly used tools for bioclimatic modelling, how these data influence the model output and how they ought to be prepared in order to obtain accurate results. Special focus is put on the usage of remote sensing data in this context. This way, we intend to make a contribution to a standardised procedure for the assessment of thermal stress in urban environments.

Most of the findings presented in this paper emerged from the first phase of the REGKLAM project, which is aimed at the development and testing of an integrated regional climate change adaptation programme for the model region of Dresden (Germany).

Besides that, some aspects that we refer to in this paper are embedded in the project "Urban nature and green space development under climate change", funded by the German Federal Ministry for the Environment, Nature Conservation and Nuclear Safety.

2 Human Bioclimatology – Background, Application and Linkage to Geospatial Data

Human bioclimatology deals with the interactions between the human body and the atmospheric environment. In general, these interactions are described by thermal, air quality and actinic factors (VDI 1998). The experimental and model applications used in this study are related to the thermal factors of human bioclimatology. These factors comprise all meteorological parameters which are needed to describe the heat balance of the human body. Complex energy balance models such as the so called "Klima-Michel" (JENDRITZKY et al. 1979) have been established in order to calculate quantitative measures of human thermal well-being, such as the Predicted Mean Vote (PMV) and the Physiological Equivalent Temperature (PET).

The meteorological factors controlling the thermal exchange between the body and the surroundings are: air temperature, mean radiation temperature (being an output of the energy balance model), air humidity (vapour pressure) and wind velocity. These parameters determine the so called comfort equation after FANGER (1973), which describes the interaction between the non-closed energy balance of a person and the subjective perception on a psycho-physical scale. PMV as a measure for this phenomenon was determined empirically and depends on the above mentioned atmospheric factors as well as on the internal heat production and the thermal insulation of clothing. It takes values between –4 (very cold) and +4 (very hot).

Another measure for describing the thermal comfort of humans is PET. It includes data on skin temperature and sweat evaporation. PET is a measure for the perceived temperature in a specific atmospheric situation. It is defined as the air temperature at which, in a standardised indoor setting, the heat budget of the body (same skin temperature assumed) is in balance with the real conditions to be assessed (HÖPPE 1999). PET is directly influenced by radiation and temperature conditions in the atmospheric environment. For example, during a summer day with an air temperature of 30 C and a wind velocity of 0.5 m/s, PET may range from 30°C in the shadow to 45°C in the sun.

Different simulation programmes for modelling human bioclimate have been developed over the years, two of which are used here. ENVI-met is a three-dimensional microclimate model used to simulate surface-plant-air interactions in urban environments. The typical resolution of ENVI-met ranges between 0.5 and 10m in space and 10s in time. Fields of application are urban climatology and environmental planning. ENVI-met is a prognostic model based on the fundamental laws of fluid dynamics and thermodynamics. The model includes the simulation of flow around buildings, exchange processes of heat and vapour at the ground surface and at walls, as well as the exchange at vegetation and bioclimatology (BRUSE & FLEER 1998). The main model input parameters are wind speed and wind direction, roughness length, temperature of the atmosphere, specific humidity, relative humidity, specific plant and soil types present, cloud cover, background CO_2 concentration, as well as soil temperature and relative soil water storage in three layers. We use ENVI-met for modelling the cooling effects of green spaces and for medium-scale simulation approaches.

The models RayMan and RayMan Pro respectively (MATZARAKIS & RUTZ 2006) are further options for carrying out bioclimatic simulations. This software particularly emphasises the importance of the radiation budget acting upon the human body. By taking complex urban structures into account, which are presumed to control the relevant short- and longwave radiation fluxes, this model offers a potential for several applications, such as urban planning or street design (MATZARAKIS et al. 2007).

Important outputs of the model are the thermal indices PMV, PET and standard effective temperature (SET). These indices are calculated for a selected point site and require personal data (age, weight, etc.), meteorological data (air temperature, relative humidity), and structural parameters (topography, building, vegetation structure) as input parameters. Especially the built environment and the vegetation have a large impact on the outcome since they cause shading effects and control the radiation budget. Therefore, accurate and updated data on the physical urban structure are necessary. Within the scope of the present study, RayMan Pro is used for providing the foundation for area-wide modelling approaches as well as for fine-scaled modelling of single streets and squares (see chapter 3).

3 Modelling Approaches and Data Basis

In order to reflect the complex interrelations between the thermal conditions in the city of Dresden (Germany) and the settlement and green space structure on different spatial scales, different modelling approaches are pursued within the scope of the study at hand. Figure 1 provides an overview of these modelling levels, the necessary input data and the software we use. From this compilation of methods and data it becomes obvious that the output of the various approaches is highly dependent on the quality and availability of data.

Meteorological data were mainly retrieved from the local meteorological stations and were measured directly by a field campaign carried out in several central districts of Dresden by the Institute of Hydrology and Meteorology (Technische Universität Dresden) in summer 2009. Besides that, measured and simulated raster data sets of different meteorological parameters with a horizontal resolution of $500 \times 500m^2$ obtained from the web-based Ra-K-liDa data portal were used (BERNHOFER et al. 2009).

Land surface temperature (LST) derived from remote sensing data is a particularly important input for an area-wide modelling approach. Various satellites carry thermal sensors which allow for the recording of LST data. Landsat 4, 5, and 7 satellites, for example, capture thermal information in the 10.4–12.5µm band with a spatial resolution of 120m for the Thematic Mapper (TM Band 6, Landsat 4, 5) and 60 m for the Enhanced Thematic Mapper+ (ETM+ Band 6, Landsat 7). This data is widely used for surface thermal analyses (e.g. WLOCZYK et al. 2006). Besides that, we utilise these data for the energetic characterisation of land surface types in order to generate area-wide maps of PET using selective simulations and statistical modelling approaches.

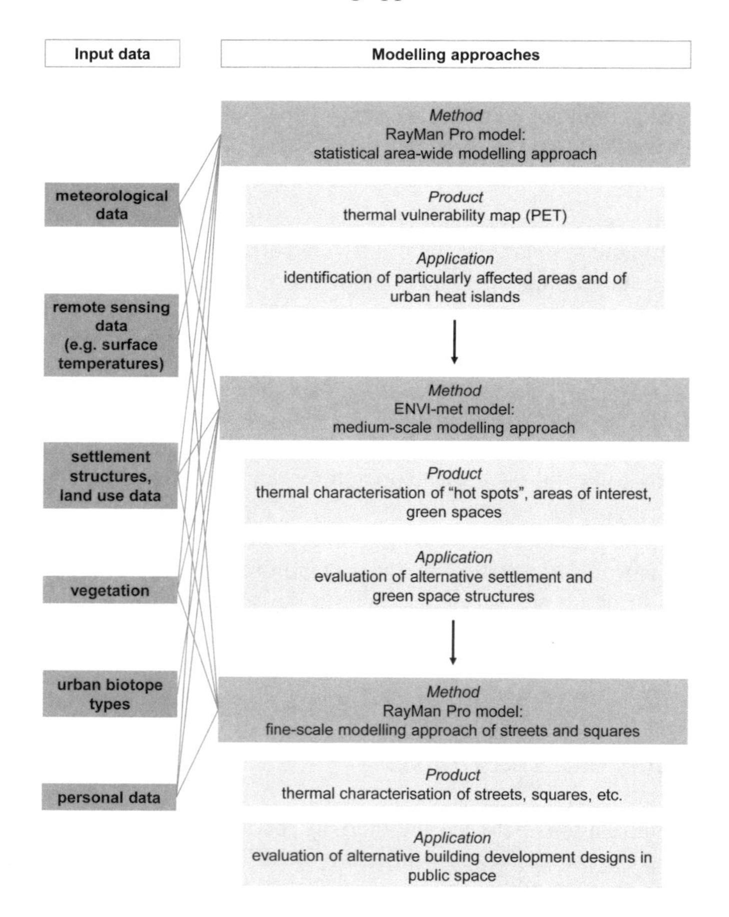

Fig. 1: Overview of the various stages of modelling approaches on different spatial scales, including the most important input data used

In order to get values of land surface temperature, the image pixel values of the Landsat data have to be converted into thermal radiance from which temperatures can be derived. This process implies several steps, starting with the rescaling of the retrieved 8-bit integer values into at-sensor spectral radiance values (CHANDER et al. 2009). The resulting radiation can be converted into the so-called at-sensor brightness temperature, which differs from LST because of atmospheric influences on the signal and of the emissivity of the surface. The effects of these have to be accounted for in order to get reliable information on the actual land surface temperature. For the estimation of atmospheric correction parameters we used an online calculator by BARSI et al. (2003) which allows for the input of the meteorological situation (regarding air temperature, atmospheric pressure and relative humidity) during the data acquisition. For emissivity estimation we employed several approaches which are mainly based on the discrimination between water bodies and vegetation and non-vegetation (urban built-up areas). As a result, we obtained area-wide LST maps for the city of Dresden which can be used as input for further analysis.

For describing and analysing the physical characteristics of *settlement structures*, the urban structural type (UST) approach is used. A UST is a basic spatial unit of homogeneous character on a neighbourhood scale which can be used to differentiate the urban area. Physical aspects can be characterised by typical values and indicators. There are different kinds of methods and nomenclatures to get USTs and their characteristic values. We used the city structure type map of the city of Dresden in order to make detailed analyses in the inner city, and the software SEMENTA® developed by MEINEL et al. (2009) for area-wide analyses in the greater Dresden area.

Especially building volume and building height are important input parameters of the climate models we used. The building volume calculated by SEMENTA® is based on empirical studies and not detailed enough for our purpose. Another possibility to get data about building volumes is a digital 3D-city-model, which is also available for parts of the Dresden area. In general, the data structure of such a 3D-model is rather complex and the amount of detail present in these data cannot be processed anyway by most of the climate models. Therefore, in our analysis we used airborne laser scanner data in combination with the geometry of the buildings from the German authoritative digital map ATKIS-Basis-DLM to calculate the building volume and the building heights. These data are easy to use, widely available and the method is thus easily transferable into other study areas.

Urban green space influences microclimate, soil water content, biodiversity of flora and fauna, atmospheric composition, soil erosion, noise and exposure to sunlight (ARLT et al. 2005). Thus, data on the *distribution and the elevation of vegetation* and the spatial pattern of greenery in the cityscape is also crucial for determining the thermal situation. It has been shown that increasing the canopy cover along streets or in settlement areas can reduce the air temperature considerably (O'NEILL et al. 2009). Therefore, in bioclimatic analyses data on vegetation needs to be considered. The RayMan Pro model uses point data determining the position, height and geometry of single trees. We obtained these data from the respective departments of the Dresden municipality. Similar data sources are available in most German municipalities. For most analyses, this level of detail is sufficient.

For more complex examinations regarding the actual effect of different types of green spaces and for using the ENVI-met simulation model, however, more detailed information about the vegetation is needed. For determining the proportion of green space and vegetation volume in urban districts, the *urban biotope type approach* is applied. This offers a basic definition of vegetation structures and their assessment by type, dimension, and location. Cities exhibit a mosaic-like pattern of biotopes. As a general presupposition, these biotopes are spatially confined and internally homogeneous. Urban biotope mapping provides a good overview of the biotope types and vegetation structures in a city. In our case, 52 biotope types were identified, on the basis of which vegetation structures and volumes were assessed. The vegetation structure of urban biotope types was analysed in representative areas. The analysis included the physiognomic identification of areas with different vegetation heights and areas without any vegetation (built-up land, other kinds of sealed and open ground, water bodies). Results of the analysis in representative areas are specific attributes such as surface proportion of different vegetation layers as well as the proportions of surfaces without vegetation and the specific green volume, differentiated by vegetation layers and as a total of all vegetation layers as arithmetic mean values. Also, for each biotope type, the degree of soil sealing, the coefficients of runoff and the proportions of the different kinds of sealed soil surfaces have been determined.

Personal data are needed in order to determine the biothermal stress of different groups of persons. For example, the heat balance of the body is dependent on the person's weight, age or sex, as well as on the insulation properties of the clothing or the physical activity. In order to simulate the thermal stress of different age groups, we are able to use socio-demographic data provided by the municipality of Dresden. Some of the analyses carried out are focused on districts with a large proportion of elderly people, who are particularly affected by heat events. However, due to reasons of data security, such data are frequently not available with a high spatial resolution.

4 Results

Several Landsat scenes have been acquired and prepared for analysis. There are two night scenes amongst them. Figure 2 shows the situation on June 20th, 2000 (11:48 a.m.) which was a so called "tropical day" with a maximum temperature T_{max}>30°C. It becomes obvious that areas especially in the city centre are heating up very fast in the morning. The surfaces of the large green areas of the Dresdner Heide (a forest in the northeast of the city) and of the Großer Garten (a large park near the city centre) are cooler compared to the built-up areas. This way, besides serving as a means for the energetic characterisation of surfaces, such data sets can provide valuable large-scale information on the general pattern of the temperature distribution of a certain region of interest. Moreover, they are used as input parameters for the area-wide analysis of thermal stress.

Interactions of different urban structures with the microclimate of Dresden were determined by calculating the temperature differences of various structure types to a reference area and by the distribution of PMV. The ENVI-met model output of air temperature and PMV clearly showed the effects of green areas in Dresden and the impact on the thermal environment of humans. For example, PMV inside the park Großer Garten showed distinctly lower values (implying a pleasant thermal environment for humans) compared to the sealed areas in the vicinity. The above-mentioned field measurement campaign confirmed the model results and verified the cooling effects of green areas.

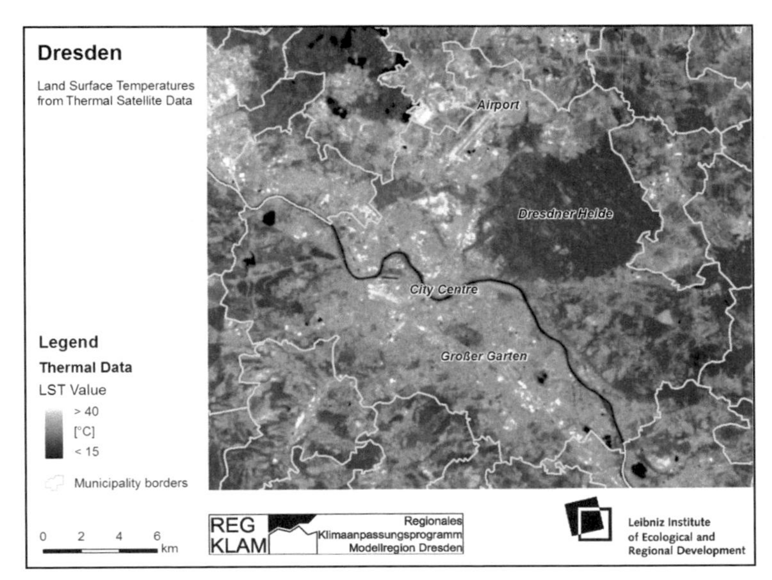

Fig. 2: Land surface temperatures derived from Landsat data (acquisition date: June 20th, 2000; 11:48 a.m.)

Green and open spaces have a large potential for cooling down the urban area (OKE 1987). This effect can be demonstrated with the ENVI-met model output: The interaction between typical urban vegetation structure types and the microclimate was simulated for the conditions on a summer day in the city of Dresden. Figure 3 shows the air temperature differences and thus the cooling effects of several urban structures compared to sealed vicinity for July 16th 2009 at 02:00 p.m. The given difference of the mean temperature of an urban structure and mean temperature of the sealed surrounding illustrates the importance of the vegetation inventory and the shading effects of trees for the microclimate during a sunny day. Because of these shading effects combined with a good ventilation, mixed green and urban structures are cooler (dark bars) than the sealed vicinity. Sealed urban structures with less shadowed areas and ventilation, e.g. traffic areas, showed the highest temperature. This example shows the positive climatic effects of open and shadowed areas and vegetation structures on the urban microclimate of Dresden.

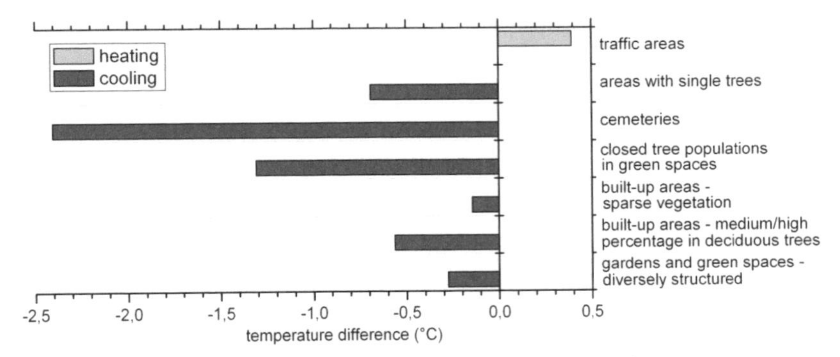

Fig. 3: Temperature effects of different urban vegetation structure types for July 16th, 2009 at 02:00 p.m.

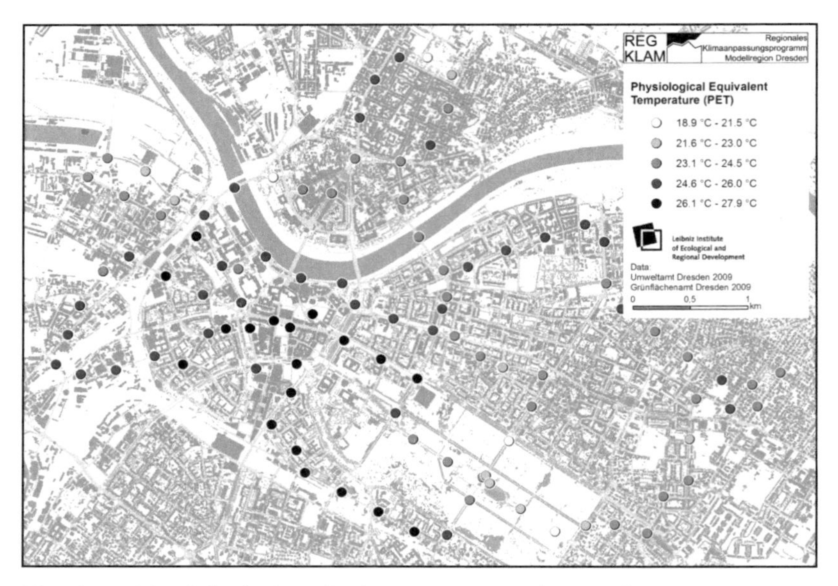

Fig. 4: Physiological equivalent temperature in Dresden (reference date: July 16th, 2009; 07:30 p.m.)

As another analysis step, we carried out a series of more than 100 representative simulations of PET using the RayMan Pro model. The locations for these point measurements were selected along the route of the field campaign mentioned in the previous chapter. This way, different urban areas regarding settlement structure and density can be characterised concerning their thermal comfort for humans on a summer day. Figure 4 provides an illustration of the results, calculated for 07:30 p.m. in the inner city of Dresden. Even on this spatial scale, the cooling effect of green spaces can be demonstrated, which is not only reflected by lower surface temperatures, but also by a lower thermal stress as expressed by the PET value. The high density areas of the city centre exhibit high values of PET even in the evening hours. Planting of trees, creating shaded areas or increasing the reflectivity of the surface could be appropriate steps for improving the living conditions in these districts.

5 Discussion

The findings emerging from our study so far reveal that the combination of different simulation approaches offers a large potential for serving as monitoring scheme for the thermal situation in cities.

We argue that the examination of the city climate over a range of scales is necessary in order to reflect the complexity and heterogeneity of the cityscape. The area-wide approach using Landsat data is appropriate for identifying critical parts of the city and areas where urban heat islands are likely to occur. Medium-scale simulations are suitable for assessing the cooling effects of different types of green spaces on their surroundings. The fine-scale observations finally can be used to evaluate different development scenarios of the built environment and to elaborate concrete measures for the mitigation of heat effects. Since we mainly relied on widely available data sources, this procedure can be adapted to other urban areas. Moreover, the products obtained during the several steps can be used effectively for visualising the potential consequences of planning alternatives.

The next steps comprise the production of area-wide PET maps according to the outlined procedure. Concerning the method of deriving reasonable results for land surface temperature we want to find possibilities to approximate emissivity values by calculating several indices from the visible and near infrared spectra. The use of NDVI for finding a relation between vegetation cover and emissivity and, thus, a suitable function to estimate emissivity from the visible and near/middle infrared bands of TM/ETM+ would allow to establish stable time series on LST in Dresden.

Acknowledgements

REGKLAM is funded by the German Federal Ministry of Education and Research as a part of the KLIMZUG initiative. We would like to thank Peter Teichmann (Environmental Office, city of Dresden), Anne Bräuer and Christian Hoyer (Leibniz Institute of Ecological and Regional Development), and Anna Westbeld (formerly Technische Universität Dresden) for assisting in data collection and preparation. Special thanks also go to Andreas Matzarakis (University of Freiburg) and Michael Bruse (University of Mainz) for permission to use the models RayMan Pro and ENVI-met respectively.

References

ARLT, G., HENNERSDORF, J., LEHMANN, I. & THINH, N. X. (2005), Auswirkungen städtischer Nutzungsstrukturen auf Grünflächen und Grünvolumen. IÖR-Schriften 47. IÖR, Dresden, 136 p.

BARSI, J. A., BARKER, J. L. & SCHOTT, J. R. (2003), An atmospheric correction parameter calculator for a single thermal band earth-sensing instrument. IGARSS03, 21th-25th July 2003, Centre de Congres Pierre Baudis, Toulouse, France. Proceedings.

BEG, N., CORFEE MORLOT, J., DAVIDSON, O., AFRANE-OKESSE, Y., TYANI, L., DENTON, F., SOKONA, Y., THOMASC, J. P., LÈBRE LA ROVERE, E., PARIKH, J. K., PARIKH, K. & RAHMANF, A. A. (2002), Linkages between climate change and sustainable development. Climate Policy, 2, pp. 129-144.

BERNHOFER, C., FRANKE, J., KURBJUHN, C. & HUPE, F. (2009), Ableitung von Rasterdaten aus gemessenen und projizierten Klimazeitreihen für den Freistaat Sachsen. Abschlussbericht zum Forschungs- und Entwicklungsvorhaben des Sächsischen Landesamtes für Umwelt, Landwirtschaft und Geologie (AZ: 24-8802.26/22/1). – http://www.umwelt.sachsen.de – 95 p.

BRUSE, M. & FLEER, H. (1998), Simulating surface-plant-air interactions inside urban environments with a three dimensional numerical model. Environmental Modelling & Software, 13, pp. 373-384.

CHANDER, G., MARKHAM, B. L. & HELDER, D. L. (2009), Summary of current radiometric calibration coefficients for Landsat MSS, TM, ETM+, and EO-1 ALI sensors. Remote Sensing of Environment, 113, pp. 893-903.

DEFRA (2007), Climate change and sustainability: The crucial role of the new local performance framework. Department for Communities and Local Government – Strategies and action plans, Department for Environment, Food and Rural Affairs. Product code 07LGSR05025/a.

FANGER, P. O. (1973), Thermal Comfort. McGraw-Hill, New York, 265 p.

HÖPPE, P. (1999), The physiological equivalent temperature – a universal index for the biometeorological assessment of the thermal environment. International Journal of Biometeorology, 43, pp. 71-75.

JENDRITZKY, G., SÖNNING, G. W. & SWANTES, H. J. (1979), Ein objektives Bewertungsverfahren zur Beschreibung des thermischen Milieus in der Stadt- und Landschaftsplanung („Klima-Michel-Modell"). Beiträge der Akademie für Raumforschung und Landesplanung 28, 85 p.

KOSATSKY, T. (2005), The 2003 European heat waves. Eurosurveillance, 10 (7-9), pp. 148-149.

MATZARAKIS, A. & RUTZ, F. (2006), RayMan: A tool for research and education in applied climatology. 8th Conference on Meteorology-Climatology-Atmospheric Physics, Athens, May 24th-26th, 2006.

MATZARAKIS, A., RUTZ, F. & MAYER, H. (2007), Modelling radiation fluxes in simple and complex environments – application of the RayMan model. International Journal of Biometeorology, 51, pp. 323-334.

MEINEL, G., HECHT, R. & HEROLD, H. (2009), Analyzing building stock using topographic maps and GIS. Building Research & Information, 37 (5-6), pp. 468-482.

O'NEILL, M. S., CARTER, R., KISH, J. K., GRONLUND, C. J., WHITE-NEWSOME, J. L., MANAROLLA, X., ZANOBETTE, A. & SCHWARTZ, J. D. (2009), Preventing heat-related morbidity and mortality: New approaches in a changing climate. Maturitas, 64 (2), pp. 98-103.

OKE, T. R. (1987), Boundary layer climates. Second edition. Routledge, London, 435 p.

PATZ, J. A., CAMPBELL-LENDRUM, D., HOLLOWAY, T. & FOLEY, J. A. (2005), Impact of regional climate change on human health. Nature, 438 (17), pp. 310-317.

VDI (1998), Human-biometeorologische Bewertung von Klima und Luft für die Stadt- und Regionalplanung. Verein Deutscher Ingenieure. VDI-Richtlinie 3787, Bl. 2, 29 p.

WLOCZYK, C., RICHTER, R., BORG, E. & NEUBERT, W. (2006), Sea and lake surface temperature retrieval from Landsat thermal data in Northern Germany. International Journal of Remote Sensing, 27 (12), pp. 2489-2502.

Using Mobile GIS for Water Networks' Management – The Open Source Approach

Iraklis KARAMPOURNIOTIS and Ioannis PARASCHAKIS

The GI_Forum Program Committee accepted this paper as reviewed full paper.

Abstract

Being able to move freely and get on the spot information about a water network's condition is one of the most important factors in the work of a utilities technician. Mobility in GIS and GPS can provide such a solution to everyday work and support actions such as online and near real time database updates regarding the condition of the network, real time updates from sensors deployed throughout the water network, pin point guidance of the technician to the point of malfunction and other. In this paper such a research is presented, regarding water network management specific applications. By using a PDA or mobile phone equipped with a GPS and a custom Mobile GIS application that connects over the air to a database server, the utilities' technician is able to access a huge amount of data regarding the area he is examining and act accordingly. The focus of this paper is not a new application. There are many commercially available applications for Mobile GIS. It is about providing a better understanding of the technology behind the scenes and how it can be put to use in the case of water network's management. This is the main reason why Open Source software is used apart from the well known benefits of free use, low or no cost, professional and community support. Everything from the DBMS to the client software is entirely based on Open Source packages and development.

1 Introduction

Networks in one or another form are an integral part of our everyday lives. We see networks every single day and in most of the cases we don't even recognize them, although we use them to make our life easier. Commonwealth or utility networks are part of the general networks' family. These networks are used to supply people with water, energy, heat, transportation and other services and goods and are utilized for the common good hence the term commonwealth networks as we refer to them in Greece.

A network is an organized set of lines called edges or links and junction points where lines meet that are called nodes (RAVINDRA et al. 1993). According to the above statement a node is the point where two or more edges meet and this signifies the change in the flow of the network theme. A link is the line that connects two nodes and defines the flow of the network theme from and to the nodes. Finally a network is the set of nodes and the interconnected links with a specific theme.

This paper uses a water network as an example for the application of Mobile GIS. In this case a node could be a pressure valve junction, a T-Junction, a pump, a water tank, a reservoir or other. The links of the water network would be water pipes of various diameters, materials and shapes and the theme would be the water flowing inside the pipes between nodes.

2 Mobile Computing Power and Large Databases

A spatial database is the most significant part of any GIS application. In the majority of the cases these databases tend to become huge and require quite an amount of processing power in order to provide results. With today's systems the power hungry databases can be satisfied quite easily but for small handheld devices in the form of PDAs or mobile phones it is quite a strenuous process. Hence it is of great importance to be able to use either a small footprint DBMS/Database, only with data for a specific job and running on the handheld device, or be able to process all the data on a server system and then provide the device with the required result set of the processed query. The communication between the server and the handheld device is accomplished via a wireless connection be it 3G, GPRS or Wi-Fi. The device runs a custom application that communicates with the spatial database server over the aforementioned connection methods sending the query to the server and receiving the result set back in order to view it. In the following figure a schematic representation of the system is given (see Figure 1).

The query is formed on the handheld device from the selections and choices of the user via the User Interface of the application. After the formulation, the query is sent (light grey arrow) to the database server for processing and the result set is sent back (dark grey arrow) to the device for viewing. No processing is done on the handheld relieving it from this burden. Since the data is in an open format WKB/WKT (Well Known Binary/Well Known Text) (OGC INC. 2006) which is prescribed in the OGC's specifications, the application which is compliant to the standard can parse the data and view it in an application window on the device.

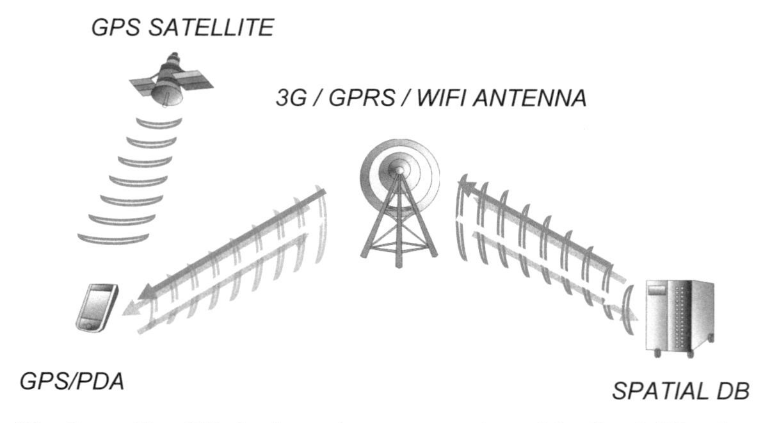

Fig. 1: Simplified schematic representation of the Spatial Database Server and handheld device communication

It is important to understand that even only with the processed result set there still could be a great amount of data to be viewed. But the only processing to be held on the device is the viewing part which is not power hungry since it is only 2D vector data. Apart from the data transfer the device also packs a GPS receiver and is able to know its current position thus aiding the technician on his work. The communication with the server enables the device to handle large amounts of data since all of the processing is outsourced and only the viewing is done on the device. Since today's mobile phones and PDA devices are offered with fast graphics processors and CPUs, the graphics' viewing is not so much of a power problem.

Because of the database being handled by a dedicated server or a service running on a server machine there is no need to customize the database for mobile use. Of course the database can provide information on either mobile or desktop systems and there is no need for the server device to be synchronized with the handheld after field work since they are always on-line. The only case where a synchronization would be required is if for any reason the connection between the two devices (handheld and server) would be lost. In that case the handheld would not be able to process any data.

There is a solution for this problem and it is to keep a backup of the working area on the device which would be synchronized every time with the result set from the server. Keeping a very small part of the database on the handheld is not much of a processing burden since it is only the work area. If for any reason the connection is lost, the device reverts to the internal database and when the connection to the server is re-established; synchronization between the two is accomplished thus updating the server with the current database status.

3 Water Network Spatial Database

Designing the spatial database to support network topology is one of the most important parts in constructing a GIS application for Utility Networks. A great number of parameters has to be taken into account in order to satisfy all the constraints of the network. For example, the flow in a water network does not follow the physical terrain according to gravity. Pumps are used in various positions of the network in order to push water higher or pull water from a well countering the effect of gravity. Obviously the database contains a table that refers to the nodes of the network and another one that refers to the links.

The nodes are of various types. Each type is described by one table and all of them are connected by means of foreign keys with the main nodes' table. The links are contained in a table and as far as topology in concerned they are single direction defined by a FROM and a TO node attribute. These are foreign keys referencing the primary key of the nodes' table (Node ID). In the following diagram (Figure 2) a part of the spatial database is shown covering the main structure of the water network. Although not clearly shown on the diagram for space economy reasons, each of the node type tables (Stop and Pump) has extra attributes making it impossible for the various types to be implemented in a single table. This is the main reason why each type is represented by a single table. All the attributes not common to every type have been removed from the main table and transferred to peripheral tables creating the various node type tables. In the diagram below eight tables can be seen. The main network tables as mentioned previously are the Water Pipes and Water Nodes. They contain all the necessary data attributes and data with which the network is implemented. There are other tables related to the main ones. As far as the Water Pipes one is concerned there are three tables related to it. The Pipe Diameter contains all the data about the used pipe diameters in the network. All measurements are in mm. The Pipe Material table contains the necessary data about the material used to build the pipe such as PVC, steel and other. The Pipe Shape table contains the data about the various shapes the pipes have like circular, trapezoid, rectangular etc. The three attributes combined together provide a full description of the pipe in a manner like *Ø100 PVC* which means Circular Shape Pipe of 100mm Diameter made from PVC.

The Water Nodes table is related to the Node Type which holds the data for the various types of water nodes such as T-Junctions, Pressure Valves, Stop Junctions, Pumps, Reservoirs, etc. As describe in previous paragraphs, every type has its own table containing the relevant node type data.

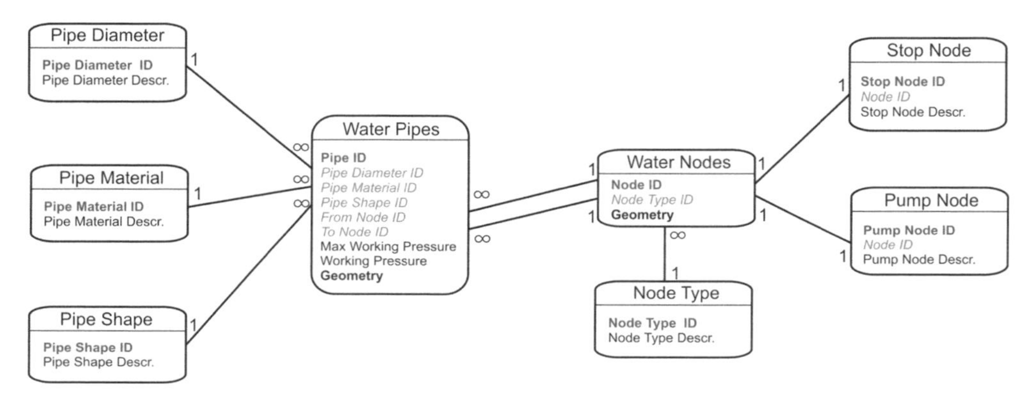

Fig. 2: Part of the Spatial Database ERD – Network

3.1 Geometry Columns and Types

The spatial database follows the Simple Features for SQL (OGC INC. 2006) standard as well as the ISO/SQLMM Part 3 (ISO. 2000) standard in order to support the WKB/WKT (Well Known Binary/Well Known Text) spatial data format and the various data types. The two main tables of the network system (Water Pipes and Water Nodes), in order to accommodate spatial data, incorporate a geometry column of a standard type describing the spatial format of the table. The Water Pipes table has a geometry column of *LINESTRING* (REFRACTIONS RESEARCH 2010) type according to the SFS standard and the Water Nodes table a geometry column of *POINT* (REFRACTIONS RESEARCH 2010) type according to the same standard. These are both WKB/WKT (Well Known Binary/Well Known Text) compatible thus enabling the programmer to use or develop any parser for this format in order to read and consequently render the data with a graphics engine.

3.2 Spatial Indexes and Routing

Spatial indexes are used to increase the search speed of the database. The indexes used are R-Tree indexes implemented thru GiST (HELLERSTEIN et al. 1995, MANOLOPOULOS et al.

2005, Refractions Research 2010). The scheme used for the database was to index with the above method the two main tables of the network part of the database. Using the spatial indexes with the routing algorithm applied to the search there is a great increase in the speed of the search and the provided results.

It is important that the flow of water is not interrupted for any reason. Of course there is always the possibility of malfunctions, repairs and other works and so the flow should be routed via other pipes leaving only a small part of the network affected; meaning without water. The algorithm selected for the application is the shortest path shooting star algorithm used in pgRouting (Philipona & Kastl 2008) which is Open Source Software. Shooting star can be used in any type of network with almost no modifications since it provides a great set of tools for the algorithm application. Also because it is Open Source it is easy to modify and apply to any type of situation and network. PgRouting is an extension to PostGIS providing network routing capabilities. PostGIS (Refractions Research 2010) is part of PostgreSQL (PostgreSQL Global Development Group 2009) which is also the DBMS used for the database of the GIS application presented in this paper.

4 The handheld device

The device used to run the client GIS application could be a mobile phone with Wi-Fi and 3G connection abilities and an internal GPS receiver. The above configuration is proposed because it is an all round configuration with no incompatibilities among the hardware used. The consumer grade GPS receiver of the device provides up to 5 meters of accuracy but with the appropriate incorporation of new technologies like RTCM broadcasters (Ntrip) (Fotiou et al. 2008), there is great improvement on this issue (Fotiou et al. 2009).

Since the application is a GIS one mapping, drawing and viewing require a large screen. There are many devices in the market with screens reasonably larger than the 2.4 inch mainstream standard. Of course if the touch screen ability is available then this is an even greater asset for the end user. Last but not least is the storage capacity of the device. Any type of memory cards should be supported and the higher the supported capacity the more data to be available or stored on the device. In the following table (see Table 1) a summary of the device's characteristics is provided.

Table 1: Device characteristics summary

Component	Speed, Capacity, etc
CPU	>500MHz
Memory, Storage Capacity	>128MB, >4GB of storage space
Screen	>3 inch touch screen
Connectivity	At least 3G

The device used to test the application was a PDA with a 624MHz CPU, 128MB memory of which 90MB available to the applications, 4 inch touch screen display with 640 × 480px resolution and 8GB memory card. Since it was not equipped with an internal GPS receiver, a Compact Flash GPS was used and the PDA was also coupled via Bluetooth with a mobile phone for 3G access when not using the Wi-Fi connectivity.

5 The Application

The client part of the application was developed for use with Microsoft Windows Mobile. It references the DBMS's API and uses a connection string to connect to the database and then calls upon the API again to issue various SQL Commands. All the commands are processed on the server and the result set is send back to the device for viewing.

5.1 Server Side

The server applications, mainly the DBMS, is Open Source. The server side part of the application is nothing more than the database running on a PostgreSQL Database Server backed by PostGIS Spatial Extensions and pgRouting routing capabilities All of the above mentioned software is Open Source and of course freely available. Since they have been under heavy development and testing for quite a few years they can be safely used in a business environment fully supported by the developers and not only.

5.2 The Client Side

The client application runs on the handheld device. A connector class is instantiated and provides the means to connect to the database server. The viewer object is used to create and show the map on the device from the result set. The Query Formulation Class is used to create queries that are sent to the server. Finally the GPS positioning object is used to provide access to the GPS receiver on the device. The viewing is made possible by using another Open Source package which is SharpMap CF. The engine of SharpMap CF is used in order to read WKB (Well Known Binary) or WKT (Well Known Text) data and render it on the device's screen.

5.2.1 Classes

The classes used (see Figure 3) in the application development provide access to the database and the GPS of the device. The connector class is used in order to communicate with the database and provide a layer of information transportation between the database and the handheld device. When connecting to the DB a set of information, such as username, password, port, IP, etc, is used to create a connection string. This connection string is formulated by the Connector Class and so the connection to the DB is established.

After the connection, the Viewer Class receives from the Connector Class all the available data tables of the database. The user is informed and prompted by the application to select the tables with spatial data which he wants to view and/or edit. Also using the GPS's signal, the application is able to load the specific area within 200 meters from the user's position. The GPS Positioning Class is used to provide assistance to the user and the application

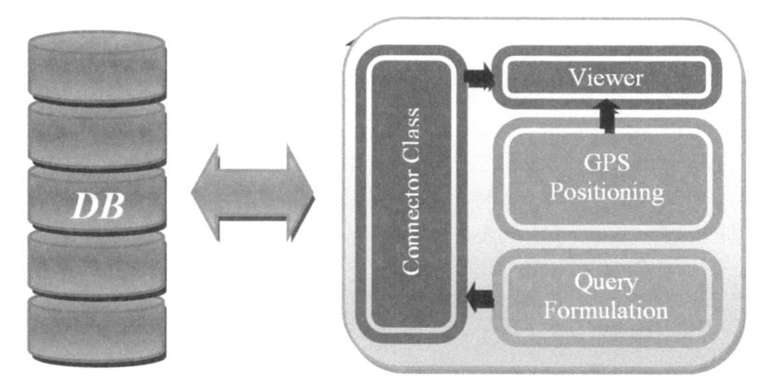

Fig. 3: Class – Communication Diagram

itself with positioning data and pinpoint guidance to the required area as requested by the user. Last but not least the Query Formulation Class is used to form queries and send them to the database via the Connector Class.

5.2.2 Querying the database

The most important part of the database is the part where queries are executed and then sent back to the handheld device. The Query Formulation Class in conjunction with the User Interface of the application is used in order to accommodate the creation of queries and their forwarding, by means of the Connector Class, to the database server for execution.

The composition of queries is done through menus and dialogs presented to the user on the U.I. After the user has completed the query form, a query string is sent as a SQL Command to the database. Consequently the database server executes the command and produces an appropriate result set. The connector class, using the provided query notification service of the PostgreSQL Database server, is informed for the result, picks the data and then feeds them to the Viewer Class which in turn shows the data on the device's screen.

6 A simple example

Starting the application the only available data is a base map which can be preloaded on the device and the position blip of the user as shown in the figure below (see Figure 4). By clicking the Connect button, the user requests for the available data around his current position (see Figure 5).

Since digital camera pictures of the device running the application could not produce clear results of the various screens, an emulator was used to run the application and get the screenshots. The emulator run through the development environment connected to a virtual GPS device which fed the client interface with positioning results in the same way as the real one would. The emulator also had internet access via the PC's internet access and in this way it was able to communicate with the database server listening on a specific IP range.

The connection to the database using the available network connections is established and the available data tables are shown in a selection screen for the user to pick the ones to view directly on screen (see Figure 6).

From here on the user can form queries and view the appropriate data. After having requested the needed data and viewed on the screen the user can select by taping, for example, a pipe, and view any additional information regarding the selected object and/or edit them (see Figure 7 and Figure 8).

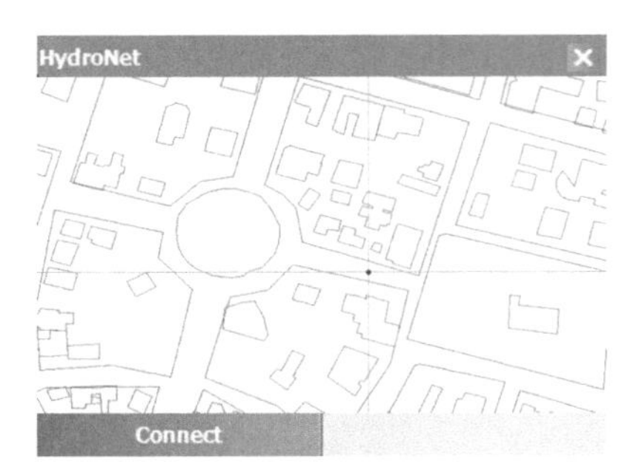

Fig. 4: Application Initialization

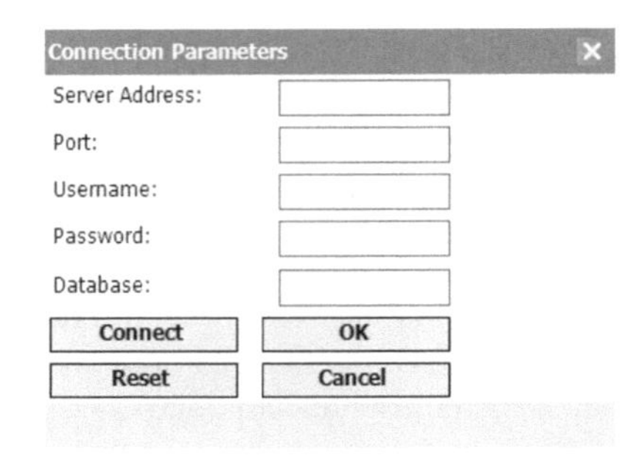

Fig. 5: Connecting to the Database

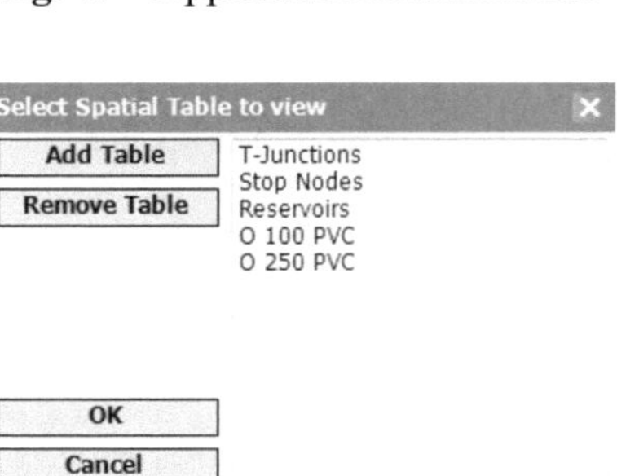

Fig. 6: Table Selection

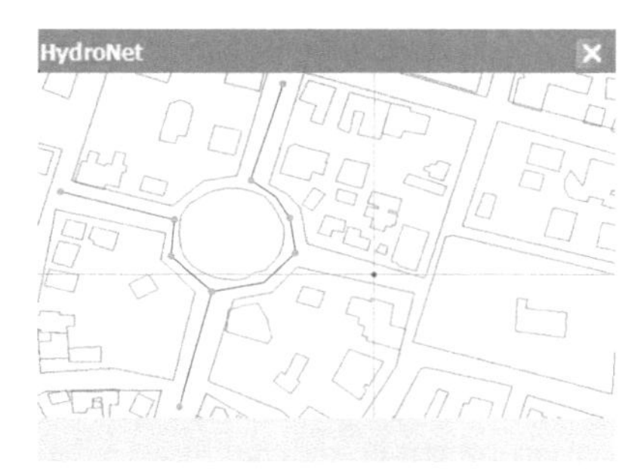

Fig. 7: Selecting an object

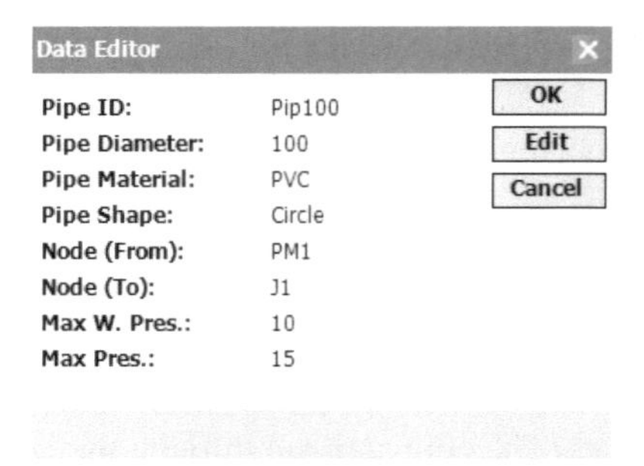

Fig. 8: Additional Information

7 Benefits and Conclusions

By putting Mobile GIS to use in the field work of water network management there is great improvement in flexibility and productivity. The field crew can be anywhere and still be able to access mission critical data, maps and schematics thus spending less time travelling between the office and the various job sites. The crew is also able to respond and react faster to emergencies like breakdowns, leaks and others. This is possible because the one that is closer to the emergency can be routed faster and with great accuracy.

Direct, over the air access to data concerning the network and real time sensor updates decrease the risk of accidents that could prove costly to the service and also prevent risking the worker's life and well being. Also by using the GPS's positioning ability the device can retrieve data relative to the area of work and not bother the technician with data that can prove to be an obstruction. It also improves on the spot productivity by handling irrelevant data automatically and providing the technician with job specific information. The GPS is also used for navigation and guiding the crew to the exact location of the job site with accuracy and above all with safety. Using GIS in the field means real time or near real time data updates which in turn provides the enterprise part of the water management service to work with more accurate and up to date information thus leading to better decisions and decision support.

Viewing all of the above under the health and safety scope, Mobile GIS can prove to be an exceptional tool in minimizing risks. Of course there is always the economic and administrative point of view. By increasing productivity, response speed and assisting decision making, Mobile GIS helps reduce running costs, maximize the crew's and service's efficiency which in turn produces greater value for the enterprise because it can provide its customers with higher quality of services and products.

Using Open Source in an application of this magnitude surely requires skills both in the technical and in the managerial department. From an economic and legal point of view there is almost no cost in acquiring the software to be used and the license management is more simplified than in proprietary software since the software can be obtained once and installed as many times and in as many places needed. It is quite an economical burden released from the budget of enterprises especially the ones run by the state. There is also no mandatory maintenance fee although there will always be a situation when support is required and will have to be purchased if there is no other immediate solution from the community support mechanism.

In general, less hardware power is required to run Open Source applications because of their programming and this also adds to the low or no cost benefit since the frequency of hardware upgrades is lessened. Also support is widely available free of charge by the community involved in the software's development or under a price tag from professionals. In the GIS area under the watchful eye of OSGeo and OGC the applications provided, at least the ones coming from the incubation process of the first, are well maintained and with excellent quality. Also the fact that the source code is publicly and openly available creates a drive for excellence.

Another situation that can rise to become a problem is training. Nowadays training for Open Source Software in the GIS area is widely available in many forms. Universities and other institutes have training programs on Open Source GIS software, companies provide training in such matters and there are many on line and off line schools and numerous tutorials that can be accessed.

Coupling a Mobile GIS application with Open Source Spatial technology can provide a new way out to enterprises and public services and help them maintain their usual if not higher level of provided quality in products and services and in parallel reduce the running and working cost. The use of mobile GIS applications provides a powerful tool for field operations. Mobile GIS accompanied by a GPS receiver can lead to a very simplified and friendly system for data management in the field. The usability of current mobile device technologies especially with the new types of touch screens can create a user friendly environment and in combination with the high availability of communication methods such as Wi-Fi, Wi-MAX and 3G can produce a very adaptive solution to any kind of GIS compatible work conditions.

References

FOTIOU A., PIKRIDAS C., BIMPISIDOU A. & PAPANIKOLAOU D. (2009), DGPS and RTK Positioning using Hermes Ntrip Caster. Proc. of the International Symposium on Modern Technologies, Education and Professional Practice in Geodesy and Related Fields, Sofia, Bulgaria, 5 – 6 November 2009.

FOTIOU A., PIKRIDAS C., ROSSIKOPOULOS D., SPATALAS S., TSIOUKAS V. & KATSOUGIANNOPOULOS S. (2009), The Hermes GNSS NtripCaster of AUTh. Proc. of the EUREF 2009 Annual Symposium, Florence, Italy, 27 – 30 May 2009.

HELLERSTEIN, J. M., NAUGHTON, J. F. & PFEFFER, A. (1995), Generalized Search Trees for Database Systems. Proc. 21st Int'l Conf. on Very Large Data Bases, Zürich, September 1995, pp. 562-573.

ISO (2000), ISO/IEC 13249-3:1999, Information technology – Database languages – SQL Multimedia and Application Packages – Part 3: Spatial. International Organization For Standardization, 2000.

MANOLOPOULOS, Y., NANOPOULOS A., PAPADOPOULOS, A. N. & THEODORIDIS, Y. (2005), R-Trees: Theory and Applications. Springer, 2005, ISBN 1-85233-977-2.

OGC INC. (2006-10-05), OpenGIS Implementation Specification for Geographic information – Simple feature access – Part 1: Common architecture. Open Geospatial Consortium Inc.

PHILIPONA, C. & KASTL, D. (2008-09-29), FOSS4G routing with pgRouting tools and OpenStreetMap road data. FOSS4G 2008 Workshops.

POSTGRESQL GLOBAL DEVELOPMENT GROUP (2009), PostgreSQL 8.4.2 Documentation. The PostgreSQL Global Development Group, 2009.

RAVINDRA, K. A., MAGNANTI, T. L & ORLIN J. B. (1993), Network Flows: Theory, Algorithms and Applications. Prentice Hall. ISBN 0-13-617549-X

REFRACTIONS RESEARCH (2010), PostGIS 1.5.1 Manual. Refractions Research, 2010.

Enabling INSPIRE for Aeronautical Information Management

Gernot KLINGER

The GI_Forum Program Committee accepted this paper as reviewed full paper.

Abstract

The scope of this work is to evaluate the draft guidelines of the INSPIRE (Infrastructure for Spatial Information in Europe) Data Specification on Transport Networks answering the question whether from aeronautical information management (AIM) perspective the INSPIRE data model concept, especially on air transport networks, is suitable as an aeronautical data exchange platform in the sense of a European Spatial Data Infrastructure (ESDI). The necessary modifications and enhancements concerning overlapping spatial data themes of the INSPIRE initiative as well as the used temporal and topology concepts are elaborated. Finally changes in the data model itself to ensure the usefulness of INSPIRE as an aeronautical spatial data infrastructure (SDI) are successfully undertaken by using the Unified Modeling Language (UML) for specification.

1 Introduction

From the beginning of civil aviation geospatial information is in common use. The navigation of aircraft and the control of air traffic demand extensive knowledge about geographic realities because they have an impact on air traffic services provided (e.g. airspace structure, air route and waypoint characteristics). Furthermore civil aviation does not stop at country boundaries and is international by its nature.

There is a proliferation of operating requirements to data quality, timeliness, provision, and exchange of digital aeronautical information due to safety reasons. So today's main goal of AIM is to provide aeronautical data to in-house or international partners or users. The importance of geospatial methods and infrastructure concepts, like an SDI, developed by geographic information science is growing steadily in AIM because the benefits are realised in the geospatial niche of aviation.

A review of future developments and imperatives concerning SDI and AIM clarifies a connection between these topics in the European initiatives of INSPIRE and SESAR (Single European Sky ATM (Air Traffic Management) Research Programme). So far the aviation community has rarely considered the INSPIRE initiative as an adequate platform for data provision and exchange.

The scope of the project consists of a review of the draft guidelines of the INSPIRE Data Specification on Air Transport Networks (INSPIRE 2008) answering the question whether from AIM perspective the INSPIRE data model concept, especially on air transport networks, is suitable as an aeronautical data exchange platform in the sense of an ESDI. This could establish INSPIRE as a counterpart to the EAD (European AIS Database) initiative. The necessary adjustments in the data model for the usefulness of INSPIRE as an aeronautical SDI are undertaken by using the Unified Modeling Language (UML) for specification (KLINGER 2009).

2 Spatial Data Infrastructures

SDI is the current keyword in the growing Geoinformation industry. RAJABIFARD and WILLIAMSON (2002) define an SDI as the interaction of the widest possible group of potential users, the network linkages and access, the political issues, the standards, and the data of different data sources for the effective flow and interoperable exchange of spatial information at different levels based on partnerships. Many organisations, administrations, and companies spend a lot of money on the establishment of an SDI. Now it is a known fact that an SDI does not have a primary advantage but is the essential precondition for a developing geospatial data exchange (MASSER 2006).

At present AIM is not a typical field of application for SDIs although the essential parameters are given. The current operational infrastructures of the air navigation service providers (ANSP) are very complex and often individually built. There is lack of interoperability, data sharing, and cooperative scheduling in investment, planning, and management concerns. Due to the awareness of resulting disadvantages the ANSPs are starting a transformation of their (spatial) data driven by the continuous progress in (information) technologies. In 2007 VAN DER STRICHT and STANDAR finally introduced the idea of SDI to the aeronautical community. A complete transition of the current infrastructures would be a difficult process because this SDI would have to fit different scales usually described as a pyramid of building blocks (RAJABIFARD et al. 2000). Normally an ANSP is a company which is in charge of an ATM for a state. So the establishment of a corporate SDI and a national (aeronautical) SDI has to be accomplished at the same time including all of their basic necessities and imperatives. Additionally attention should be paid to the future needs of the integration and provision of the spatial data of the ANSP in a regional (e.g. European) SDI.

The current draft of the INSPIRE initiative contains the management and provision of aeronautical information. But is it suitable for an AIM?

3 Aeronautical Information Management

Spatial data for AIM cannot be fully handled like "common" geospatial data. Additional to geospatially related standards many tried and tested international aeronautical regulations to be kept in mind in order to distribute matching data to the customers (KLINGER 2009). ICAO (International Civil Aviation Organization) delivers common reference systems, units of measurement, and requirements in temporality and data quality which an ANSP has to fulfil to be permitted to provide AIM to the customers (ICAO 2001a, ICAO 2001b, ICAO 2002, ICAO 2004). Data quality of aeronautical (spatial) data consists of the user requirements in terms of accuracy, resolution, and integrity.

The temporality concept of aeronautical data of a lasting character is called aeronautical information regulation and control (AIRAC) (ICAO 2004). The AIRAC system regulates the coming into effect of innovation, withdrawal, or significant change of aeronautical information. These effective dates are defined to an interval of 28 days. The information notified therein shall not be changed further for at least another 28 days after the effective date. Nevertheless it is possible to publish important ad hoc or temporary restricted announcements by notices to airmen (NOTAM) via an international fixed telecommunication network. Based on the temporality concept data traceability by the use of appropriate procedures in every stage of data production or data modification process has to be assured (ICAO 2004, EUROCONTROL 2007).

The fragmentation of the European ATM network and the development of separate national infrastructures cause substantial inefficiencies, a multiplicity of workload and costs, and a delay when introducing technical equipment on an international level. There is currently no common architectural design of a European ATM system (SESAR CONSORTIUM 2006a).

SESAR is the European ATM modernisation programme and combines technological, economic, and regulatory aspects. Finally a global, distributed aeronautical data management environment, managing aeronautical information as well as technical fundamentals, shall be established (EUROCONTROL 2006, SESAR CONSORTIUM 2006b). Due to the fact that the majority of air navigation data has a spatial dimension, a spatially enabled information management system like an SDI would lead towards a common aim.

Timely and reliable data for all phases of flight have to be made permanently and dynamically available for use in applications that perform the required tasks of any strategic or tactical ATM activity. Thereby AIM will be responsible for both the content (including formats, timeliness, collection, checking, distribution, etc.) and the proper management of the data (storage, consistency between databases, interfacing with other systems, etc.) (EUROCONTROL 2006, REID 2008).

Therefore EUROCONTROL develops an aeronautical information exchange model (AIXM) for electronic data exchange of AIM data to ensure that the data flow of aeronautical information keeps its quality, is efficient and cost effective, and supports real time (spatial) information. Another goal of this model is to support the vision of a single global data source for data-centric operations. AIXM is based on Extensible Markup Language (XML) and supports the sophisticated aeronautical temporality model (AIRAC and NOTAM), the International Organization for Standardization (ISO) standards for spatial data including the use of Geography Markup Language (GML), and the latest ICAO and user requirements (EUROCONTROL 2010).

In a nutshell, an SDI (supporting AIXM) could provide solutions for many requirements of SESAR because of its power in interoperability, scalability, and seamless transition. So far the SDI-concepts are not taken into consideration by the ANSPs and SESAR. Moreover it is important that SESAR deals with the INSPIRE subject and derives advantage from its integration. It would be an inconceivable benefit for the aviation industry to abandon its narrow perspective concerning solutions for aviation solely (KLINGER 2009).

4 Implementation

4.1 General considerations

First of all, questions concerning AIM cannot be solved by looking at the INSPIRE Air Transport Network alone. Important links and overlaps with other proposed spatial data themes elevation, buildings, atmospheric conditions, and meteorological geographical conditions are found to meet AIM topics like (natural and artificial) obstacles, aerodrome mapping, or weather and topographical impacts on air routes (KLINGER 2009).

On the basis of the draft guidelines of the INSPIRE Data Specification on Air Transport Networks (INSPIRE 2008) a revised data model is developed to meet the demands of environmental issues, the fields of application the draft proposes, and AIM. The data model shall make INSPIRE suitable as a framework for a European aeronautical SDI and set a counter movement to the EAD initiative.

The life cycle concept described in the draft guidelines of the INSPIRE General Network Model has to be adjusted to be suitable for the temporality concept of ICAO, that complies all aeronautical information. Due to the fact that the validity of aeronautical information can last several times the multiplicity factor of the validity of aeronautical information has to be adjusted. For instance a waypoint can be part of an air route network for some AIRAC cycles then the network changes and the validity of the waypoint expires. Another change in the network may lead to the reuse of a waypoint (for another air route).

The data quality issues of all attributes defined for the data model have to meet the requirements of ICAO Annex 15 (ICAO 2004) and of the European Commission Regulation (EU) No 73/2010 of 26 January 2010 laying down requirements on the quality of aeronautical data and aeronautical information for the single European sky initiative.

4.2 Topology

The topological concept has to be adapted in order to minimise distortions in the connectivity to other transport subthemes (road, rail, and water). Additionally the adaptation mirrors the realities in air navigation much more lifelike.(KLINGER 2009).

The most important change in the topology concept is done within the aerodrome area. Due to the resulting distortion between the different transport network subthemes when using the aerodrome reference point (ARP) for connectivity, the aerodrome point is displaced to the main entrance of the aerodrome. Usually the road and rail network end in its vicinity. Additionally this node is used for connectivity purposes only. The ARP is not dismissed because of its importance in daily aviation and becomes a new spatial object type which contains the main information concerning an aerodrome (Fig. 1).

The starting point of a standard instrument departure (SID) and the end point of an instrument approach procedure (IAP) are represented by a newly defined spatial object type called threshold point. This point indicates location and height of the runway threshold, which marks the point where published SIDs usually start and the touch-down of the IAPs is located (Fig. 1).

The distance between the aerodrome point and the threshold point is described by a newly introduced link, the aerodrome connector. This link feature indicates the (fictitious) distance the passenger is covering within the aerodrome area (by foot, vehicle, or aircraft).

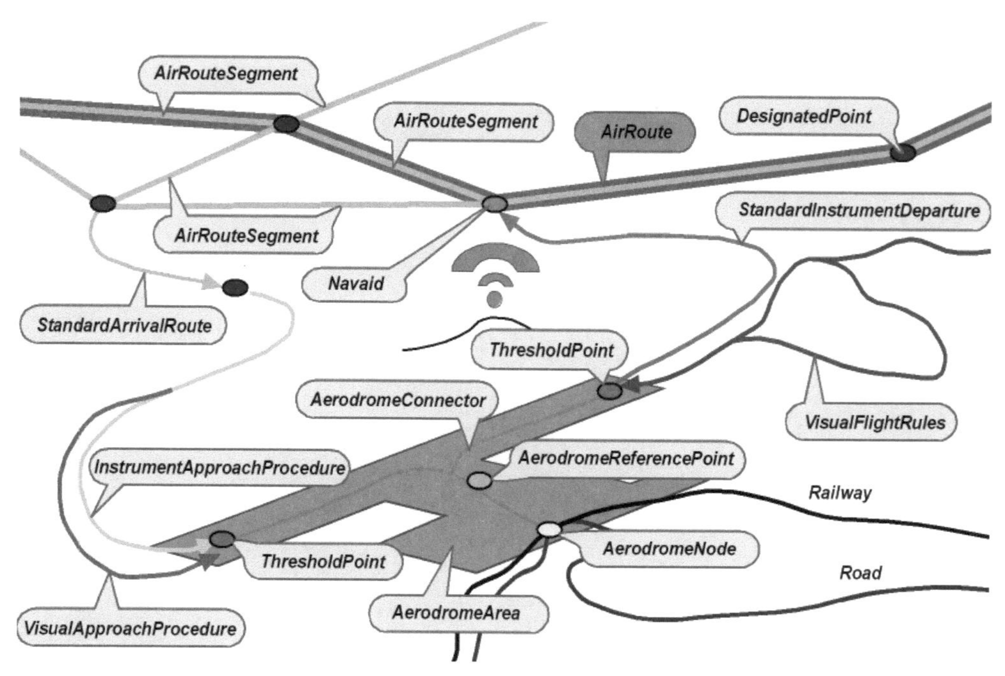

Fig. 1: Modified topology concept (KLINGER 2009)

Furthermore the transport links are simplified and all movements of an aircraft in the air are summarised in the spatial data type air route. A differentiation between the air route types is done by different attributes (departure, enroute, arrival, approach, ...).

Finally the aerodrome area has been diversified in its main parts of usage in order to provide more detailed information concerning the apron, the clearway, the infrastructure area, the runway, the stopway, and the taxiway (Fig. 2). Moreover this fits future AIM topics like the aerodrome mapping data base (AMDB).

4.3 Data model

UML is used to describe and depict the modifications and enhancements of the data model itself (Fig. 2). The design of the figures is ajar the INSPIRE draft document.

Although there is no big difference in the amount of the feature types in the UML models, major differences can be found in the number of the defined attributes, enumerations, and code lists including their values (Fig. 3, Fig. 4, Fig. 5, and Fig. 6). This indicates that the UML model in the project is much more explicit than the INSPIRE draft and more information is maintained in order to be suitable for AIM as a framework for European aeronautical data exchange. This leads to the assumption that the requirements of the fields of application (in AIM), in fact the INSPIRE draft should support, can be fulfilled better (KLINGER 2009).

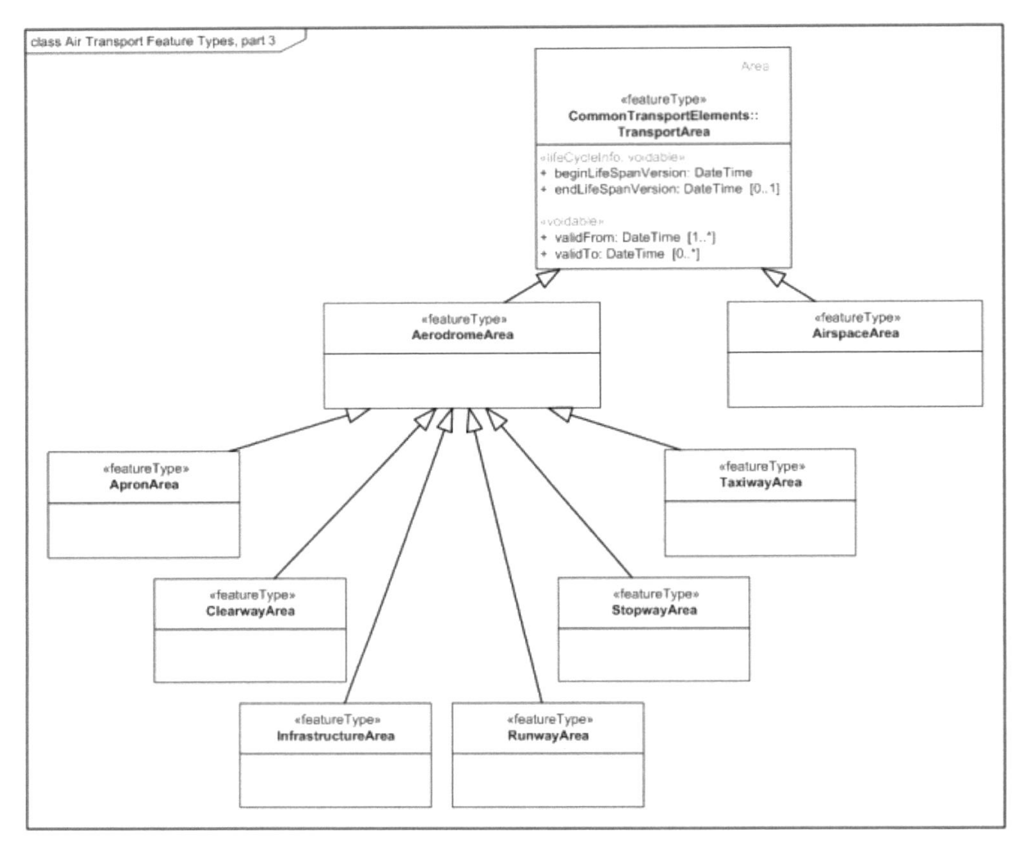

Fig. 2: Modified UML model of the aerodrome area (KLINGER 2009)

The integration of detailed information concerning the aerodrome infrastructure, runways, obstacles, air routes during take-off and landing phase, and airspaces solves the demands of AIM clearly. Moreover the defined data model is expandable which is important to integrate future AIM tasks to the data model of the aeronautical ESDI (e.g. AIXM, AMDB, electronic terrain and obstacle data (ICAO 2004)).

4.4 Example

With the help of the feature type concerning navigational aids (navaids) the modifications and enhancements of the defined attributes, enumerations, and code lists shall be demonstrated in order to enable INSPIRE for AIM. First of all the feature type was renamed from RadioBeacon (Fig. 3) to NavaidPoint (Fig. 4), because a beacon is only one type of a navaid. The INSPIRE draft only differentiates between the navaid type (e.g. NDB, DME) but no special information is delivered which is needed for air navigation and AIM. Now the definitions of all important attributes concerning navaids are procured like the identification, the frequency, or the channel where air traffic service is provided.

Fig. 3: Attributes of feature type RadioBeacon in the INSPIRE draft (INSPIRE 2008)

«featureType»
NavaidPoint

+ navaidPointClass: NavaidClassValues
+ navaidPointName: CharacterString
+ navaidPointIdent: CharacterString
+ navaidPointWorkingHoursType: WorkingHoursTypeValues
«voidable»
+ navaidPointFrequency: Measure
+ navaidPointFrequencyUom: UomFrequencyValues
+ navaidPointChannel: CharacterString
+ navaidPointWorkingHoursRemark: CharacterString
+ navaidPointMagVarValue: CharacerString
+ navaidPointMagVarDate: Measure
+ navaidPointMagVarAnnualChange: CharacterString
+ navaidPointAntennaElevation: Measure
+ navaidPointAntennaElevationUom: UomVerticalValues
+ navaidPointCollocatedNavaid: CharacterString

Fig. 4: Attributes of the feature type NavaidPoint in the data model suitable for AIM (KLINGER 2009)

«codeList»
RadioBeaconClassValues

+ DME = 1
+ NDB = 2
+ Tacan = 3
+ VOR = 4
+ VOR-DME = 5
+ Vortac = 6

Fig. 5: Code list concerning RadioBeaconClassValues in the INSPIRE draft (INSPIRE 2008)

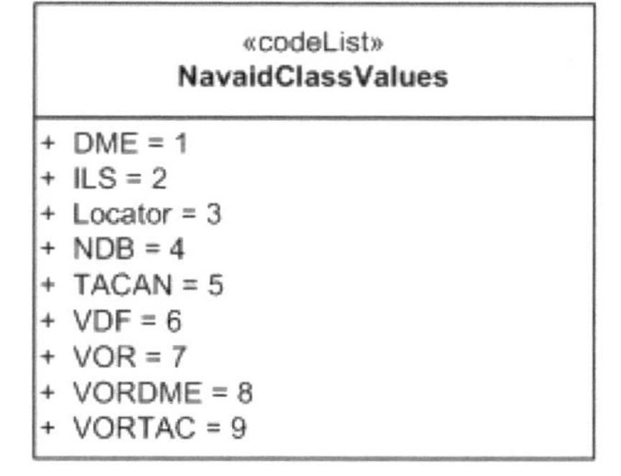

Fig. 6: Code list concerning NavaidClassValues in the data model suitable for AIM (KLINGER 2009)

Furthermore the enumerations or code lists of the defined attributes were adapted to the current needs of ATM (Fig. 5, Fig. 6).

5 Conclusion

A connection between SDI and AIM is clarified in the European initiatives of INSPIRE and SESAR. This work seizes the idea and reviews the draft guidelines of the INSPIRE Data Specification on Air Transport Networks to identify whether they are suitable as an aeronautical data exchange platform in the sense of a European aeronautical SDI (supporting AIXM). The results show that the INSPIRE data model on air transport networks needs some improvement and essential information is missing to meet the demands of AIM. Lack of information leads to problems in using the data model for considered fields of application and for AIM especially.

Based on the review results general modifications of the data model are provided with the objective of meeting the requirements of INSPIRE and AIM. The temporality concept is adapted in order to fulfil the required AIM demands concerning the concepts of AIRAC and NOTAM. The provided topological concept of connectivity to other transport subthemes like road, rail, and water is modified to suit the reality and to minimise discovered problems. Furthermore the topology within the INSPIRE Data Specification on Air Transport Networks has been changed on a big scale to meet the air navigation facts.

UML is used to describe and depict the modifications and enhancements of the data model itself. Major changes are made concerning the definitions of feature types, attributes, enumerations, and code lists. The UML model in the work is much more explicit than the INSPIRE draft. Supplementary information is maintained in order to be suitable as a framework for European aeronautical data exchange from the perspective of AIM. Moreover the defined data model is expandable which is important to integrate future AIM tasks to the data model of the aeronautical ESDI (e.g. AIXM, AMDB, electronic terrain and obstacle data).

Hopefully this work raises the awareness in the aviation community that INSPIRE and the data model defined in this work could be an important piece of a puzzle in the SESAR programme as this work is an important contribution to the design of INSPIRE from the aeronautical user perspective.

References

EUROCONTROL (2006), From AIS to AIM – A Strategic Road Map for Global Change, Version 1.0, June 2006, 56 p. – www.eurocontrol.int/aim/gallery/content/public/pdf/global_aim_strategy.pdf (accessed 1.04.2010).

EUROCONTROL (2007), Final Report for the Draft implementing rule on aeronautical data and information quality, Edition 2.0, 16 October 2007, 237 p. – http://www.eurocontrol.int/ses/gallery/content/public/docs/ru/ses_iop_adi_rep_v2.0.pdf (accessed 1.04.2010)

EUROCONTROL (2010), Aeronautical Information Exchange. – http://www.aixm.aero/public/subsite_homepage/homepage.html (accessed 1.04.2010).

ICAO (2001a), Annex 4 to the Convention on International Civil Aviation: Aeronautical Charts, Tenth Edition, July 2001, 159 p. (Incorporating Amendments 1-54; effective date: 22.11.2007).

ICAO (2001b), Annex 11 to the Convention on International Civil Aviation: Air Traffic Services, Thirteenth Edition, July 2001, 115 p. (Incorporating Amendments 1-44; effective date: 23.11.2006).

ICAO (2002), Document 9674 to the Convention on International Civil Aviation: World Geodetic System – 1984 (WGS-84) Manual, Second Edition, 2002, 125 p.

ICAO (2004), Annex 15 to the Convention on International Civil Aviation: Aeronautical Information Services, Twelfth Edition, July 2004: 132 p. (Incorporating Amendments 1-34; effective date: 22.11.2007).

INSPIRE (2008), D2.8.I.7 INSPIRE Data Specification on Transport Networks – Draft Guidelines: 142 p. – http://inspire.jrc.ec.europa.eu/reports/ImplementingRules/DataSpecifications/INSPIRE_DataSpecification_TN_v2.0.pdf (accessed 1.04.2010).

MASSER, I. (2006), What's Special about SDI Related Research?, International Journal of Spatial Data Infrastructures Research, 1, pp. 14-23. – http://ijsdir.jrc.it/editorials/masser.pdf (accessed 1.04.2010).

KLINGER, G. (2009), Enabling INSPIRE for Aeronautical Information Management, Master Thesis (unpublished), Z_GIS – Centre for Geoinformatics, Salzburg University, 143 p.

RAJABIFARD, A., WILLIAMSON, I., HOLLAND, P. & JOHNSTONE, G. (2000), From Local to Global SDI Initiatives: a pyramid of building blocks, 4th Global Spatial Data Infrastructure Conference, Cape Town, South Africa, 12 p. – http://www.sli.unimelb.edu.au/research/publications/IPW/ipw_paper41.pdf (accessed 1.04.2010).

RAJABIFARD, A. & WILLIAMSO, I. (2002), Spatial Data Infrastructures: an initiative to facilitate spatial data sharing, Global Environmental DBs – Present Situation and Future Directions, ISPRS-WG IV/8, GeoCarto International Centre, HongKong, 2, 30 p. – http://www.geom.unimelb.edu.au/research/publications/IPW/Global%20Environmental%20Book-SDI%20Chapter%206.pdf (accessed 1.04.2010).

REID, K. (2008), Towards a Common Understanding – The AIM Concept – Creating the Network-centric Environment, Presentation at the Global AIM Congress 2008, 17 p. – http://www.eurocontrol.int/aim/gallery/content/public/events/congress_2008/S2_ECTRL_Reid.pdf (accessed 1.04.2010).

SESAR CONSORTIUM (2006a), Air Transport Framework – The Current Situation, Version 3.0, July 2006, 70 p. – http://www.eurocontrol.int/sesar/gallery/content/public/docs/DLM-0602-001-03-00.pdf (accessed 1.04.2010).

SESAR CONSORTIUM (2006b), Air Transport Framework – The Performance Target, Version 2.0, December 2006, 99 p. – http://www.eurocontrol.int/sesar/gallery/content/public/docs/DLM-0607-001-02-00a.pdf (accessed 1.04.2010).

VAN DER STRICHT, S. & STANDAR, Å. (2007), Geospatial Interoperability for ATM, Presentation at the Global AIM Congress 2007, 17 p. – www.eurocontrol.int/aim/gallery/content/public/events/mini_globalais/S4-D-AIM.pdf (accessed 1.04.2010).

Algorithms for Detecting and Extracting Dikes from Digital Terrain Models

Tobias KRÜGER

The GI_Forum Program Committee accepted this paper as reviewed full paper.

Abstract

Digital Terrain Models are the main input data for flood simulations besides precipitation and/or flood gauge information. Hereby the terrain is considered as static and unchangeable in most cases. But in the flood risk management and flood simulations context it can be of interest to perceive dikes as modifiable terrain features. The consideration of dikes as separate objects makes it possible to establish a Digital Dike System as part of a dike information system.

The following study deals with the adaption and application of algorithms used for object detection and extraction modified for dikes in DTM. The presented methods allow the automated identification and mapping of dike geometries.

1 Introduction

Although flood protection has a long tradition and is well developed in the Central European region (e. g. the Rhine River has been embanked and straightened from the middle 19th century on) flooding can cause extraordinary damages as one can see in Table 1 which shows an (incomplete) list of considerable recent flood events in Central Europe.

Table 1: Major recent floods in Germany (MERZ 2006, unless indicated otherwise)

Date	**Event description**
1993 (December)	Flood on River Rhine, labelled as “Centennial Flood”
1995 (January)	Flood on River Rhine; total damage 1993/95: 5.5 billion €
1997 (July/August)	Flood on River Oder (Germany, Poland), damage: 5 billion €
1999 (May)	*Whitsun Flood* in Bavaria and Austria, dam.: 335 million €
2002 (August)	Flood on Rivers Elbe and Danube, damage: 11.8 billion €
2005 (August)	Flood in southern Bavaria, drainange > HQ_{500} (LFU 2006, 3)
2006 (March, April)	Flood on River Elbe, underflow all-time record gauges because of less headwaters dike breaches than 2002 (NLWKN 2007)
2007 (August)	Heavy-rain induced flood event in Germany (Rhine), Switzerland and Austria

River floodplains are extensively used as settlement areas and the need for flood protection is evident. However, absolute protection cannot be guaranteed and therefore there are efforts of re-naturation of floodplains. In this context and especially after the centennial flood of the Elbe River in 2002 a considerable number of research activities dealing with flood protection, flood risk, and flood risk management has emerged. The German BMBF-National Research Programme "Risk Management of Extreme Flood Events" (RIMAX) is an example for the facilitation of research in these fields (RIMAX online 2010).

1.1 Digital Terrain Models in use for flood simulations

The simulation of flood events is essential for the estimation of the potential flood hazard in a given area. The variety of algorithms used for flood simulation is very broad. It ranges from one-, two-, and three-dimensional static to non-static modelling approaches and can include lots of input parameters such as precipitation, evaporation, surface roughness, polder effects, viscosity of the water, and others.

Due to the increasing availability of laser scanning data the amount of high precision Digital Terrain Models (DTM) is growing constantly. Also in a hydrological context the use of laser scanning DTMs is essential. The *Bundesanstalt für Gewässerkunde* (BfG, engl.: German Federal Institute of Hydrology) provides gauge prediction systems for the *Bundeswasserstraßen* (federal waterways) employing high precision DTM datasets which are combined from laser scanning data and bathymetric data of the river banks.

For a research project on the change of flood risk due to climate change, a high resolution DTM of the River Elbe was assembled from various sources (SCHANZE 2008). All parts derived from laser scan DTMs and had a pixel spacing of 2 m and an absolute elevation accuracy of ±15 cm. The resulting DTM covered the German part of the River Elbe and its drain effective floodplain from the Czech border to river kilometre 586.

Flood risk research deals with the needs and perception of several scientific disciplines. Geographic data analysis allows for estimating effects of flooding for certain regions or even individual buildings. Digital Terrain Models are always necessary to perform flood simulations and, therefore, for flood risk mapping or the planning of flood protection buildings like dikes or mobile protection walls.

For I intend to understand dikes as partly non-static features built onto the earth surface, the term *Modifiable DTM* has been introduced. Indeed the dike system is intended to be a modifiable part of the terrain because it can be a part of flood protection planning which results in dike system adaptations.

1.2 Digital Dike Model

Dikes are the most important flood-protection facilities. This is why they are of outstanding interest in the public perception of the flood protection. The dike's crest level is determined by the design flow for which the flood protection is desired. Unfortunately, dikes are often not adequately surveyed or the information on crest levels in geo-datasets does not correspond with reality (KRÜGER 2009, 107).

Because of their importance, it seems to be reasonable to establish an information system dealing with dikes as objects which are semantically and geometrically independent from the earth surface. So one would be able to manage dikes and hold information of any kind for dikes or dike sections. Therefore, a Digital Dike Model (DDM) is proposed which enables to locate the footprint on dikes within a map as well as to change dike geometries or even relocate dikes in order to spare floodplain areas for natural floodings.

For the establishment of a Digital Dike Model one has to decide which kind of digital representation would be applicable. When combining the needs of mapping and modelling there are some main factors that have to be accounted for: (i) For flood modelling the DDM has to provide 3D-information about the dike bodies. (ii) For topographic mapping the DDM has to provide lineage (vector) information. (iii) The information system should be able to store attributes, e.g. crest level, slope angle, construction date, maintenance dates etc. (iv) Dikes are to be modelled as changeable objects. That means that new dikes can be erected and merged with the underlying DTM or that existing dikes can be removed, relocated, lowered, or heightened.

These conditions can be met effectively by a combination of different data types. It is useful to provide three-dimensional data as raster as well as vector data. A raster model can be merged with existing DTM while the attributes of the vector lines can hold additional information.

Geometrical dike information is implicitly contained in DTMs because they usually appear as parts of the surface (BERKHAHN & MAI 2006, KRÜGER & MEINEL 2008). So it is necessary to extract them from the digital elevation data in order to get explicit information for storage in a DDM. The separation of the dikes allows a number of analyses in the flood risk management context. These include performing flood simulations under unaffected conditions of the natural floodplain, the derivation of specific geometric information on dikes (mainly crest levels above ground), the testing of crest levels for their reference values, the simulation of dike building/deconstruction/relocation measurements, and the semi-automated geometrical correction of dike lines for mapping purposes.
Initially the base DTM has to be divided into a DTM_0 approximating the natural earth surface and a Raster DDM (DDM-R) from which vector data (DDM-V) can be derived. Vector lines can hold attributes and are accessible for editing. Modifications can be re-modelled as DDM-R which itself is in use for merging with the DTM_0 resulting in a changed DTM^* for further flood simulations (Fig. 1).

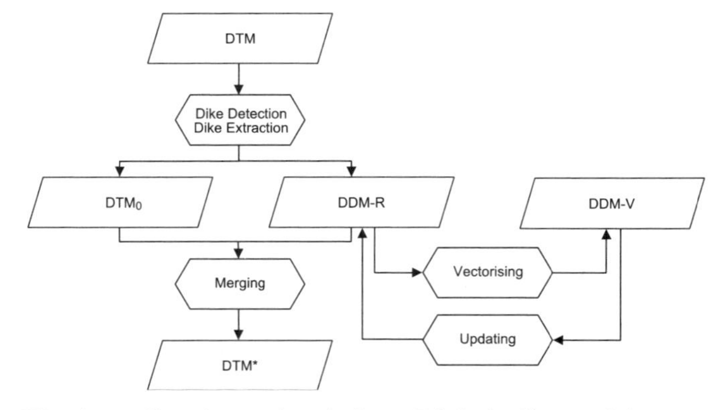

Fig. 1: Creating and updating a Digital Dike Model

2 Dike Detection in DTM

In order to establish a Digital Dike Model it is necessary to perform dike detection and extraction on existing laser scan DTMs which is an object identification task. Object detection algorithms always analyse level differences or gradients to decide if a certain point belongs to the ground or is part of an elevated object. The identification of dikes is also based on the evaluation of height differences, which means to identify sudden level leaps occurring within short distances (MAYER 2003).

2.1 Elevation difference thresholding

Basically, this method examines the difference of any cell value to its neighbourhood. The neighbouring cells are used to determine the surrounding ground level which is referred to the minimum cell value of the neighbourhood. If the difference exceeds a given threshold the evaluated pixel is marked as potentially elevated. Both, the size of the neighbourhood and the threshold value, have impact on the size of the detectable objects. Vice versa it is necessary to know about the dimensions of the objects to be identified for a reasonable dimensioning of neighbourhood and threshold.

Filter design

The neighbourhood assessment is done by a moving window. For each pixel its value difference to the minimum within the neighbourhood is calculated. If this elevation difference exceeds a predefined threshold $\Theta(\Delta h)$ the pixel is marked as elevated and recorded in a binary pixel mask $P_{elevated}$. The dimensions of the filter F have to ensure that at least one ground pixel is to be found inside the window. Further it has to be taken into account that only terrain elevations are detected which are embraced by two contrarily arranged slopes of similar angles. Otherwise any terrain edge would be marked as elevated object.

This is done by dividing the moving window into two halves covering the area left and right of the dike line respectively. Given the dike lineage is in straight north-south direction it would be possible to calculate the elevation difference sequentially for the eastern and western half of the filter matrix. Only if both level differences exceed the threshold the elevation flag is set. For the dike line direction is usually unknown the moving window is split twice into a north-south filter pair and an east-west filter pair (see Fig. 2). This way the filter design is assuring that dikes of either azimuth direction can be detected. The elevation mark would be set only if both half-window differences of a filter pair exceed the threshold:

$$P_{elevated} = 1 \text{ if } (\Delta_{h,north} \geq \Theta) \wedge \Delta_{h,south} \geq \Theta) \vee (\Delta_{h,west} \geq \Theta \wedge \Delta_{h,east} \geq \Theta)$$

The conditional filtering is done by sequentially applying the four half-windows on the raster DTM. So each pixel is evaluated four times against the cell minima within the regions indicated with “1” to get its level differences to either direction. To make calculations less time-consuming the four half-matrices can also be reduced to matrix vectors of the dimensions of [1×n] or [n×1]. This way only the cell values within the same row or column of the central pixel are evaluated. The main parameters for the application of the filtering are the dimensions of the filter matrix itself and the threshold value. To detect objects of 25m width, which can be seen as a maximum object width of major river dikes, a half filter

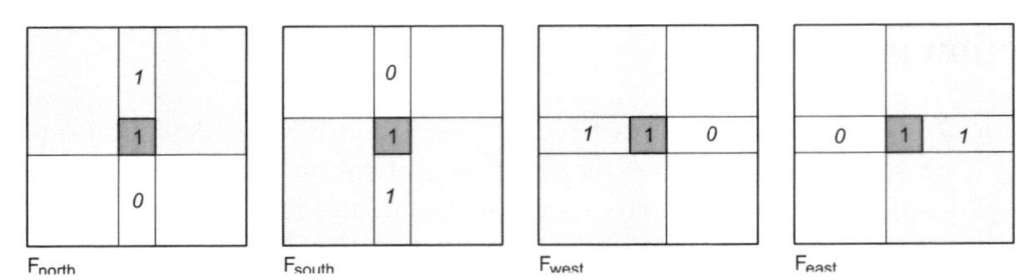

Fig. 2: Combined filter windows by half matrices and row/column matrix vectors

has to stretch at least over a distance of 12.5 m. The elevation difference threshold has to be adapted to the object height that is desired to be detected.

Evaluation of detected dike objects by form parameters

The pixels marked as potentially elevated have to be examined before they can be accepted as dike surface points. Form parameters describing geometric properties for each object can be calculated. These parameters are the *Area A* and the *Form Factor F*.

A is determined by the number n of pixels in the object and by raster width Δx:

$$A = n\,\Delta x^2$$

The *Form Factor F* is the quotient of the squared perimeter u divided by area A and describes the object's deviation from circular shape, for which is $F_{Circle} = 4\pi$:

$$F = \mathrm{u}^2\,A^{-1} \geq 4\pi$$

Only those pixel objects passing two thresholds of minimum *Area* and minimum *Form Factor* will further be treated as detected dikes. The thresholds depend on the scene investigated and have to be adapted. A minimum of 100 pixels for A and a minimum of 100 for F obtained reasonable results with the used DTM.

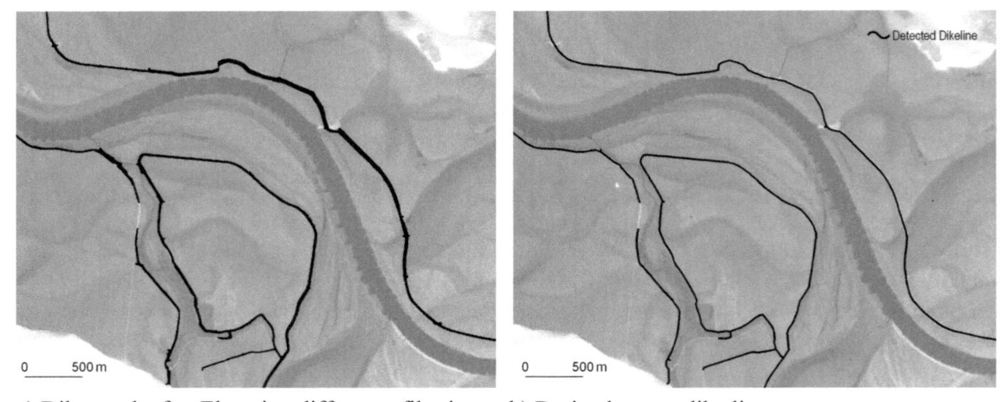

a) Dike mask after Elevation difference filtering b) Derived vector dike lines

Fig. 3: Dike detection by Elevation difference thresholding and derived vector lines (DTM data: BfG)

2.2 Slope analysis and relative maximum analysis

The detection method described above is based on absolute elevation differences. Additionally, dikes have other very distinct geometrical characteristics. This especially concerns the embankment slopes which are described by the parameters gradient (slope angle) and aspect (difference from the southern direction) and their relative extrusion from the surroundings.

Dike slopes

The *Slope* θ is defined by constructional needs. The width-to-height ratios of dike slopes range from 1:2 to 1:6 and are in most cases defined by 1:3 (PATT 2001, 255). This means that dike slope angles vary within the interval [8°; 26°]. Therefore a slope mask can be defined as:

$$P_{slope} = \begin{cases} 1 \text{ if } 8° \leq \theta \leq 26° \\ 0 \text{ otherwise} \end{cases}$$

Aspect difference

The *Aspect* ψ is defined as the angle between a straight line pointing south and a line feature on the earth surface. The crestlines of dikes divide the two dike slopes facing contrary directions. The aspect difference of the slope surfaces ideally has a value of 180°. The selection mask of DTM regions with valid aspect differences of $\Delta\psi$ is containing the possible dike crestlines. Due to the geometry constraints of the raster DTM a tolerance ε is introduced and an aspect mask is generated:

$$P_{aspect} = \begin{cases} 1 \text{ if } \Delta\psi = 180° \pm \varepsilon \\ 0 \text{ otherwise} \end{cases}$$

Relative maximum

Dikes significantly protrude from their surrounding. Given the assumption that dikes are usually situated in a quite flat and even terrain, the DTM can be analysed for significant local maxima which might be part of an elevated object. If the filter window is positioned on top of a crest level it can be assumed that the central pixel value is similar to the other crest level pixels in the window which should all be ranging around the neighbourhood maximum. To evaluate the proximity of a cell value z to the local maximum Z, an interval is defined using the standard deviation σ of the cell values within the moving window scaled with factor k. So a tolerance of the allowed relative difference from the actual neighbourhood maximum is defined and is used to establish the relative-maximum mask:

$$P_{relmax} = \begin{cases} 1 \text{ if } z > Z - k\sigma \\ 0 \text{ otherwise} \end{cases}$$

The addition of the three masks resulting from the slope interval selection, aspect difference, and relative maximum analysis enables to detect dike lines in a DTM. The summation of the three preliminary masks results in a raster with integer values ranging from 0 to 3 depending on the number of times a pixel has been marked. All pixels that have met every condition will be flagged as elevated:

$$P_{crest} = \begin{cases} 1 \text{ if } P_{slope} + P_{aspect} + P_{relmax} = 3 \\ 0 \text{ otherwise} \end{cases}$$

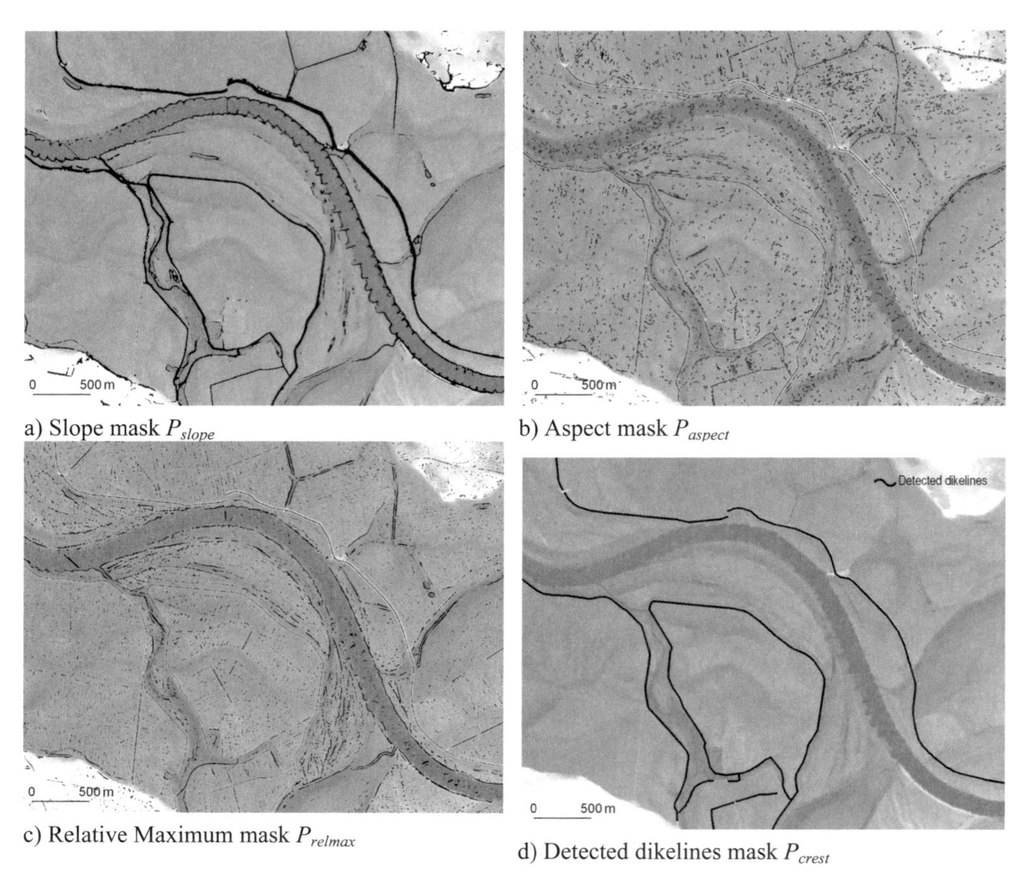

a) Slope mask P_{slope}

b) Aspect mask P_{aspect}

c) Relative Maximum mask P_{relmax}

d) Detected dikelines mask P_{crest}

Fig. 4: Dike detection through combined slope, aspect and relative maximum analysis (DTM data: BfG)

This method focuses on the crest of the dikes and therefore the result merely consists of lines rather than footprints as the dike mask resulting from the Elevation difference thresholding (Fig. 4).

2.3 Dike extraction and derivation of DDM information

After the detection of existing dike geometries it is necessary to extract the objects in order to derive geometrical information for the Digital Dike Model. Dike extraction removes the detected dike bodies from the digital terrain surface and restores the state before the dikes had been erected and approximates the natural surface level. The actual task of dike extraction therefore is to cut out the detected areas and interpolate the surface with a smooth transition to the remaining regions.

Dike base height restoration

Base height interpolation can be performed by interpolation using the cell values of the edge pixels as reference points. Since geometries of dike footprints are quite elongated and

a) DTM

b) DTM_0 after dike extraction

Fig. 5: Perspective view of a DTM section before and after dike extraction (DTM: BfG)

the surrounding terrain level is usually flat, a triangulation should achieve reasonable results. So the DTM gaps remaining after the cut-out of the dike footprints are closed by triangular irregular networks (TIN) on the basis of their edge pixel values. To get smooth edges working with small overlapping bands around the dike cut-outs is recommendable.

Derivation of geometrical dike information

Once the dikes have been detected and removed the desired geometrical information for the Digital Dike Model can be derived. The result of the subtraction of the resulting DTM_0 from the original DTM forms the DDM-R:

$$DDM\text{-}R = DTM - DTM_0$$

From this dataset we can derive a vector line representing the dike crest. The cell values directly indicate the objects heights, and so the crest levels on any position on the dike line can be extracted.

With raster-vector conversion it is possible to get lines that approximate the axis of the raster objects. In most cases the vertices of resulting lines do not lie exactly on the top of the dike crest (Fig. 3b). This can be corrected by shifting the line vertices onto near local maximum cells of the DTM. It is possible to get dike-line vectors representing the actual dike lineage very tight depending on the maximum distance of its vertices.

The graphics in Fig. 6 show dike lines of official German topographic data (ATKIS®, Authoritative Topographic-Cartographic Information System) and dike lines derived from DTM based dike detection. The main difference is that dikes with crest levels lower than 3 m are not recorded in the official dataset. Contrarily the minimum crest level for automated dike detection is dependent on the employed parameter settings.

The numbers on the dike lines in subfigure a) are indicating the stored object height. This information is not available for all dike vectors (value 0). Figure 6b shows detected dike lines with crest levels that have been extracted for the single line vertices. Therefore it is possible to selectively check dike heights or to calculate mean crest levels for larger dike sections.

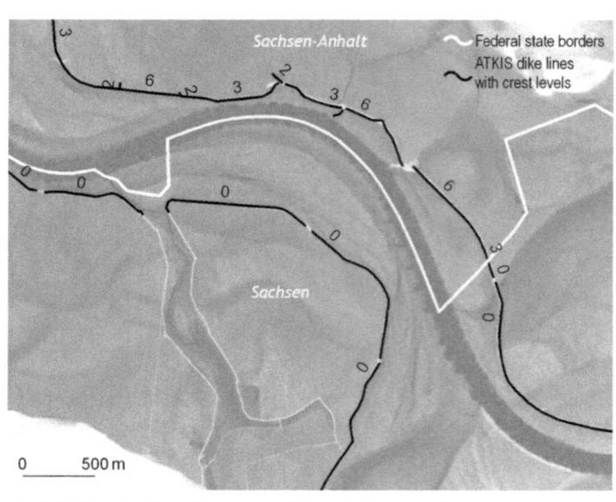

a) ATKIS dike lines with height attributes

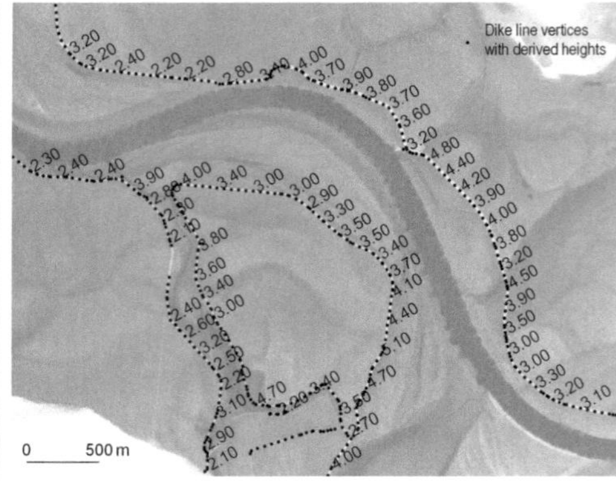

b) DTM-detected dike lines with heights

Fig. 6: Comparison of height attributes official topographic data (a) and of detected dike line vertices (b) from DTM (DTM data: BfG, Topographic data: ATKIS®)

3 Conclusions and Outlook

The presented algorithms of object identification, raster data manipulation and data conversion have been adapted and further developed to meet the special needs of a dike management and planning. The concept of a Digital Dike Model (DDM) was developed that is able to combine digital surface data in form of a laser scan DTM with semantic and object-specific information on dikes and dike sections. The DDM concept is based on the separation of the actual earth surface from the dikes which are modelled as individual objects. This allows modelling the consequences of changes in the dike system for flood events.

Once the separation of surface and embankments has been performed it is feasible to model planning alternatives for flood protection before the actual in-situ realisation. The DTM becomes a dynamic element (in a specific aspect) within the flood risk system. So the implementation of Digital Dike Models can be a useful brickstone for an integrated flood risk management.

Generally we could show that standard laserscan DTM with a resolution of 2 m as they are employed for flood simulations are suitable for all kinds of dike modelling. Furthermore the methods can also be used for crest level control and for the control of official topographic information (e. g. ATKIS). The latter turned out to be incorrect in many cases concerning the object heights of dikes. Also the geometry of existing dike lines can be corrected for an improved fitting of the actual local maxima within the DTM.

Raster widths larger than 2 m have not been tested in this study. But in general it is evident that the employed DTM has to ensure that dikes emerge significantly from the surrounding. Coarse-scale raster datasets do not contain the degree of detail necessary for the thresholding of level differences within a neighbourhood or to calculate slope angles of dikes. Another aspect is that the crest width of dikes is often about 3 m. So a larger raster resolution would result in inaccuracies of crest levels due to the resampling of pixels in the upper zone of the dikes. Therefore the derivation of crest line geometry as well as the determination dike heights is complicated.

Ongoing research and development should mainly emphasise the detection algorithms. Especially on dike crossings there are still problems which at the moment can only be overcome by sequentially applying the filtering with modified parameter settings or by manual editing of the dike footprint masks. Possibly other object identification methods could be applied that not only rely on the raster data but work immediately with point clouds of the laser scanning.

References

BERKHAHN, V. & MAI, S. (2006), Detection of Terrain Features Embedded in a Pre-Processor for Topographic Data. 7th International Conference on Hydroinformatics (HIC 2006). Nice, France, pp. 2383-2390.

KRÜGER, T. (2009), Digitale Geländemodelle im Hochwasserschutz: Detektion, Extraktion und Modellierung von Deichen und vereinfachte GIS-basierte Überflutungssimulationen. Kartographische Bausteine, 37. Dresden.

KRÜGER, T. & MEINEL, G. (2008), Using Raster DTM for Dike Modelling. Lecture Notes in Geoinformation and Cartography: Advances in 3D Geoinformation Systems. Berlin, pp. 101-113.

LFU – BAYERISCHES LANDESAMT FÜR UMWELT (2006), August-Hochwasser 2005 in Südbayern (Endbericht). Augsburg.

MAYER, S. (2003), Automatisierte Objekterkennung zur Interpretation hochauflösender Bilddaten. Phdthesis. urn:nbn:de:kobv:11-10032574. Berlin

MERZ (2006), Hochwasserrisiken. Grenzen und Möglichkeiten der Risikoabschätzung. Stuttgart

NLWKN (Niedersächsischer Landesbetrieb für Wasserwirtschaft, Küsten- und Naturschutz, 2007), Jahresbericht 2006. – http://cdl.niedersachsen.de/blob/images/C35499083_L20.pdf. Norden.

PATT, H. (Ed.) (2001), Hochwasser-Handbuch. Berlin.

RIMAX online (2010), BMBF-National Research Programme "Risk Management of Extreme Flood Events" (RIMAX). – http://www.rimax-hochwasser.de/400.html?&L=1 (last visited: 04/19/2010).

SCHANZE, J. (Ed.) (2008), Veränderung und Management der Risiken extremer Hochwasserereignisse in großen Flussgebieten – am Beispiel der Elbe (VERIS-Elbe). Endbericht, Stand Mai 2008. Unpublished. Dresden.

Performance Evaluation of Two Solution Methods for Time Dependent Bi-objective Shortest Path Problems

Senthanal Sirpi MANOHAR and Günter KIECHLE

The GI_Forum Program Committee accepted this paper as reviewed full paper.

Abstract

Path finding methods have been extensively studied by the researchers for the past three centuries. The richness of the shortest path problem has lead to the development of various solution methods. Availability of traffic information due to the revolution in the Information and Communication Technology (ICT) field for the past two decades added further richness to the problem.

The two solution methods discussed in this research work are a method with hierarchization of objective functions and a method with a meta-criterion function. The objective of this research work is to extend these two existing solution methods for time dependent bi-objective shortest path problems and compare the performance of both. The two solution methods discussed in this research work use a modified Gandibleux revision of Martin's algorithm and Dijkstra's algorithm with Analytic Hierarchy Process (AHP) in the time dependent routing environment. The two objectives discussed in this research work are minimization of trip travel time (travel time objective) and trustworthiness of the computed trip travel time (reliability of travel time). Both solution methods use travel time as the first objective and reliability of travel time as the second objective. The two solution methods use historical travel time data and Floating Car Data (FCD) as a data source for dynamic traffic information and compute distance matrices, which should be used later as input for Vehicle Routing and Location problems (VRP).

The aim of the performance comparison is to find which Multi Objective Shortest Path (MOSP) solution method fits better in the realistic VRP environment. The two solution methods discussed in this research work yield same results for smaller path lengths and different results for larger path lengths. Performance of the method with a meta-criterion function is better than the method with hierarchization of objective functions. Both solution methods can be extended to work with further potential objectives. Hence, this research work concludes that the method with meta-criterion fits better in realistic VRPs.

1 Introduction

The shortest path problem deals with the geographic questions like "*which is the shortest/fastest/safest/cheapest route from one/more place(s) to one/more destination(s)*". Various solution methods have been continuously developed to handle such geographical questions. The geographic question which is addressed in this research work is "*which is the reliable faster route from one/more place(s) to one/more destination(s)*". The time dependent bi-objective fastest path problem discussed in this research work is visualized in the figure 1, where S is the start location and T is the destination location. Two fastest paths (*a* and *b*) were computed with the corresponding trip travel time and reliability of trip travel time as shown in the figure 1. According to the geographic question addressed in this research work, *path a* is one pareto-optimal path, since *path a* has a better realiability parameter than *path b* although travel time is slightly longer than for *path b*. In scenarios, where high reliability is desired by the user, *path a* would be considered as the optimal path.

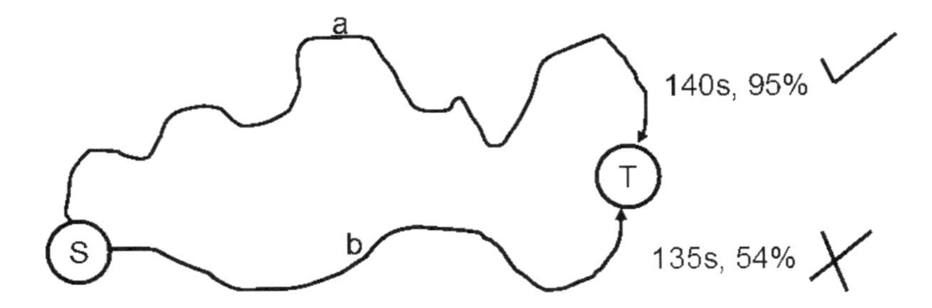

Fig. 1: Problem visualization

This kind of geographic question may be considered as the bi-objective or multi-objective fastest path problem and can be dealt like multi-objective routing problems. The following subsections discuss relevant issues about estimating the reliability of the travel time and time dependent bi-objective routing process model.

1.1 Street Networks & Travel Time Reliability

The formal definition of a bi-objective fastest path problem can be described as follows: Let G = (A, V) be a street network, where A is the number of edges and V is the number of nodes contained in the street network. The edges of the street network have two non-negative weights T (i, j, t) and R (i, j, t), where T (i, j, t) is the forecasted travel time required to travel between the nodes i and j at the time of day t and R (i, j, t) is the reliability of the forecasted travel time between the nodes i and j at the time of day t. The travel time parameter T is expressed in seconds and the reliability parameter R is expressed in percentage.

The estimation of reliability on Floating Car Data (FCD) travel time largely depends on probe vehicle sample size. The research by QIANG (2007) discusses two methods which estimate the reliability of FCD data. Standard Deviation Formulation and Confidence Interval (CI) (QIANG 2007) are those methods. The CI method is more appropriate when the probe vehicle sample size is small (QIANG 2007). The reliability parameter used in this research work is computed using CI method. The CI method defines a travel time interval range for FCD and observes if the historical travel time falls within this interval range or not. If the historical travel time falls within this CI interval range then the reliability is 95% and if not, the further the differences, the lower the reliability. Figure 2 represents a street link with historical travel time (20s), FCD travel time (22s) expressed in seconds along with its reliability (90%) parameter expressed in percentage.

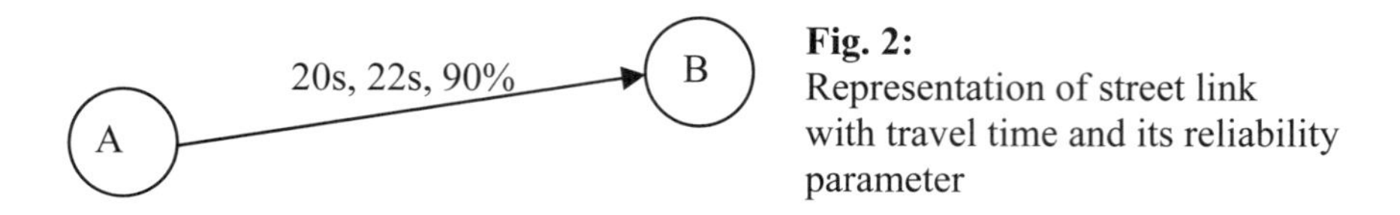

Fig. 2: Representation of street link with travel time and its reliability parameter

This research work emphasizes the comparison of performance metrics of two solution methods and assumes the availability of the travel time reliability data.

1.2 Bi-objective routing process model

Figure 3 shows the bi-objective fastest path routing process model. The historical travel time and reliability of travel time along with the street network are the input to the bi-objective fastest path routing process model. The routing module computes multi-period distance matrices which can be later used in vehicle routing and location optimization procedures.

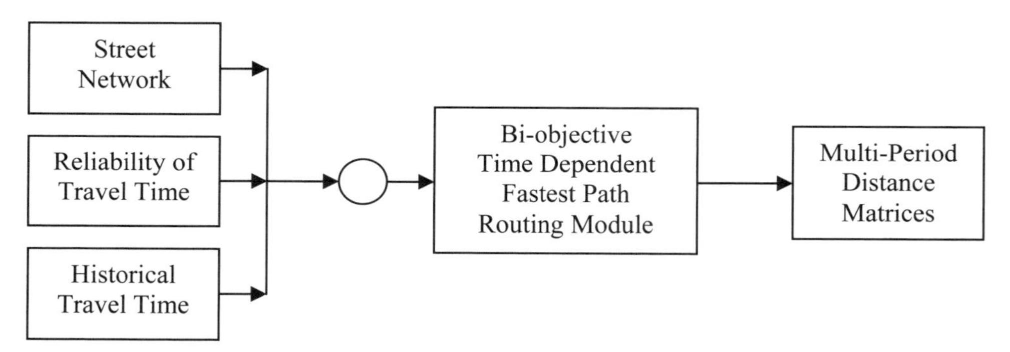

Fig. 3: Process model for bi-objective fastest path routing

1.3 Motivation & Literature Review

Traditionally, shortest path problems were assumed to be single objective in nature. In practice, several real-life situations require characterization of multiple objectives (TARAPATA 2007). Shortest path problems with multiple objectives in nature are addressed as multiobjective (multicriteria) shortest path (MOSP) problems. Due to the revolution in ICT live traffic congestion data becomes available. Solution methods which can utilize these data in multiobjective shortest path problems are still underdeveloped. The main goal of this research work is to extend the two existing multi-objective solution methods to accommodate live traffic congestion data produced by FCD. This research work also focuses on performance evaluation of the both solution methods.

Shortest path problems have been one of the interesting research topics for the scientific community for the past three centuries. Hence the literature for the shortest path problems is huge and spreads over a large variety of problems. Despite abundant literature, dealing with time dependent bi-objective shortest path problems is rather limited. The research work by TARAPATA (2007) provides an overview of categorization of several MOSP solution methods and categorized the MOSP into five categories. The following research work discusses two of those MOSP solution methods. The method with the hierarchization of objective functions utilizes the revised Martin's algorithm provided by GANDIBLEUX et al. (2006) which generates k-shortest paths in the first stage of optimization. In the second stage of optimization an optimal path is chosen using an Analytical Hierarchy Process (AHP) based decision rule (XIANG et al. 2007). The method with a meta-criterion function uses the Dijkstra's algorithm with an arc meta-function. The First-in-first-out (non-overtaking property) condition is considered in both solution methods. Non-overtaking is achieved using piecewise constant speeds for edges proposed by ICHOUA et al. (2003).

2 Discussion of Solution Methods

Traditionally path finding methods are used to find, either shortest path or fastest path assuming the shortest path problem is single-objective in nature. Most of the routing situations require considering multiple criteria to find an optimal path. In such cases MOSP solution methods are used (TARAPATA 2007). The choice of a MOSP solution method depends on the type of problem and the number of criteria involved in the routing problem. This research work discusses two of these MOSP solution methods described in the work of TARAPATA (2007). The solution methods are as follows:

1. Method with hierarchization of objective functions.
2. Method with a meta-criterion function.

Both the solution methods extend Dijkstra's algorithm for single objective shortest paths calculation. The following sections discuss the original Dijkstra's algorithm and the solution methods in detail.

2.1 Dijkstra's Algorithm – An Overview

Dijkstra's algorithm is a label-setting procedure which marks only one label permanent at every iteration and finally converges towards the fastest path. In ist limited to problems with non-negative costs (here, travel time of each arc). The performance and the adaptability of the Dijkstra's algorithm largely depends on the representation of the street network. The original Dijkstra's algorithm is based on node labeling where only one node was marked at every iteration. Representation of turn restrictions and turn costs are inefficient in the node labeling approach. However, the arc labeling based representation of street network provides efficient integration of turn restrictions and turn costs and also consumes less memory (GUTIERREZ & MEDAGLIA 2008). This research work utilizes arc labeling instead of node labeling in Dijkstra's algorithm.

Original Dijkstra's Algorithm

The basic idea behind the Dijkstra's algorithm is to maintain a list of arcs with travel time from the source arc location and choosing the least cost arc during each iteration. This list is a data structure. The runtime performance of Dijkstra's algorithm heavily depends on the process of selecting an arc with least travel time cost in the list. This list can be a special data structure like a heap or a bucket. This research work utilizes the bucket data structure as means to store arcs with its travel time.

Time Dependent Dijkstra's Algorithm

In time dependent routing problems, travel time depends on the time of day. The original Dijkstra's algorithm has to be extended to address these time dependent issues. Figure 4 shows a street representation with time dependent travel time, in this case with values for 96 time intervals.

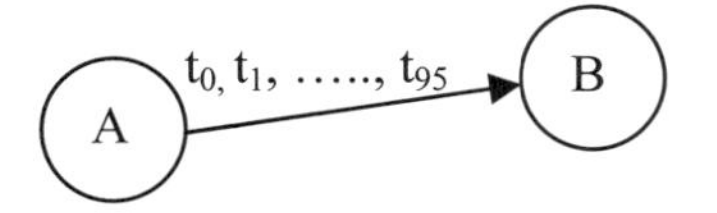

Fig. 4: Representation of street link with time dependent travel time

This research work uses floating car data as source for time dependent travel time data. The floating car data is collected for every 15 minutes and based on time intervals of 15 minutes length, 24 hours of a day contain a number of 96 time intervals.

2.2 The Analytic Hierarchy Process

The AHP is used to construct a meta-criterion function which computes the comprehensive cost for each arc. The decision making process uses this cost value to find an optimal path. The AHP process starts with computing criterion weights. Techniques proposed in the literature to compute the criterion weights are as follows:

1. Ranking
2. Rating
3. Pairwise Comparison
4. Trade Analysis Method

This research work uses the Pairwise Comparison Method (PCM) to compute the criterion weights which define relative importance of objectives according to the user defined preferences. PCM is easy to implement and only two criteria are compared at a time (SATTY 2008). Preferences over the objectives are defined in a scale from 1 to 9. Table 1 shows the description of each scale (SATTY 2008).

Table 1: PCM scale (SATTY 2008)

Scale	Description
1	Equal Importance
2	Equal Plus
3	Moderately
4	Moderately Plus

Scale	Description
5	Strongly
6	Strongly Plus
7	Very Strongly
8	Very Strongly Plus
9	Extreme Importance

After constructing the comparison matrix, the right principle of eigenvector is applied to the comparison matrix to extract the criterion weights (TRIANTAPHYLLOU & MANN, 1995). But realistic computation of criterion weights depends on consistency between the criterion comparisons. Estimation of consistency ratio guarantees the validity of the computed criterion weights. The next step is to compute the comprehensive cost. Comprehensive cost is calculated as follows:

$$\text{Comprehensive cost} = [(\eta\,(TT) / TT_{max})\ \ W_{TT}] + [1- ((\mu(R) / R_{max})\ \ W_R)]$$

Eqn 1: Comprehensive cost formulation

Where,

η (TT): Travel time of an arc

TT_{max}: Maximum travel time spent on an arc in the street network

W_{TT}: Weight of the travel time criterion computed from the PCM.

$\mu(R)$: Reliability of an arc

R_{max}: Maximum reliability of an arc in the street network

W_R: Weight of the reliability criterion computed from the PCM.

Initially, the comprehensive cost for each arc is computed. Later, bi-objective fastest path query is sent to the routing module to compute the multi-period distance matrices.

2.3 Method 1: Method with Hierarchization of Objective Functions

In this solution method, objectives are optimized in orderly fashion starting from most important to least important objective. Initially Martin's algorithm is used to generate k-fastest paths by optimizing the travel time objective while the reliability parameter is aggregated. K-fastest path generation is achieved by extending Dijkstra's algorithm by allowing multiple labels for each arc where each label represents a path (GANDIBLEUX 2006). Later, an optimal path is chosen by applying Analytical Hierarchy Process (AHP) technique in the decision making process.

Phase 1: Generating k-shortest paths

The Martin's algorithm based on the revision provided by Gandibleux allows multiple labels added to arcs and each label potentially represents a path to an arc. The algorithm terminates once the number of labels added to the destination arc reaches the maximum number of paths or all the labels are made permanent. During this phase, the reliability parameter is aggregated and used along with the travel time to find an optimal path. The steps involved in the Martin's algorithm are described as follows (GANDIBLEUX 2006):

1. Initialize the source arc with a temporary label.
2. At each iteration the lexicographically minimum label is selected from the list and marked as a permanent label. Then, propagate the information contained in this label to all its successor temporary labels.
3. The step 2 is iterated until there are no temporary labels left.
4. Paths are generated by moving backwards from the target arc to the source arc while considering the label number.

Each permanent label represents a unique path. K-shortest paths are generated by moving backwards till it reaches the source arc. Figure 5 shows a street network with three non-dominated paths generated using Martin's algorithm.

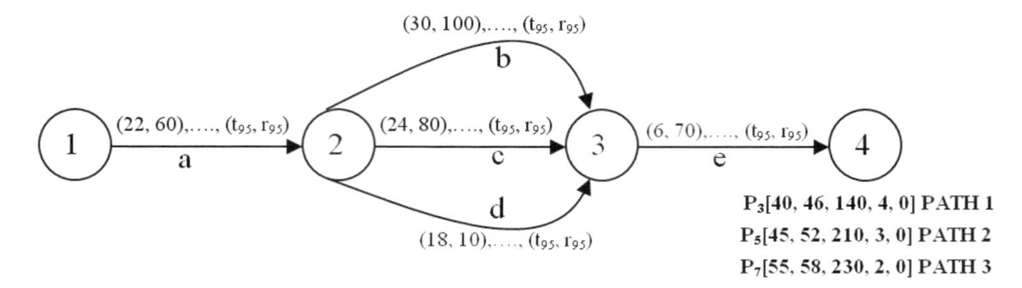

Fig. 5: Sample street network with multiple shortest paths

Phase 2: Finding optimal path using AHP

In this phase, a meta-criterion function based on AHP theory is used to choose an optimal path from Pareto (non-dominated) paths. The first step towards computing the comprehensive cost is to define the relative preference of objectives through the comparison matrix and use PCM to compute the criterion weights. The criterion weights vary based on the relative preference scale.

Example: Travel Time is highly important

PCM		TT	R
	TT	0	9
	R	1/9	0

After computing Eigen Vector:

$W_{TT} = 0.900$; $W_R = 0.100$; $TT_{max} = 30$; $R_{max} = 100$

Apply these values in Eq.1 yields: **Path1 = 2.33**; Path2 = 2.49; Path3 = 2.66

Least cost path is Path 1. Hence, Path 1 is the optimal path.

2.4 Method 2: Method with a Meta Criterion Function

In this solution method, a meta-criterion function computes the comprehensive cost for each arc by merging travel time and reliability criteria (TARAPATA 2007). Later, Dijkstra's algorithm is used to optimize the computed comprehensive cost. Since the problem is a time dependent shortest path problem, the comprehensive cost is computed for every time period.

Example: Travel Time is highly important

PCM		TT	R
	TT	0	9
	R	1/9	0

$W_{TT} = 0.900$; $W_R = 0.100$; $TT_{max} = 30$; $R_{max} = 100$

Apply these values in Eq.1 for each arc to compute the comprehensive cost:

Arc(a) = 1.6; Arc(b) = 1.8; Arc(c) = 1.64; Arc(d) = 1.53; Arc(e) = 1.11

After computing Eigenvector:

Now, Dijkstra's algorithm computes the optimal path. The optimal path is 1→ 4 → 5 which is path 1 in the earlier method.

3 Performance Comparison

The experiment discusses the quantitative (CPU computational time) and qualitative (trip duration & trip reliability) performance requirements of both solution methods. The experiment carried out with the user preference "reliability is highly important compared to travel time". The street network of Vienna is used for the experiments with FCD as the source for dynamic travel time. This street network contains 70775 arcs. The computer used for the experiments is a desktop PC equipped with an Intel core 2 quad 2.83GHz Processor, 3GB of RAM installed, running under Windows XP operating system. The algorithms were implemented in C++ without any compiler optimization.

Experiments were carried out on qualitative and quantitative aspects. The quantitative experiment compares the CPU computational time of both solution methods for a shortest path query with different path lengths (various test cases). The outcome of this experiment shows that the method with meta criterion has better runtime performance over the method with hierarchization of objective functions. Figure 6 shows that the higher path lengths yields the better the runtime performance for the meta-criterion method over hierarchization of objectives method. The runtime performance of hierarchization of objective functions is influenced by the number of potential fastest path available between source location and destination location. The multiple label-setting procedure is the reason for the increase in the runtime and memory consumption of the method with hierarchization of objective functions. On the other hand, the method with meta-criterion merges all the criteria into one criterion and works with single-objective Dijkstra's algorithm. Hence, the meta-criterion method is significantly faster.

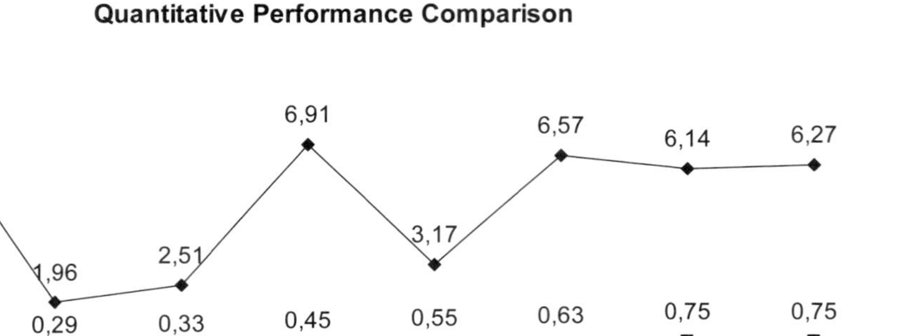

Fig. 6: Quantitative performance comparison of solution methods

The outcome of qualitative experiment shows that method with hierarchization of objective functions produces in general better performance over the method with meta-criterion in terms of both trip travel time and trip travel time reliability. Figure 7 shows that for most test cases the method with hierarchization of objective functions produces better qualitative performance. On average, travel time reliability is 2.2% higher and travel time is 12,7% shorter compared to solutions of the method with meta criterion.

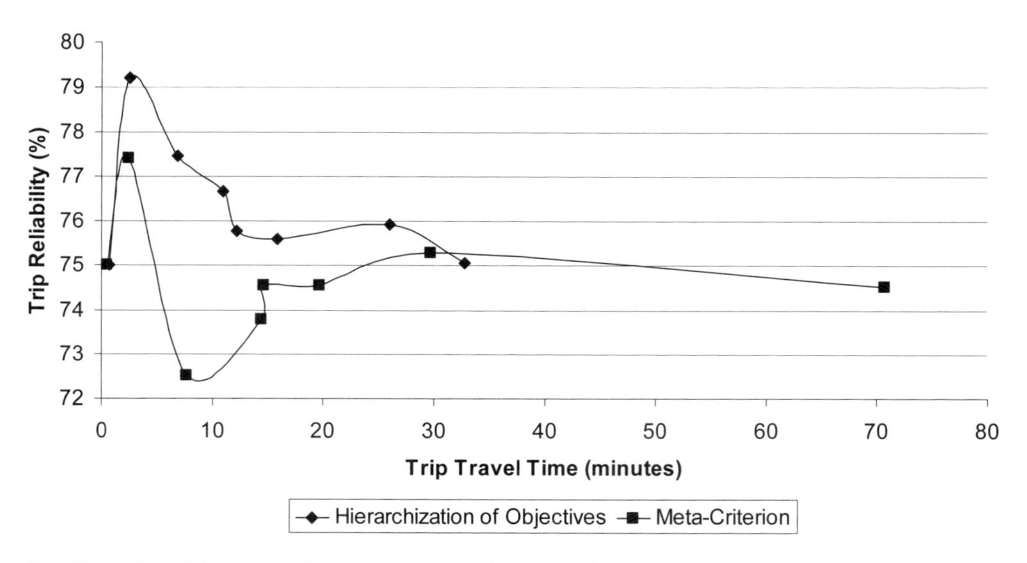

Fig. 7: Qualitative performance comparison of solution methods

4 Conclusion

This research work extended two existing solution methods for the multi objective shortest path problem in time dependent scenarios with bi-objective configurations and presented a quantitative and qualitative performance comparison of these. The MOSP solution methods utilize the Martin's algorithm and extended Dijkstra's algorithm to achieve the optimization. The multi-objective optimization process uses the AHP based decision rule to find an optimal path. The AHP based decision rule provides a better framework to integrate several real-world objectives into the realistic routing environment. The experiment shows that the method with meta-criterion function produces better runtime performance over the method with hierarchization of objective functions. Due to better runtime performance the method with meta-criterion produces lower quality results compared to method with hierarchization of objective functions. Thus, the decision which method is suitable for a specific task yields to another bi-objective optimization problem.

Acknowledgments. This work was supported, in part, by the Austrian Science Fund (FWF) Grant no. L510-N13. Floating car data was provided by the Austrian Institute of Technology.

References

GANDIBLEUX, X., BEUGNIES, F. & RANDIRAMASY, S. (2006), Martin's algorithm revisited for multi-objective shortest path problems with a MaxMin cost function. 4OR 4, pp. 47-59.

GUTIERREZ, E. & MEDAGLIA, A. (2008), Labeling algorithm for the shortest path problem with turn prohibitions with application to large scale road networks, Annals of Operational Research, 157 (1), pp. 169-182.

ICHOUA, S., MICHEL, G. & JEAN-YVES, P. (2003), Vehicle Dispatching with Time Dependent Travel Times. European Journal of Operations Research, 2003, pp. 379-396.

QIANG, L. I. (2007), Arterial Road Travel Time Study using Probe Vehicle Data. – http://ir.nul.nagoya-u.ac.jp/dspace/bitstream/2237/8888/1/Liqiang.Thesis.pdf (last visited: 19/01/2010), 1 p.

SAATY, T. L. (2008), Decision making with the analytic hierarchy process. International Journal of Services Sciences, 1 (1), pp. 83-98.

TARAPATA, Z. (2007), Selected Multicriteria Shortest Path Problems: An Analysis of Complexity, Models and Adaption of Standard Algorithms. International Journal of Applied Mathematics and Computer Science, 17, pp. 269-287.

TRIANTAPHYLLOU, E. & MANN, S., H. (1995), Using the Analytic Hierarchy Process for Decision Making in Engineering Applications: Some Challenges. International Journal of Services Sciences, 2 (1), pp. 35-44.

XIANG, Q. J., MA, Y. F. & LU, J. (2007), Optimal Route Selection in Highway Network Based on Travel Decision Making. IEEE Intelligent Vehicles Symposium, 2007, pp. 1266-1270.

Landslide Susceptibility Mapping with Support Vector Machine Algorithm[1]

Miloš MARJANOVIĆ

The GI_Forum Program Committee accepted this paper as reviewed full paper.

Abstract

This paper introduces one current machine learning approach for solving spatial modeling problems in domain of landslide susceptibility assessment. The case study addresses NW slopes of Fruška Gora Mountain in Serbia, where landslide activity has been quite substantial, but not inspected in detail. Regarding this lack of precise landslide inventory, an expert-driven zoning of landslide susceptibility was created as the referent model. The Support Vector Machines (SVM) as the last generation machine learning classifier was used to device the susceptibility model after criteria used in the expert-driven model. Training and testing of the SVM algorithm was performed over an assembly of spatial attributes, which included elevation, slope angle, aspect, distance from flows, vegetation, lithology, and rainfall with respect of the referent model. The algorithm was optimized for its learning capacity and kernel dimension parameters. In addition, the autocorrelation precaution was undertaken by specifying the sampling strategy. The results show that high accuracies (87,1%) and agreement coefficients (0,77) between the referent model and the SVM model were feasible with small sample sizes (10% of original instances).

1 Introduction

Landslide susceptibility stands for likelihood of landslide occurrence over a specified area or volume of the terrain (VARNES 1978). It had been illustrated in versatile techniques in various case studies, yielding more or less reliable results depending on the complexity of the terrain and suitability of the approach (ALEOTTI & CHOWDHURY 1999, CHACON et al. 2006, FERNANDEZ et al. 2003, GUZZETTI et al. 1999). The central idea of all those studies implies the processing of input spatial attributes into a single final model through the various weighting and interpolating methods (heuristic techniques, deterministic techniques, statistical and probabilistic analyses, fuzzy logics, artificial intelligence algorithms, and others) usually generalized as knowledge-driven or data-driven approaches (ALEOTTI & CHOWDHURY 1999). There are disagreements on which one to prefer, but it is widely believed that certain subjectivity expressed through the knowledge and experience of experts turns to be useful, and that the combination of two techniques is the most desirable (ERCANOGLU et al. 2008, VAN WESTEN et al. 2003). Namely, landslide assessment tends to be an ambiguous (non-linear) problem due to the complexity of the examined

[1] This paper is written in the framework of Methods of artificial intelligence in GIS, a project of Czech Republic Grant Agency (CR GA 205/09/079).

geological environment and forces that trigger the sliding process, and it had been traditionally handled by favoring site-specific investigations, empirical models and personal practice (CHACON et al. 2006). This kind of investigation requires considerable expenses and time to present solutions, which ends up with discontinuity and incompleteness of data (CARRARA & PIKE 2008), but provides very reliable information. On the other hand, rapid development of the computer techniques and Geographic Information System (GIS) empowered application of complex mathematical and geostatistical tools. These cheaper and faster techniques retrieved more complete and continual results, offering new perspectives on landslide assessment (BRENNING 2005). Thus, it seems inevitable to combine the virtues of both approaches, i.e. to train data-driven techniques over limited expert-driven analysis and surveys. This idea was one of the foci in the present paper in terms of implementing artificial intelligence algorithms with the expert-driven zoning for mid-scale research. Mapping the expert-driven interpretation proved the algorithm's ability to replicate the expert's criteria, which was the main goal of this research. A particular machine learning technique, called the Support Vector Machine (SVM) was used for this task. Since the SVM optimizes its capacity during the learning process (BURGES 1998), our motif was also to show how accurate mapping is feasible with sparse inputs, as suggested in some earlier studies (YAO et al. 2008).

The practice of SVM in spatial modeling has quite recent history. Pioneering the application in landslide susceptibility (YAO et al. 2008) compared single-class (binary) vs. two-class SVM in the Hong Kong area affected by the landslides. The authors demonstrated how the latter provides better conditions for algorithm training and testing, since it is clearly favorable to know both, where landslides should and should not be expected. This brings about another dimension to the problem, since geotechnical engineering practice turns more reliable in specifying landslide occurrence sites than the stable ground. Another study (YUAN et al. 2006) regarded only one aspect of the landslide phenomena, which included the debris flows analysis by using SVM and fuzzy approach. Since it outperformed fuzzy method in the testing mode, SVM was considered appropriate and more convenient for these kinds of assessments in the area of interest (Yunnan Province, China). SVM was also proven suitable for another site-specific geological study involving 3D modeling of geological bodies from the borehole samples (SMIRNOFF et al. 2008). Nevertheless, similar philosophy could equally hold true in the case of geotechnical 3D models, for advanced surface interpolation, be it particular geological strata, groundwater table or stress/strain distribution. Research of the landslide susceptibility in Ecuadorian Andes (BRENNING 2005) was performed by using the-state-of-the-art techniques. Beside SVM, the author used logistic regression and decision trees algorithms over very cautiously prepared input datasets, and compared their performance. Shown results encouraged the use of SVM or logistic regression for the landslide susceptibility zoning . Very recent research (YILMAZ 2009) appeared with the most complete perspective on the landslide assessment methodology so far. Various modeling methods have been considered and compared, including SVM learning. The study shows, that several methods turned very precise, and efficient. However, it also underlines the GIS compatibility issue as a serious drawback of advanced methods (ANN, SVM, logistic regression and so forth).

The gravity of the accurate landslide susceptibility assessment is especially emphasized if the ultimate aim is the landslide risk assessment (KOMAC 2006). The latter involves hazard analysis (derived after susceptibility model) and vulnerability analysis for the area of

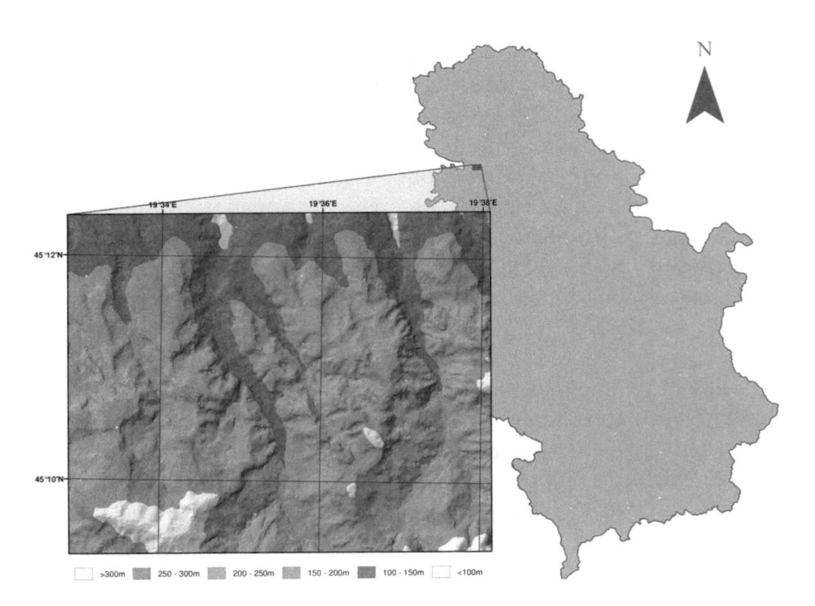

Fig. 1: Location of the study area, NW slopes of the Fruška Gora Mountain in Serbia

interest (VARNES 1978). It has been shown (KOMAC 2006) that successful hazard and risk evaluations are possible by appending necessary thematic data (temporal frequency, vulnerability of elements at risk) upon the initial landslide susceptibility map.

2 The Case Study Area

Study area encompasses NW slopes of the Fruška Gora Mountain, in vicinity of Novi Sad, Serbia (Fig. 1). The sight (N 45°09'20", E 19°32'53" – N 45°12'25", E 19°38'05") spreads over approximately 15 km^2 of hilly landscape, but yet with an interesting dynamics, since these remote slopes have been chosen by abundance in landslide occurrences and their indications. Geotechnical practice in this area implies that the superficial dynamics directly depend on geological background, meaning that rocks exhibit diverse geodynamical behaviors under the agencies of different geological processes (JANJIĆ 1962). In terms of stability, the most problematic are Tertiary terrains with marlstones and clays, especially in the zones of seismically active structures (PAVLOVIĆ et al. 2005). According to the international typology (VARNES 1978), the most of the landslide phenomena belongs to the slips and shallow translational slides.

3 Methods of Landslide Susceptibility Mapping

Methodology applied in this research involved knowledge-driven modeling and machine learning techniques. Former was performed in order to supply class reference map (knowledge-driven landslide susceptibility model) for training/testing stage of machine learning algorithm. Prior to those stages input data, represented by different spatial attributes, needed to be gathered and prepared. ArcGIS 9.+ package was used for data preparation, referent model calculation and display, while an open-source package Weka 3.6 (HALL et al. 2009) served for placing the SVM experiments.

3.1 Input spatial attributes

Input data were gathered from different sources and could be categorized as geological, morphometric and environmental attributes. Sources included elevation data from 1:25000 contour line maps, table datasets of rainfall measuring stations, Landsat ETM imagery with 30 m resolution (bands 3 and 4) and digital geological map compiled at 1:50000 scale. Availability of data and specificity of present mid-scale research limited the choice to the following attributes (S_i): elevation S_7, slope angle S_2, slope aspect S_6, distance from flows S_4, vegetation S_5, lithology S_1 and rainfall distribution S_3. Most of these attributes are among the ones that were proven significant for mid-scale type of studies, and could be regarded as standard (FERNANDEZ et al. 2003, GUZZETTI et al. 1999, VAN WESTEN et al. 2006). The attributes were adapted for the raster modeling approach, which is recommended for any spatial multi-criteria (multiple inputs) analysis in GIS platforms (BONHAM-CARTER 1994). Later on, those initial input sets were refashioned and adapted (converted to ASCII and CSV formats) to meet the format requirements of machine learning software package.

3.2 Referent landslide susceptibility model

Given the required inputs, a simple knowledge-driven analysis of landslide susceptibility took place. In ideal circumstances (CHACON et al. 2006, FERNANDEZ et al. 2003) this stage would be performed by referring to a thematic survey, a geomorphological and remote sensing research or the engineering geology research conducted at site-specific scale and then up-scaled to the desired level. Unfortunately, the conditions concerning the availability and formats of the data were far from ideal throughout this research, but since the main goal of the study was to replicate an expert-driven (subjective) criterion by machine learning algorithm, the problem only affects the reliability and usability of the final model. Referent model was designed by multi-criteria analysis which involved weighting of the input attributes (in accordance to their importance), ending up with an additive model. The weights were knowledge-driven, i.e. judged by the prevailing opinions (averaged votes of the experts) among several researchers (PAVLOVIĆ et al. 2005) familiar with the landslide problematic over the specified or similar terrains. In order to quantify the attributes weights more accurately, the Analytical Hierarchy Process (AHP) was employed. AHP initially relates the relative weights by pair-wising, i.e. by comparing the parameters to each other and creating normalized (0-1 or 0-100% range of values) eigenvector (a sequence of eigen-values). The eigen-value of each parameter represents its final weight. The susceptibility model N was then calculated as linear additive function in a GIS environment (1). In (1) S_i corresponds to the value of the i-th attribute, while index W_i represents its weight.

$$N = \sum_{i=1}^{7} W_i S_i = 0{,}29S_1 + 0{,}27S_2 + 0{,}15S_3 + 0{,}14S_4 + 0{,}08S_5 + 0{,}05S_6 + 0{,}02S_7 \quad (1)$$

Furthermore, the obtained float (continual) raster model (Fig. 3a) was to be somehow discretized and information gain was chosen as the classification criterion (PÁSZTO et al. 2009). Namely, an entropy model tracked the optimal information gain for this particular raster, yielding four classes of susceptibility, denoted as Very high, High, Moderate and Low (MARJANOVIĆ 2008). Subsequently, the model of susceptibility was used as class reference map in the SVM learning algorithm.

3.3 Support Vector Machine classifier

Support Vector Machine (SVM) stands for the last generation machine learning classifier, that deals with the binary classification tasks (i.e. pixel belongs to only one of the two classes of the referent raster), but one can easily transform n-class problem into the sequence of n binary classification tasks – one-versus-all (BELOUSOV et al. 2002). The algorithm initially attempts to generate a linear separating plane h in the original space of a set of n-attributes – vector of n-coordinates (Fig. 2a), and accepts the rule which classifies the coordinates well enough (KECMAN 2005). The latter means that the rule provides maximal margin (minimal **w**) between the classes (2), where the separating hyper-plane takes geometrical center of the margin. If linear separation is not feasible, the method introduces complex hyper-plane h' (Fig. 2b), by appending slack variables ε_i (slacks of misclassified instances) and corresponding penalty factor C (3). If the classification turns unfeasible even with the complex h' in the original space (which is inevitable in the case of spatial modeling) the procedure goes one step further and transforms coordinates (attributes) into a higher feature spaces by kernel functions (polynomial, radial basis functions or some special functions) until the coordinates become separable (ABE 2005). The problem is then brought down to the initial case and the classification is performed.

$$\min_{\mathbf{w},b} \frac{1}{2}\|\mathbf{w}\|^2 + C\sum_i \varepsilon_i \tag{2}$$

$$\mathbf{w}^* = \sum_{i=1}^{m} \alpha_i y_i \mathbf{x}_i , \quad C \geq \alpha_i \geq 0, \quad i = 1,...m \tag{3}$$

$$f(\mathbf{x}) = \operatorname{sgn} \sum_{i=1}^{m} \alpha_i y_i (\mathbf{x}_i \cdot \mathbf{x}) + b^* \tag{4}$$

The optimization problem (maximizing the margin) of (2) is usually solved in its dual form and the solution is (3). Variable **w*** is a linear combination of training examples for an optimal hyper-plane (Fig. 2b). It can be shown that **w*** depicts linear combination of support vectors x_i for which the corresponding (*Lagrangian coefficients*) α_i are non-zero values. Support vectors for which $C>\alpha_i>0$ condition holds, belong either to h_1 or h_2 (Fig. 2a). In that case, x_a and x_b are two support vectors for which holds $y_a=1$, $y_b=-1$ and $b^*=-0{,}5\boldsymbol{w}^*\cdot(\boldsymbol{x}_a+\boldsymbol{x}_b)$, as the classification function (rule) finally becomes (4). After removing all training data that are not support vectors and retraining the classifier we would obtain the same solution. The ability to distinguish between support vectors and noisy data points enables SVM to increase its generalization capacity in the learning process (MICHELL 1997).

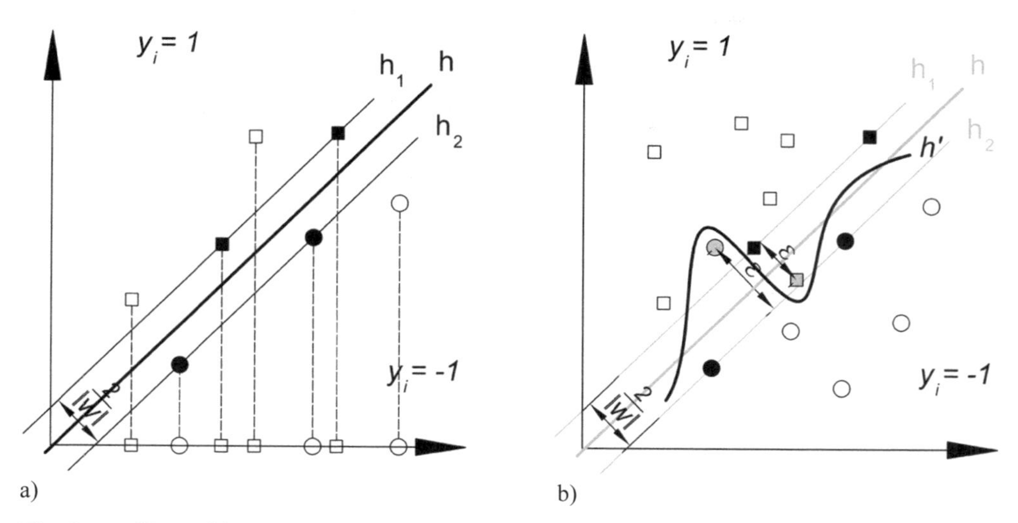

Fig. 2: SVM illustration (bold squares and circles are support vectors): a) linearly inseparable in 1D (x-axis) became separable binary case in a higher (2D) feature space, b) SVM margin and penalty relation for the complex hyper-plane *h'*

4 Results and Discussion

Initial hypothesis of this research suggests that accurate mapping of expert criterion is possible by SVM classifier, and that reducing the size (amount of attribute and reference data) of input samples will not significantly affect the result. In addition, it was necessary to estimate the influence of data autocorrelation, so that the classifier would not face the overfitting problem. Contextually, the sampling strategy was designed in such fashion to gradually increase the size of the training/testing samples, and to use their different configurations in terms of spatial reference. The latter regarded creation of attribute and complete sets. It is also important to mention that the sampling had been performed by randomly selected and evenly distributed instances throughout the input datasets.

The SVM experiments were executed through two-folded cross-validation, meaning that the training mode was performed over the first split of the sample, while the remaining half was used for testing. In the next iteration, the procedure was repeated by swapping the roles of training and testing splits. Samples containing 5%, 10% and 50% of original instances (2000, 4000 and 20.000 of original 40.000 instances of attribute rasters) were run in ten iterations in both sets. The arithmetic mean of obtained values was used as the representative result. Since it was expected that the problem would not be linearly resolvable, all experiments in both sets used a radial basis function – Gaussian kernel, to map the instances to a higher feature-space. Optimized parameters of this kernel were set after 20 initial attempts, where consistency of result was tracked while the parameters were increased successively and with various steps. The optimal parameters equaled C=100 for the penalty factor, and γ=2 for the kernel dimension. The algorithm's sensitivity to multi-fold cross-validations was also tested, and similar results were obtained with two, five, and ten-fold cases.

Performance of the algorithm (Table 1) was expressed as the accuracy, and particularly for the complete set through the κ-index statistics (which was not possible for the attribute set due to the lack of spatial reference in retrieved result). Accuracy is an estimate based on the proportion of correctly classified and total instances (0-100%) calculated directly in Weka 3.6 package, while κ-statistics measures exact agreement between cross-tabulated datasets (FIELDING & BELL 1997) and it was calculated after the cross-tabulation in ArcGIS 9.+ package. This applies only to data with the same number of classes (BONHAM-CARTER 1994), which was the case with comparison of the referent map (landslide susceptibility) and resulting SVM map. Substantial agreement is achieved if κ-index is in the 0,61-0,81 range (VAPNIK 1995).

Table 1: Accuracy of the SVM algorithm

sample size (% of original instances)	accuracy (%)	
	attribute set	complete set
50	86,4	87,6 (κ=0,86)
10	85,9	87,1 (κ=0,77)
5	84,5	86,0 (κ=0,73)

4.1 Attribute set

This training/testing set involved only samples of attribute values (without geo-coordinates) and the class reference values. The idea was to compare the performance of the algorithm with and without spatial reference, in order to reveal the influence of autocorrelation in the particular sample size. The SVM algorithm reached notably high accuracy ranging in a very subtle 84,5%-86,4% span (Table 1), which implies that high accuracy was reached with sparse inputs (only 5% of original instances were sufficient to reach approximately 85% of the accuracy, and there were no significant improvements when the sample size increased to 10% and 50%). The result also suggests that the attribute set due to the reached accuracy level, gained some autonomy over spatial reference, meaning that the autocorrelation overfit could be surpassed, and that the sampling strategy seemed proper.

4.2 Complete set

In this set the SVM's capacity optimization and the autocorrelation overfit were finally challenged. The accuracy differences in relation to the sample size (Table 1) were only 1-2%, which confirms that learning capacity was optimal, and that sample size could be substantially reduced with minor consequences. Furthermore, there was a very small increase of accuracy in comparison to the attribute set, which once again approves the applied sampling strategy. Values of κ-index falls into the range of substantial agreement between cross-tabulated models, in this case – knowledge-driven referent model and SVM model of landslide susceptibility. Since the instances were spatially referenced, appropriate SVM-driven susceptibility map was retrieved (Fig. 3b)

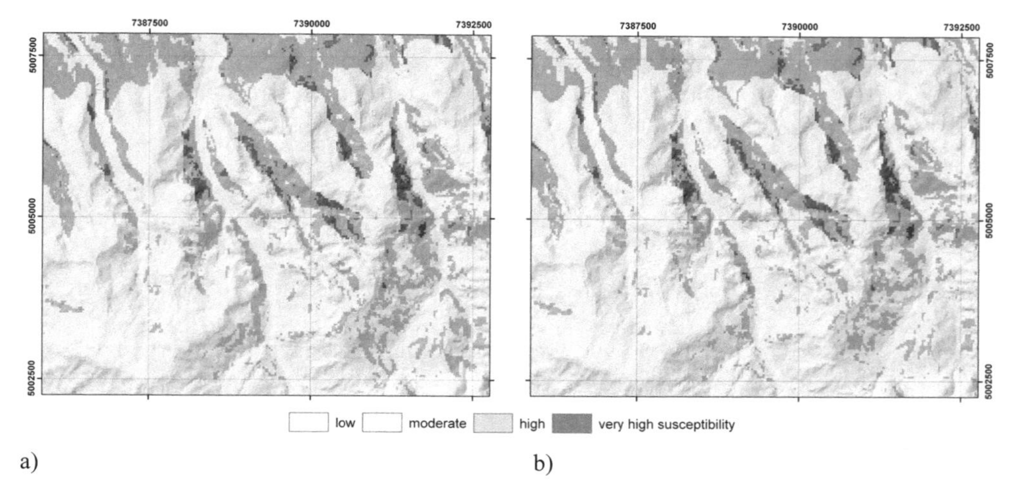

a) b)

Fig. 3: a) AHP expert-driven model of landslide susceptibility, b) 15% sample SVM model of landslide susceptibility

5 Conclusion

The SVM classification problem succeeds a knowledge-driven approach that dealt with landslide assessment by weighting of particular attributes and creating an additive model of landslide susceptibility. Subjective criteria that drove the model were tracked by the SVM classifier, whose performance was optimized by both, optimizing the algorithm's capacity and adjusting the input datasets. Speaking of former, it is significant to mention that optimization, seemed less demanding in comparison with other machine learning techniques (such as ANN), since it is handled by fitting only two parameters (C and γ). The shortcoming of the method is certainly incompatibility with existing GIS platforms, resulting in time-consuming manipulations over data preparation and data exchange between GIS software and machine learning simulator. Thus, potential SVM modules could be interesting for integration with GIS platforms.

The SVM classifier, with suggested sampling strategy turns quite convenient for the chosen non-linear problem, gaining very consistent and precise results. The most important result is the one revealing that small training/testing sets were sufficient to reach very high accuracy and that attribute set achieved high precision, not too different from the precision of the complete set. These facts could be exploited two-sidedly, that usually expensive expert-based surveys could be substantially reduced, but that the sampling needs to be cautiously strategized. According to the result of this study, only 5% of the overall area would require full data coverage, with proper sampling. This particularly regards geological and environmental attributes, as well as the referent landslide model (detailed, site-specific investigations could be performed over much smaller area). However, this optimistic result needs to be tested in detail, with accurate and up-to-date referent model, with additional input attributes, and including of the adjacent terrains. In addition, the threshold for the sample size versus learning capacity and desired accuracy should be determined and put to the test.

This approach and its result could serve for extending the analysis toward landslide hazard and risk assessment, even the spatial vulnerability, since they all cope with the similar problems in terms of relating the useful expert knowledge/estimation and particular phenomenon (landslides for instance).

References

ABE, S. (Ed.) (2005), Support Vector Machines for pattern classification. Springer, pp. 25-33.

ALEOTTI, P. & CHOWDHURY, R. (1999), Landslide hazard assessment: summary review and new perspectives. Springer-Verlag, Bull. Eng. Geol. Environ., 58, pp.21-44.

BELOUSOV, A. I., VERZAKOV, S. A. & VON FRESE, J. (2002), Applicational aspects of support vector machines. J. Chemom., 16, pp. 482-489.

BONHAM-CARTER, G. (Ed.) (1994), Geographic information system for geosciences – Modeling with GIS. Pergamon, pp. 51-81, pp. 177-334.

BRENNING, A. (2005), Spatial prediction models for landslide hazards: review, comparison and evaluation. European Geoscience Union, Nat. Haz. Earth Sys. Sci., 5, pp. 853-862.

BURGES, C. J. C. (1998), A tutorial on support vector machines for pattern recognition. Data Mining and Knowledge Discovery, 2 (1), pp. 121-167.

CARRARA, A. & PIKE, A. (2008), GIS technology and models for assessing landslide hazard and risk. Geomorphology, 94, pp. 257-260.

CHACON, J. et al. (2006), Engineering geology maps: landslides and geographical information systems. Bull. Eng. Geol. Environ., 65 (3), pp. 341-411.

ERCANOGLU, M. et al. (2008), Adaptation and comparison of expert opinion to analytical hierarchy process for landslide susceptibility mapping. Bull. Eng. Geol. Environ., 67, pp. 565-578.

FERNANDEZ, T., IRIGARAY, C., EL HAMDOUNI, R. & CHACON, J. (2003), Methodology of landslide susceptibility mapping by means of a GIS. Application to the Contraviesa area (Granada, Spain). Natural Hazards, 30, pp. 297-308.

FIELDING, A. H. & BELL J. F. (1997), A review of methods for the assessment of prediction errors in conservation presence/absence models. Environ. Conserv., 24, pp. 38-49.

GUZZETTI, F., CARRARA, A., CARDINALI, M. & REICHENBACH, P. (1999), Landslide hazard evaluation: a review of current techniques and their application in a multi-scale study, Central Italy. Geomorphology, 31, pp. 181-216.

HALL, M., FRANK, E., HOLMES, G., PFAHRINGER, B., REUTEMANN, P. & WITTEN, I. (2009), The WEKA Data Mining Software: An Update. University of Trier, SIGKDD Explorations, 11 (1), pp. 10-18.

JANJIĆ, M. (1962), Inženjerskogeološke karakteristike terena Narodne Republike Srbije. Nučna Knjiga, Belgrade (Serbia), pp. 19-189.

KECMAN, V. (2005), Support Vector Machines – an introduction. In: WANG, L. (Ed.) Support Vector Machines: Theory and applications, Springer, pp. 1-47.

KOMAC, M. (2006), A landslide susceptibility model using the Analytical Hierarchy Process method and multivariate statistics in perialpine Slovenia. Geomorphology, 74, pp. 17-28.

MARJANOVIĆ, M. (2008), Landslide susceptibility modeling: A case study on Fruška Gora Mountain, Serbia. Geomorphological Society of Czech Republic, Geomorphologia Slovaca et Bohemica, 10 (1), pp. 29-42.

MICHELL, T. M. (Ed.) (1997), Machine learning. McGraw Hill. pp. 20-50, 231-236.

PÁSZTO, V., TUČEK, P. & VOŽENÍLEK, V. (2009), On spatial entropy in geographical data. Proc. GIS Ostrava Technical University of Ostrava, Czech Republic, 25-28 January.

PAVLOVIĆ, R., LOKIN, P., TRIVIĆ, B. & RADOVANOVIĆ, S. (2005), Geološki uslovi racionalnog korišćenja i zaštite prostora Fruške gore. Specialist study, University of Belgrade, unpublished.

SMIRNOFF, A., BOISVERT, E. & PARADIS, S. J. (2008), Support vector machines for 3D modelling from sparse geological information of various origins. Computers & Geosciences, 34, pp. 127-143.

VAN WESTEN, C. J., RIGERS, N. & SOETERS, R. (2003), Use of geomorphological information in indirect landslide susceptibility assessment. Nat. Haz., 30, pp. 399-419.

VAN WESTEN, C. J., VAN ASCH, T. W. J. & SOETERS, R. (2006), Landslide hazard and risk zoning – why is it still so difficult? Bull. Eng. Geol. Environ., 65 (2), pp. 167-184.

VAPNIK, V. (Ed.) (1995), The Nature of Statistical Learning Theory, Springer-Verlag, 2nd edition, pp. 138-167.

VARNES, D. J. (1978), Slope movements types and processes. In: Landslides: Analysis and control, National Academy of Sciences, Special report, 176 (2), pp. 11-33.

YAO, X., THAM, L. & DAI, F.C. (2008), Landslide susceptibility mapping based on support vector machine: A case study on natural slopes of Hong Kong, China. Geomorphology, 101, pp. 572-582.

YILMAZ, I. (2009), Comparison of landslide susceptibility mapping methodologies for Koyulhisar, Turkey: conditional probability, logistic regression, artificial neural networks, and support vector machine. Springer-Verlag, Environ. Earth. Sci., Published on-line DOI 10.1007/s12665-009-0394-9.

YUAN, L. & ZHANG, Y. (2006), Debris flow hazard assessment based on support vector machine. Wuhan University Press, Wuhan University Journal of Natural Sciences, 14 (4), pp. 897-900.

THERMOMAP – Mapping Subsurface Thermal Potential for Selected Test Sites in the EC

Matthias MÖLLER, David BERTERMANN and Reinhold ROSSNER

Abstract

Global warming has become a big issue over recent years. The main reason for raising temperatures is the increasing CO_2 concentration in the atmosphere caused by burning of mineral resources, e.g. petroleum, fuel, coal, gas. A big source of CO_2 emissions is related to the warming of buildings and household water. This amount of energy might be partly replaced by subsurface heat energy. Two different kind of subsurface energy sources can be differentiated: shallow energy potential and energy from deep drillings. We are presenting a frontend that enables the estimation of the potential of shallow subsurface energy. "ThermoMap" makes advantage of existing maps and geo-data and it will establish a unique database for geothermal potential as a prototype. Therefore maps have to be collected, converted and analysed in a GIS environment. Finally the results will be presented to users with a specially adapted, interactive WebGIS interface.

1 Introduction

Sub-surface thermal potential is a major source for future energy supply on Earth, e.g. for heating water for daily needs and for warming buildings. The method is already well proofed: drilling a hole, fitting a hose in it and extracting the heat energy from surrounding earth layers; consequently this method is offered as a standard product all over the EC. Some specially designed federal programs support the installation in terms of saving energy from mineral resources. This leads to the reduction of CO_2 emissions, it reduces the global warming and the effects of climate change. Consequently many buildings have been equipped with this technology so far. This kind of thermal power can be separated into two different methods depending on the depth where the energy is generated: deep drillings and subsurface shallow drilling. ThermoMap focuses on the subsurface energy potential.

2 Energy from the Earth

2.1 Deep Drilling for Heat Energy

Deep drilling reaches a depth of about 300 m depending mainly on the geological structures. In these geological layers the environment reaches temperatures up to 40 °C. Two hoses are fitted into the drilling hole and a fluid circulates down to the bottom of the drilling, where the pump driven fluid is heated from the surrounding layers. Heat exchangers transfer the energy and warm up water.

2.2 Subsurface Drilling

A shallow drilling into a depth of about 20 m sub-surface reaches cooler heat levels compared to a deep drilling. To extract the energy from the so called "subsurface" layer a longer hose for heat collection has to be used. The heat energy has to be treated with heat compressors, heat pumps, to reach suitable temperature levels comparable to those of [2.1]. Finally the energy is extracted with heat exchangers similar to 2.1. This method is cheaper compared to [2.1] at construction, but it becomes more expansive during operation because more electricity power is needed for the pumps.

3 ThermoMap

ThermoMap is a EC funded project (EC-Program: CIP-ICT-PSP-2009-3, PILOT TYPE B, No. 250446 and it integrates all the influencing factors of subsurface energy together in a unique database. This enables the estimation of the potential of subsurface thermal power. ThermoMap is primarily a demonstrator; however, it will be transferable to the entire administrative area of the EC. Specially selected test sites have been chosen to develop an online prototype. Existing data, e.g. maps, analog and digital databases will be digitally merged in one, spatially referenced geo-database (fig. 1). Participating partners are from Austria, Belgium, France, Germany, Greece, Hungary, Iceland, Romania and the United Kingdom. Embraced are research organisations, national authorities and authorities of EC as well as private companies. Advanced GIS technology has been found a reliable method for database development and analysis. Results of the GIS analysis will presented through a WebGIS frontend. This frontend is running in a common web-browser environment, it is accessible for the public and it will also enable different user groups to obtain information of subsurface thermal power potential for their private parcel.

3.1 Test Sites

ThermoMap test sites are distributed over the EC, interesting test sites are located in Germany and Austria: the region around Freilassing, Germany and Salzburg, Austria. This area is interesting, because it bridges a border area between two EC member states. Challenges for this area will be especially the geo-data harmonization.

3.2 Database Design and Data Collection

A big topic is the identification of the data needed for ThermoMap. Therefore a Metadata Catalog (MDC) is designed that contains all the information about data needed in the project. Important questions are: where is the data located, where is it stored and who is the owner of the data? In a next step we are defining standards for the geo-data acquisition, digitizing, scanning rules, geo-data file formats (either vector or raster based), projection parameters, data storing and handling tasks. ThermoMap may use additional data from several sources that are mostly accessible for free by the public. This data is not connected to any intellectual property rights (IPR) and it is provided as an OGC (Open Geospatial Consortium) conformal WMS (Web Mapping Service). The data needed for ThermoMap will be available on several scale levels. After defining the rules for the structure of the

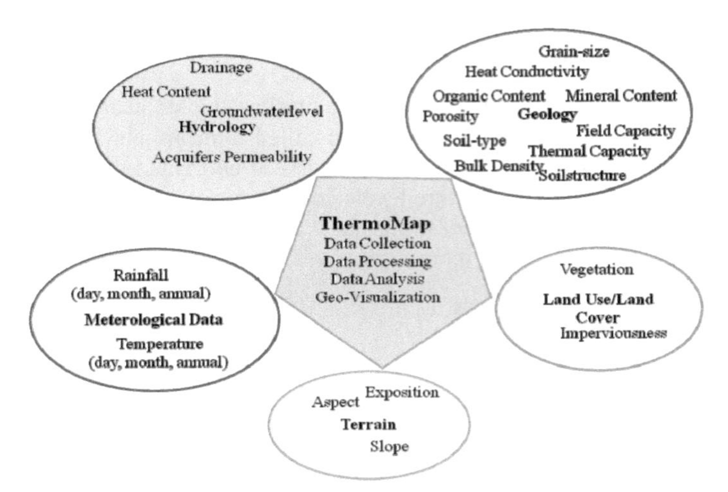

Fig. 1: ThermoMap input data and derived information

desired database each partner will transform his data according to the unique guidelines. Analog datasets need to be digitalized using various techniques to receive either vector or raster data as output. Analogue-digital conversion can be applied, for example, to obtain raster data from hard copy maps or survey plans using scanning devices. To obtain spatially referenced data which are ready to be integrated with other data sets, all data need to be properly geo-referenced. To receive vector data from referenced raster data, conversion tools implemented in a GIS can be used for a semi-automated conversion or rasterized data can be digitized manually. Fig. 1 gives an overview of ThermoMap input data sources.

3.3 Data Presentation

ThermoMap develops a user friendly target group specific Web-based GIS-System for superficial geothermal resources that will clearly reflect to the needs of the different target groups. These are the benefits of ThermoMap: (1) non-experts will have access to an easy to use WebGIS providing access to just the information and showing the potential of superficial geothermal resources. (2) Experts from science, business and administration will have access to a complex GIS that enables the use of all underlying data for scientific work and all necessary calculation of the resource. The combination of different geo-data sets in a GIS environment will lead to explicit small and detailed regions, each attached with its specific value of thermal potential. The final application scale of ThermoMap is about 1:10.000. This final Web based GI-system consists of two components, a GIS server for the geo-data storage and a visualization front-end running in a common Web-browser. The user may be enabled to define the spatial extend of the view and interactively acquire information of the geothermal potential for a specific region. The smallest object for the information of geothermal potential or value will is the parcel. An underlying grid value (containing the results of the analysis of the input data) will be clipped based on the parcel and detailed report for the selected parcels will finally be generated from the system.

4 Conclusions

ThermoMap starts its operational phase in the second half of 2010; however, the outlet and working construction is already designed and a future exploitation plan of ThermoMap has been established and it can be seen in Fig. 2.

After the realization of ThermoMap the frontend is available as a prototype for several users. Those are mainly energy supplying companies, industrial partners and administrative authorities End users are municipal and national authorities who want to use the information for planning purposes.

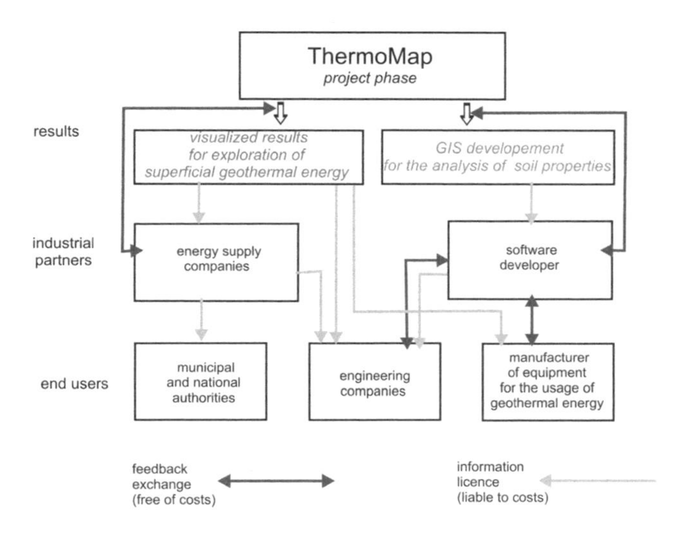

Fig. 2: Exploitation plan of ThermoMap

References

FAROUKI, O. T. (1986), Thermal Properties of Soils. Series on Rock and Soil Mechanics, 11, 136 p. Clausthal-Zellerfeld (Trans Tech Publications).

GEOTHERMAL – http://www.geo-t.de/de/index.html (6.4.2010).

KERSTEN, M. (1949), Thermal Properties of Soils. Univ. Minnesota Bull, 28, pp. L11-L21. Minnesota.

SALOMONE, L. A. (1987), Soil property and external factor effects on soil thermal resistivity. Proc.WS on GSHP Albany, Rep. HPC-WR-2, pp. 81-102. Karlsruhe.

SALOMONE, L. A. & KOVACS, W. D. (1984), Thermal resistivity of Soils. J. Geotechnical Engineering ASCE, 110/3, pp. 375-389.

SMITH, W. O. (1939), Thermal Conductivities in Moist Soils. SoilSciSoc. Amer. Proc., 4, pp. 32-40.

GIS in Planetary Geology – Standardised Moduls for Planetary Mapping

Andrea NASS, Stephan VAN GASSELT, Ralf JAUMANN and Hartmut ASCHE

1 Introduction

The last decades showed steadily growing international interest in the exploration of the planets in our solar system. The rapid technical development, the need of fundamental research and the curiosity to discover unexplored regions get more and more missions off the ground to far planets like Mars, Venus and Titan. Instruments are carried along this missions which image a variety of data in different wavelength range (multi spectral-/hyper spectral sensors, radar mode, stereo camera experiments, roentgen detectors) and further enable the derivation of data (e.g. high resolution digital terrain models).

With the aid of these data the planetary science community approaches the questions about the evolution and development of surface, atmosphere and the interior structure of planetary bodies. Therefore, the European Planetary Network (EUROPLANET 2009) for instance is concerned with the comprehensive utilisation of planetary data. The participants of the Helmholtz Alliance "Planetary Evolution of Life" (HELMHOLTZ ALLIANCE 2009) analyse the fluvial activities of several terrestrial objects and determine the potential landing sites for future robotic missions.

To explore the evolution and development of the planetary bodies, the surface structure is analysed and geo-scientifically interpreted. The results are represented in thematic, mostly geological and geomorphological maps. To enable an efficient collaboration among the different planetary disciplines all the mapping results have to be uniformly prepared, described, managed and archived. Thus the research community has recourse to already elaborated results as secondary base data for further studies.

2 Motivation and Objectives

For handling the issues mentioned above spatial information systems have been employed in the field of Planetary Research in the last years, e.g. presented by International Cartographic Association (ICA 2009), International Society for Photogrammetry and Remote Sensing (ISPRS 2009) and United State Geological Survey (USGS 2009). The necessary exploratory work concerning the data integration, analysis and visualisation are similar to the work on terrestrial geoscientific analysis. However, in contrast to terrestrial analysis the work with planetary data has specific problems caused by the missing "ground truth" that complicate the mapping process.

Therefore, in order to simplify the planetary mapping, we currently design a Geo Information System (GIS) environment, that arranges the subsequent steps of data integration, management, processing and analysis in one most efficient schema. This schema is based on a database model which is developed in the commercial software

ArcGIS and provides the layout of the data infrastructure including relations and topologies. In addition it provides a predefined, modular GIS environment according to international standards with individual tools and representation facilities for planetary applications. By the conversion of the modules into open exchange formats we will ensure the portability.

3 GIS-based Implementation and Portability

The physical implementation of the mapping schema will be performed step by step – according to the modular overall concept of the GIS environment. This will be shown here with two exemplary modules (NASS 2009).

3.1 Standardised Representation in the Planetary Geology

This module aims at the uniformed symbolisation[1] for the GIS-based preparation of thematic maps. For this purpose we used the digital cartographic standard for geologic map symbolisation (FGDC 2006) and generated the defined symbols for point, linear and area features in the GIS software. Therefore existing symbols were modified and new symbols were generated. After that all representations were stored in a predefined symbolisation catalogue (see Fig. 1). Such a symbolisation basis supports the simplification of the mapping procedure as the geologists can use the environment without having to deal with technical issues. It also provides the comparability of different mapping results. Furthermore the usage of these symbols will be simplified by the connection to the underlying data model. Therefore the option of the symbol key will be supported by an on-topic pre-selection of layer classification, i.e. is the feature volcanic or tectonic.

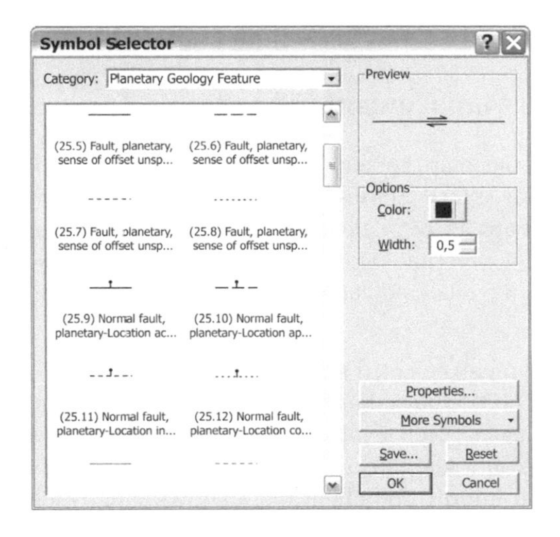

Fig. 1: Symbolisation catalogue for the planetary geology

[1] In this work the terms symbolisation and representation are applied as synonym.

In order to avoid the software dependences regarding utilisation of symbol sets, but yet to ensure the correct representation of the mapping products, the single symbols will be converted to the exchange format Scalable Vector Graphics (SVG) (W3C 2008a) (see Fig. 2). This language is based on the Extended Markup Language (XML) (W3C 2008b) and describes two-dimensional objects like vector graphics, images and text. By using these, the transparency, portability and interoperability between different software systems are possible.

```
<?xml version="1.0" encoding="utf-8"?>
<!DOCTYPE svg PUBLIC "-//W3C//DTD SVG 1.1//EN"
        "http://www.w3.org/Graphics/SVG/1.1/DTD/svg11.dtd">
<svg version="1.1" id="Ebene_1" xmlns="http://www.w3.org/2000/svg"
        xmlns:xlink="http://www.w3.org/1999/xlink" x="0px" y="0px" width="595.28px"
        height="841.89px" viewBox="0 0 595.28 841.89"
        enable-background="new 0 0 595.28 841.89" xml:space="preserve">
<g>
  <g>
    <defs>
      <rect id="SVGID_1_" x="215.144" y="358.627" width="26.88" height="25.44"/>
    </defs>
      <clipPath id="SVGID_2_">
        <use xlink:href="#SVGID_1_"  overflow="visible"/>
      </clipPath>
        <g clip-path="url(#SVGID_2_)">
          <text transform="matrix(1 0 0 1 242.5098 384.0605)"
            font-family="'TimesNewRomanPSMT'" font-size="12">
          </text>
        </g>
  </g>
    <circle fill="none" stroke="#000000" cx="229.075" cy="371.004" r="3.992"/>
    <line fill="none" stroke="#000000" x1="229.075" y1="360.596" x2="229.075" y2="367.355"/>
    <line fill="none" stroke="#000000" x1="229.075" y1="374.944" x2="229.075" y2="381.704"/>
    <line fill="none" stroke="#000000" x1="240.263" y1="371.348" x2="233.503" y2="371.348"/>
    <line fill="none" stroke="#000000" x1="225.083" y1="371.004" x2="218.323" y2="371.004"/>
</g>
</svg>
```

a) b)

Fig. 2: a) Symbol for knob or central peak, planetary geology (FGDC 2006)
b) Generated document for the representation in *.svg-format

3.2 Standardised Description of the Map Content (Metadata)

This module approaches the description of the map content through the definition of metadata. Such meta-information provides the legend entries and keys for digital maps comparable to what is known for analogues maps. A detailed specification is essential for the understanding of individual geological/geomorphological maps. Nowadays these interpretations exclusively rely on remote sensing data and can only be comprehended and understood by using this "data about data".

In order to guarantee such metadata description we generated a metadata template which is based on existing standards for spatial data (FGDC 2000), but is adapted to the individual needs of the geologist in the planetary science. This template inquires all necessary information directly after the mapping process in a Graphical User Interface (GUI) which must be filled. This contains on the one hand information about the total mapping result like basis data used for the map, on the other hand metadata for every individual mapping layer like minimum mapping scale. Through this GUI the map declaration will be stored in a predefined XML document and consequently enables the software independent exchange of the (meta) data.

4 Conclusion and Outlook

With the help of implementing the modules mentioned above all maps and single map layers generated in ArcGIS can be displayed, managed and edited independently of any (established) GIS software environment, but also in vector-based graphic software and Web Map Services (WMS) (OGC 2010). It further facilitates the efficient and traceable storage of data on a local server.

The next step will be the implementation of the numerous modules for data representation and management within the database model. The handling during the mapping and analysis process will be further improved by editing modules, e.g. the definition of a planetary map projection and reference systems.

Acknowledgment

This work is partly supported by DLR, Europlanet and the Helmholtz Alliance "Planetary Evolution and Life".

References

EUROPLANET (2009) – Web: http://www.europlanet-eu.org/demo/ (03.12.2009).

FGDC – FEDERAL GEOGRAPHIC DATA COMMITTEE (2000), Content Standard for Digital Geospatial Metadata Workbook, FGDC-STD-001-1998.

FGDC – FEDERAL GEOGRAPHIC DATA COMMITTEE (2006), Digital Cartographic Standard for Geologic Map Symbolization, FGDC-STD-013 2006.

HELMHOLTZ ALLIANCE (2009) – http://www.dlr.de/pf/en/desktopdefault.aspx/tabid-4843/ (03.12.2009).

ICA – International Cartographic Association (2009), Web: http://icaci.org/ (03.04.2010).

ISPRS – International Society for Photogrammetry and Remote Sensing (2009) – http://www.isprs.org/, (03.04.2010).

NASS, A., JAUMANN, R. & ASCHE, H. (2009), Concept of Data Archiving and Management for Geological Mapping of Martian Surface illustrated by an "Analysis and Interpretation of Valley Networks on Mars", ICC, Santiago, Chile.

OGC – OPEN GEOSPATIAL CONSORTIUM (2010), OpenGIS Web Map Service (WMS). – http://www.opengeospatial.org/standards/wms, (28.01.2010).

USGS – United State Geological Survey (2009) – http://astrogeology.usgs.gov/Projects/webgis/ (03.04.2010).

VAN GASSELT, S. & A. Nass (2010), Challenges in Planetary Mapping: Application of Data Models and Geo-Information-Systems for Planetary Mapping (in this volume).

W3C RECOMMENDATION (2008a), Scalable Vector Graphics (SVG) Tiny 1.2 Specification. – http://www.w3.org/TR/2008/REC-SVGTiny12-20081222/ (04.01.2010).

W3C RECOMMENDATION (2008b), Extensible Markup Language (XML) 1.0. – http://www.w3.org/TR/2008/REC-xml-20081126/REC-xml-20081126.xml (14.01.2010).

Applied GIS in Forest Mapping and Inventory in Rwanda

Jean NDUWAMUNGU

1 Introduction

The Geographic Information System and Remote Sensing Training and Research Centre of the National University of Rwanda (CGIS-NUR) has been created en 2001 to support the development of a GIS based education and research program at the National University of Rwanda (NUR). The centre works closely with the University Faculties and Schools to integrate Geo-Information Science courses in the different bachelor and master programs offered at NUR. However, its main focus is to undertake applied research in the various fields using GIS and remote sensing technology. This paper discusses one of the recent projects in which forested areas were mapped and forest cover changes since 1988 were assessed in Rwanda. The output forest maps served as base maps for planning and conducting the first national forest inventory. A stratified systematic cluster sampling design was adopted to select the sample plots for forest inventory. The forest maps were used as sampling frames.

2 Methodology

For the purpose of this work, the FAO definition was used to define a forest. Thus, a forest was defined as "*a land with 0.5 ha or more covered with trees higher than 7 meters and a canopy cover of more than 10%, or tree stands which have the potential to reach these thresholds in situ*" (FAO 1999). A multiple view remote sensing approach was used combining multi-stage, multi-spectral and multi-temporal sensing to establish adequate forest cover classes (LILLESAND & KIEFER 2000). The procedure has first involved the selection and procurement of satellite imageries. The imageries used include ASTER (15 m), SPOT Xs (20 m) and Landsat TM (30 m) dated from 2002 to 2005. The selection of imageries was based on compromise between spatial resolution, swath and availability for the country. The acquired imageries were then processed through geocoding, orthorectification and radiometric enhancement. The image processing software used is ERDAS Imagine 9.1. After these pre-processing operations, ground truthing followed using mobile GIS and printouts of satellite images. Basing on FAO (2005), a forest legend was further developed and a preliminary classification done. After thorough analysis of the results, forest classes were refined before running the final forest classification. It was not easy to differentiate tree species because of the low resolution of imageries used and therefore many classes were merged to reduce classification errors. In total 7 forest classes were retained including Bamboo forest, Natural Degraded Forest, Eucalyptus Forest Plantation, Humid Natural Forest, Pine Forest Plantation, Savannah and Young or Open Plantation or Coppice.

The first forest map printouts (at 1:25,000 scale) were used for planning field work for the forest inventory. The classification results were revalidated using field data collected during forest inventory field work to improve the preliminary classification. This revalidation clearly improved the reclassification result. The map accuracy was 98.6% for location and between 30 to 57% for species. After revalidation, forest classes were further refined before final classification and forest mapping. The GIS software used is ArcGIS 9.3. The tiling system based on the MININFRA Topographic Map system at 1:50,000 and the Coordinate System Rwanda_92 were adopted for map production and printout. The information on forest areas extracted from forest mapping was further used during processing forest inventory data to obtain average values on hectare basis by forest class. Moreover, in order to estimate forest cover change since 1988, forest maps of 1988 were produced through digitization of topographic maps of 1988 and aerial photographs of 1970-1980s.

3 Results and Discussion

The results of the forest mapping based on the forest definition above estimated that in 2007 forests covered about 240,747 ha (2,407 km^2) in Rwanda. This represents about 10% of the physical area of the country of which *Eucalyptus* forest plantations account for 4.3% (including Young or Open forest plantation or Coppice because most of them are made up of *Eucalyptus species*) while the humid natural forest makes up 3.4% (Fig. 1). The forest cover is also unevenly distributed in the country (MINIRENA 2007a). The Western Province had the highest percentage of forest cover (42%) while the Eastern Province accounted for only 5% of forest cover. Moreover, while artificial plantations are scattered throughout the country, the majority of natural forests remain restricted to protected areas in the western and northern parts of the country.

On the basis of the produced forest maps, in 1988 the forest covered about 265,314 ha (which is about 11 % of the total country area) of which 140,044 ha were natural forests

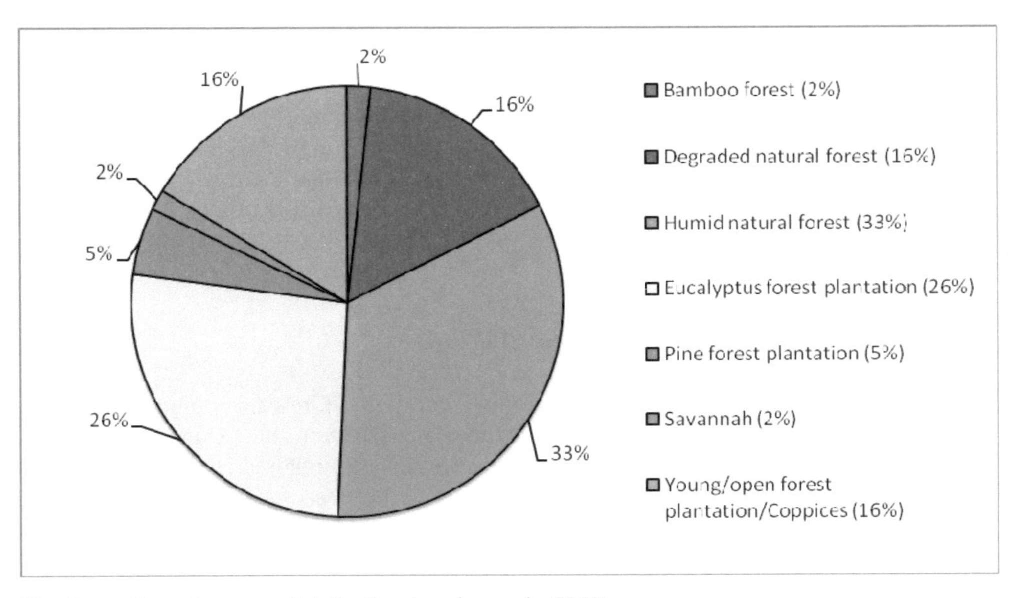

Fig. 1: Forest covers distribution by classes in 2007

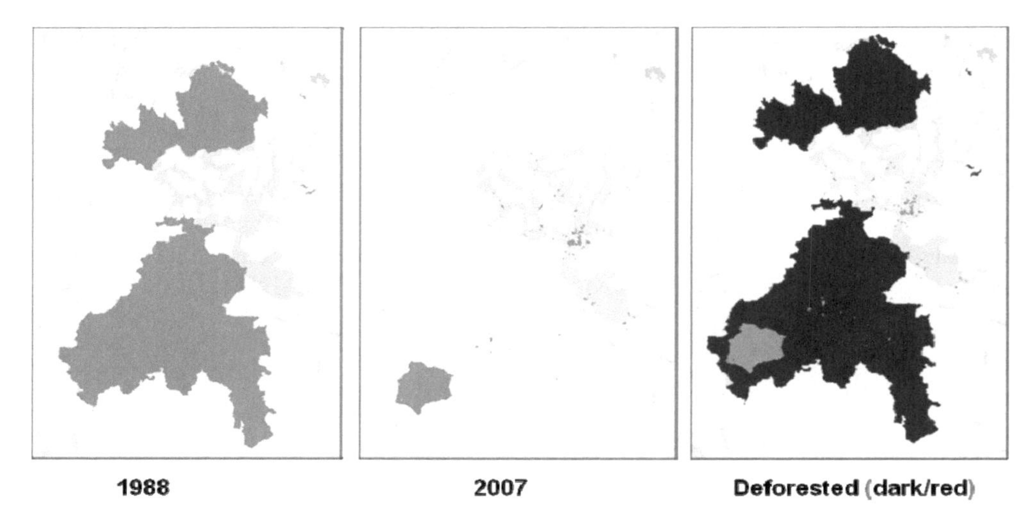

Fig. 2: Deforestation of Gishwati natural forest between 1988 and 2007

and 125,270 ha artificial plantations (MINIRENA 2007a). Using GIS functions and tools, deforested and reforested areas were identified (an example of change detection of extent of deforestation is shown in Fig. 2). A comparison of the forest cover in 1988 and in 2007 revealed that deforestation has affected about 25,441 ha of natural forests (18.2%) and 64,367 ha of forest plantations (51.4%). In other words, more than half of the forest plantations were destroyed in less than 20 years. Nevertheless, thanks to the massive reforestation campaigns undertaken during the late 90's, about 52,904 ha have been planted between 1988 and 2007 (MINIRENA 2007a). Moreover, even in natural forests, the increased control measures after 1996 have enabled rapid regeneration of patches of deforested or degraded areas of about 12,337 ha which makes up about 8% of total natural forest cover in1988. The deforestation processes in early 90's could be explained by the importance of population which were displaced by the war and the Genocide of Tutsi in 1994 but also by massive return of refugees in need of areas for resettlement and farming (MUND & CHRIST 2002). The ground truthing and forest inventory field work revealed that the remaining forests are highly degraded and are in dire need of silvicultural and management interventions to improve their yield (MINIRENA 2007b).

4 Conclusion and Recommendation

The use of GIS technologies has enabled to carry out digital forest mapping and to implement the first comprehensive forest inventory in Rwanda. Even though in the context of Rwanda, this forest inventory is still incomplete because it has ignored all small forest plots less than 0.5 ha (due to low resolution of the satellite images used), this is a step ahead towards knowing the forest resources of the country in terms of quantity and quality. The fragmentation and level of degradation of the remaining forests in Rwanda is so high that the forest definition should be adapted and new methods devised for assessing even tiny forest areas. As a result of high level fragmentation of farms, products from these tiny forest plantations are very critical to the livelihoods of the majority of rural populations. Further research is therefore needed to provide with methods, tools and techniques adapted to Rwanda specificity allowing for increased data accuracy to monitor small woodlot or trees on farm (agroforestry) and forest cover change overtime.

References

FAO (1999), FRA 2000, Global Forest Cover Map, Working Paper 19, Rome.

FAO (2005), Global Forest Resources Assessment Update 2005, Terms and Definitions (Final Version), Working Paper 83, Rome, 2004.

LILLESAND & KIEFER (2000), Remote sensing and image interpretation. 4th edition. John Wiley &Sons, New York.

MINIRENA (2007a), Projet d'Inventaire des Ressources Ligneuses du Rwanda: Cartographie des Forêts du Rwanda. Rapport Final, Volume 1 (CGIS-NUR), Kigali, Rwanda.

MINIRENA (2007b), Inventaire des Ressources Ligneuses du Rwanda. Rapport Final, Volume 2 (ISAR) Kigali, Rwanda.

MUND, D. & CHRIST, T. (2002), Land Degradation in the Akagera National Park (former and actual ANP) in Rwanda: Estimating the rate and processes, ORTPN, PRORENA & GTZ.

Modeling Flood Hazard Due to Climate Change in Small Mountainous Catchments

Stoyan NEDKOV

Introduction

Climate change is often represented by increase of extreme phenomena, like storms, torrential rains, and floods. Their regional and local dimensions vary from one area to another. This is especially valid for the regime and distribution of precipitation. Even for a country with relatively small territory, like Bulgaria, precipitation in some areas has increased during the last decades, while in others it has decreased. The only precipitation characteristic with undoubted trend of increase for most of the territory is the heavy and torrential rains (VELEV 2005, BOCHEVA et al. 2007). This means that, even in cases with no increase in the annual precipitation, there are less storm events but their quantity and intensity are higher. Torrential rains are one of the most important factors for flood formation and there is a significant increase in the number of this hazardous phenomenon during the last few years. The causal relationship between the increasing number of torrential rains and flood events for the last decades in Bulgaria can be illustrated with an example from Yantra River basin (fig. 1).

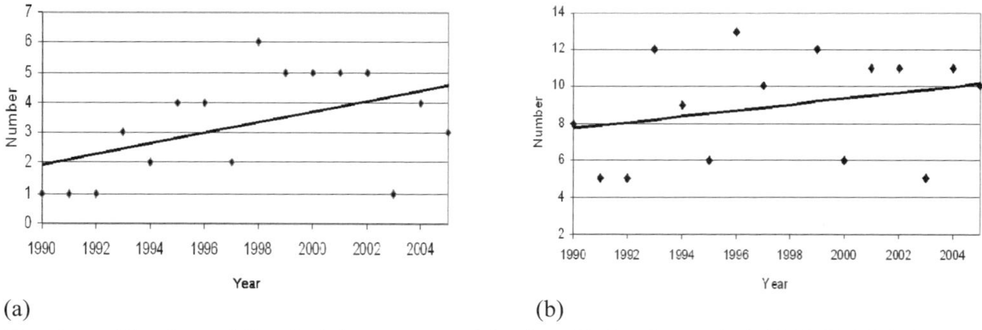

(a) (b)

Fig. 1: Number of floods (a) and torrential rains (b) for the period 1990-2005 in Yantra River basin (NEDKOV & NIKOLOVA 2006)

The problem is more serious in the mountainous areas, where the amount of precipitations is normally larger and the topography facilitates a rapid increase of surface runoff and formation of disastrous peak flows. Spatial distribution of precipitation in the mountains is usually discontinuous. This is even more typical for the torrential rains, the influence of which is often concentrated in some parts of the river basin. They seldom exert simultaneous effect on all the tributaries of the main river (NIKOLOVA 2007). Therefore, it is necessary to investigate in detail the influence of the small catchments on the formation of peak flow in the mountainous areas and, further, the impact of the increased number and quantity of the heavy and torrential rains. This study presents an approach that utilizes GIS tools for modeling the impact of increasing incidence of torrential rains on the formation of peak flows and flood hazards, as well as the response of small catchments with different landscape structure.

1 GIS Modeling

GIS is a powerful tool for modeling the processes taking place within catchment basins which affect the formation of peak flows. The Automated Geospatial Assessment Tool (AGWA), designed to facilitate all phases of the modeling for two widely used hydrologic models (SWAT and KINEROS) in GIS environment, has been used to delineate and parameterize the catchments. Kinematic Runoff and Erosion Model (KINEROS) has been used to simulate the influence of rainfalls with different intensity and quantity on the peak flow during particular storm events. It is a distributed, physically based, event model describing the processes of interception, dynamic infiltration, surface runoff, and erosion from catchments characterized by predominantly overland flow. The catchment is conceptualized as a cascade and channels, over which flow is routed in a top-down approach using a finite difference solution of the one-dimensional kinematic wave equations (SEMMENS et al. 2005).

The investigation was carried out in the case study area of Malki Iskar River before it enters the town of Etropole (Bulgaria). Situated in the northern slopes of Stara Planina Mountain the area spreads over 54.5 km^2 and has an average altitude of 1164 m. The observations carried out during the last decades show that there is a significant increase in the number and quantity of the torrential rains, which corresponds to a similar trend observed in the rest of the country. The model has been used in four small catchments within the area. The initial investigation and model calibration was carried out in the catchment of Ravna River (a tributary of Malki Iskar), because there is a hydrometric station for measurement of the river flow there. Simulated data was calibrated against runoff and precipitation data for five measured storm events. The relationship between the increasing rainfalls intensity and the peak flow was analyzed. The model has been adjusted for every event, in order to present the particular conditions of the state of the landscape during the observations, including initial soil moisture, saturated hydraulic conductivity, channel and plane roughness. Numerous rainfall files were prepared to present the increase of precipitation's amount using 2 mm step. The model has been run with everyone of them and the simulated quantities of the peak flow have been put on a graph that shows the relationship between the increasing intensity of the rainfalls and the peak flow change.

2 Results

The resulting curve (fig. 2) shows that the increasing intensity in the beginning causes almost no response in the river discharge. It means that the whole amount of water from the rainfall goes for interception and infiltration. The peak flow of the river starts to increase gradually after a particular intensity of the rainfall is reached. Its value is strongly dependent on the available moisture before the start of the storm event. More moisture

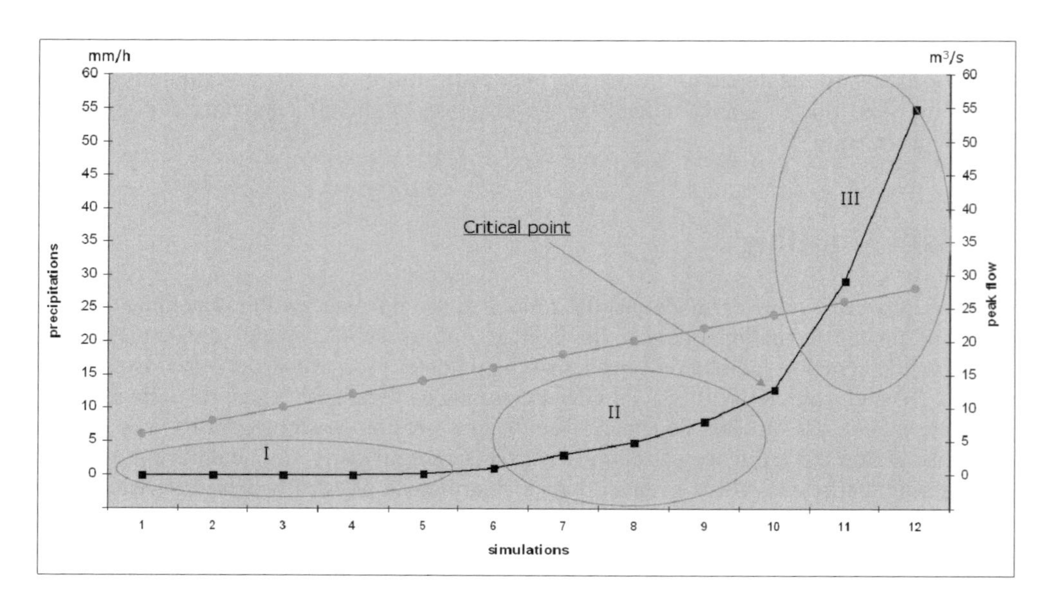

Fig. 2: Change of the peak flow in the Ravna River as a result of the increase in the rainfall intensity: I – phase of no increase; II – phase of gradual increase; III – phase of rapid increase

causes faster and earlier increase of the peak flow. As the rainfall intensity increases further, at a particular level the peak flow displays rapid growth and reaches disastrous quantities that cause floods. This level can be marked as a "critical point", after which the flood risk rises enormously. Therefore, the peak flow response to the increasing rainfall intensity can be divided into three phases (fig. 2): 1) phase of no increase, when the landscape "absorbs" almost the whole amount of water; 2) phase of gradual increase, when part of the water is transformed into surface runoff; 3) phase of rapid growth, when the ability of the landscape to "absorb" water is close to zero and almost the whole amount of rainfall goes to the surface runoff.

Next, the same procedure was implemented for the rivers Kobilya, Suha, and Jablanitsa using the same five events in order to investigate how the different catchment conditions affect the amount of peak flow. In this case, the three phases are also easily identified on the curve but their shape varies among the different catchments. The Ravna River shows the most rapid increase of the peak flow, which exceeds the other rivers at every point after the initial increase. The Suha River has slower increase in the second phase, but faster in the third. The situation is opposite for Kobilya and Jablanitsa, which have faster increase (than Suha) in the first phase, but slower in the third one. The rapid growth in these two rivers is not so expressive as Ravna and Suha. This results show that the "reaction" of the peak flow, as a result of the increasing precipitation intensity, can differ among small catchments even within relatively small area, due to the large heterogeneity of the mountainous landscapes. The most important characteristics causing these differences are topography (especially slope), land cover, and soil properties.

3 Conclusion

The main conclusion is that, climate change causes an increase of torrential rains, which can lead to bigger and more disastrous floods. The model shows significant increase of the peak flow when rainfall quantity exceeds a particular "critical" level, after which the flood risk rises enormously. This level depends on the moisture conditions before the storm and seasonal state of the land cover. It also differs among catchments with different landscape features. The implementation of the presented approach provides an opportunity to assess the flood hazard in mountainous catchments and contribute to the development of early warning systems.

References

BOCHEVA, L. BOCHAVA, L. MARINOVA, T. & GOSPODINOV, I. (2007), Variability and trends of extreme precipitation events over Bulgaria. Proceedings of 4th European Conference on Severe Storms, Trieste, Italy, 10 – 14 September.

NEDKOV, S. & NIKOLOVA, M. (2006), Modeling flood hazard in Yantra river basin. In: Proceedings from Balwois conference, Ohrid, May 23 – 26.

NIKOLOVA, M. (2007), Climatic conditions for high waves and floods in the basin of Malki Iskar River above town of Etropole. Proceedings of the Second National Research Conference on Emergency Management and Protection of the Population, Sofia, November, 9 (in Bulgarian).

SEMMENS, D., GOODRICH, D., UNCRICH, C., SMITH, R., WOOLISER D. & MILLER S. (2005), KINEROS2 and the AGWA Modeling Framework. International G-WADI Modeling Workshop.

VELEV, S. (2005), Torrential rain in Bulgaria during XX century. Problems of Geography 1-2, pp. 169-172 (in Bulgarian).

Assessing Spatio-Temporal Land Cover Changes Within the Nyando River Basin of Kenya Using Landsat Satellite Data Aided by Community Based Mapping – A Case Study

Luke O. OLANG, Peter KUNDU, Thomas BAUER and Josef FÜRST

The GI_Forum Program Committee accepted this paper as reviewed full paper.

Abstract

Spatio-temporal land cover changes witnessed within the Nyando River Basin of Kenya were assessed in this study. The land cover changes were mapped by classifying the predominant land cover classes on selected Landsat satellite images. The accuracy of the classifications were assessed using reference datasets developed and processed in a GIS with the help ground based information obtained through participatory community based mapping techniques. The results of the analysis indicated significant deforestation in the headwaters of the basin. Obviously apparent from the land cover conversion matrices was that the majority of the forest decline was a consequence of agricultural expansion. Despite the haphazard land use patterns and uncertainties related to poor data quality for environmental change assessment, the study successfully exposed the vast degradation and hence the dire need for both sustainable landuse planning and catchment management strategies.

1 Introduction

Land cover changes due to anthropogenic interventions remain a major environmental challenge in most river basins in Kenya. One of the basins that epitomize this degradation is the Nyando River Basin. With its headwaters located within the vulnerable Mau Forest currently being threatened by depletion, such changes have led to amplified flood flows during storm events and reduced stream flows during low flows (OLANG & FÜRST 2010). It is hence imperative that such changes are monitored and their sociological, economic and ecological consequences accurately quantified in order to gain valuable information important for management and future restoration efforts (BALDYGA et al. 2007). A common way to map land cover changes is through satellite image classifications. However, satellite classifications require validations through accuracy assessments using historical datasets; a factor which further hinders their applications in areas with limited data. The Nyando Basin lacks consistent and detailed land cover databases. The few existing datasets are generalized maps with inadequate thematic and statistical information essential for authentication purposes (CONGALTON & GREEN 1999).

Furthermore, the existence of amorphous land use patterns and hence lack of a clear definition between areas that intermittently vary between pasture, small scale subsistence agriculture and settlements further complicates accurate monitoring of the changes using satellite images. Previous studies have also shown that spectral signatures acquired from satellite imagery for regions within the tropics display minimal band separabilities amongst the various vegetation types (TATEM et al. 2005, TOTTRUP 2004). With these constraints, it is inevitable to apply an integrated approach that exploits every source of information available to detect and support the classification process. In this contribution, a participatory procedure that integrates rigorous community based information with the commonly used scientific tools involving GIS and RS applications were used to discern historical land cover changes (FÜRST 2004, PELLIKKA et al. 2009). Such a procedure, commonly known as community based GIS, is gaining popularity in most regions in Kenya not only due to data scarcity but also because of the fact that the riparian communities understand and do have some vital information about the environmental changes taking place within their immediate surroundings (RAMBALDI et al. 2007).

2 Study Area

The Nyando basin is located in western Kenya between 0° 25′ S – 0°10′ N and 34° 50′E – 35° 50′ E. It covers an area of about 3550 km^2 within the scarps of the Kavirondo Gulf. The basin is drained by River Nyando with its major tributaries originating from the upland Nandi and Mau Hills. The river drains into the transboundary Lake Victoria at altitudes of about 1300 m *a.m.s.l.* The climate of the basin is largely influenced by the Equatorial Convergence Zone modified by local orographic effects. The land cover types vary principally from forests in the uplands to mixed-type subsistent agriculture in the mid to lowland parts. The human population in the basin currently stands at about 0.8 million and is largely responsible for the majority of the land cover dynamics (Figure 1).

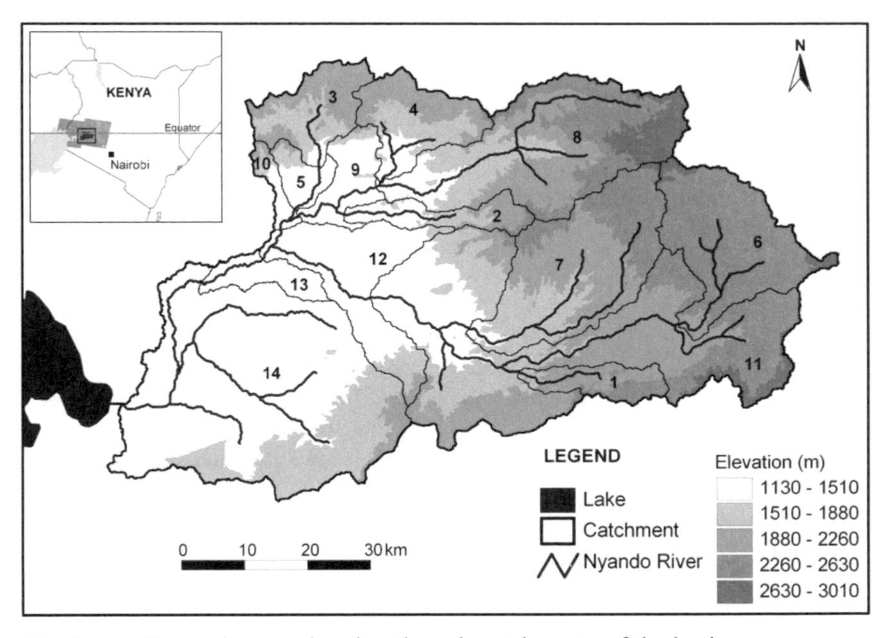

Fig. 1: The study area showing the sub catchments of the basin

3 Datasets

The global Shuttle Radar Topographic Mission (SRTM) digital elevation model (approximately 90m) was processed and used to derive the sub-catchments and their relevant geophysical characteristics. Seven major land cover classes including agriculture, grasslands, forests, wetlands, tea/coffee, shrublands and water were selected for mapping. These classes were selected based on their hydrological significance to allow for subsequent estimation of the effects of the detected land cover change on the hydrological response of the regions. Six Landsat images, two for each year, were acquired for the periods between 1973 and 2000 (Table 1).

Table 1: Characteristics of the selected Landsat images

Date	Landsat sensor	Path/Row	Approx. spatial resolution (m)
27th Jan. 2000	ETM+	169/60	30
05th Feb. 2001	ETM+	170/60	30
28th Jan. 1986	TM	169/60	30
08th Mar. 1986	TM	170/60	30
01st Feb. 1973	MSS	182/60	80
31st Jan. 1973	MSS	181/60	80

The acquired images were already orthorectified and topographically normalized (KUNDU et al. 2008). The images were resampled using the nearest neighbor technique to a common resolution of about 80 m (JENSEN 2005). Ancillary land cover datasets to support the image classifications were also obtained from various sources including the FAO-Africover data available at scale of 1:100000. This database provides good quality geo-referenced land cover data represented by unique attributes based on the FAO/UNEP land cover classification system. The dataset was acquired as vector coverage for Kenya and later processed in conformity with the study demands.

A vectorized land cover dataset for 1987 at a scale of about 1:5M was also obtained from the Ministry of Natural Resources, Kenya. Twelve topographic maps for 1972 at a scale of 1:50000 were also acquired for supporting the image classifications. The maps were digitally processed and geo-referenced based on the third order polynomial transformation (TOA & HU 2001). A general land cover map representing the visible land uses was subsequently delineated from the maps. In overall, the ancillary dataset for 1973 and 1986 did not provide sufficient thematic and statistical detail required for comprehensive authentication purposes. Nonetheless, the datasets were equally important for plausibility checks, understanding the general classification trends and for supporting the adopted proposed community based approach.

4 Methodology

Considering the quality of the available data, the procedure called guided clustering was used to discern the changes (YUAN et al. 2005). This procedure is normally favorable for complex ecological regimes of diverse composition. In principle, guided clustering encompasses the application of unsupervised and supervised classification approaches. In this study, the unsupervised technique was used strictly to identify appropriate clusters to be used as signatures in the subsequent supervised classification process (JENSEN 2005). Due to its relative good data quality of the Africover reference dataset, the study also opted for developing a consistent classification procedure that could be replicated for the other periods.

The Iterative Self-Organizing Data Analysis Technique (ISODATA) unsupervised technique was used to generate about 30 clusters. The generated clusters were carefully identified and labeled, where possible, with the help of the reference datasets. Selected regions with dependable but unidentifiable clusters were highlighted and delineated for auxiliary verification purposes. This was achieved through a comprehensive community based GIS mapping performed through participatory interactions and discussions with the local communities about the historical land cover states of the delineated polygons (RAMBALDI et al. 2007). This rigorous procedure facilitated identification and labeling of selected highlighted areas of uncertainties.

The labeled polygons were compared and carefully evaluated, especially within regions of similar spectral characteristics, and later used as signatures for the supervised classification. The optimal bands for the classification were identified using the statistical transformed divergence technique (ERDAS 2002) due to the amorphous and random land cover patterns in the basin. To discriminate and reproduce the land cover patterns of the area, different classifiers, non-parametric and the parametric, were examined. A 3×3 majority filter was then applied to the classified maps to isolate small pixels arising from the classification, and the individual classified maps mosaicked and clipped to the study area. This approach was preferred to minimize spectral distortions associated with mosaicking multi-date images before classification.

The classification accuracies were later assessed through error matrices developed using auxiliary points generated through stratified random sampling technique. The total numbers of sampling points were established using the Binomial theory, with 5% allowable error being assumed. The expected accuracy of the change maps was also estimated based on a procedure that involved multiplying the individual image classifications accuracy. The procedure is generally based on the premise that when two change maps are overlaid, an accumulation of the overall error occurs leading to the reduction of the estimated accuracy by a factor equal to the product of the individual accuracies (YUAN et al. 2005).

5 Results and Discussion

Generally, the community based approach of identifying and labeling training sites through ground truthing proved sufficient and reliable in assessing the historical land cover states. The best reproducibility of the land cover patterns of the basin was obtained when the maximum-likelihood classifier was used with *a priori* probability weights of cover classes approximated from the reference dataset. Forest and water land cover classes indicated the highest average separability. Grassland and agriculture on the other hand showed the lowest average separability due to their random occurrence and close spectral characteristics. Since, the reference dataset for the year 2000 was the most reliable; it was consequently used to test the accuracy of the classification procedure adopted. Results obtained from the error matrices revealed overall classification accuracy and *Kappa* index of about 84% and 80% respectively. More specifically, higher producers and users accuracies were noted between forests, wetlands and water land cover classes (Table 2).

Table 2: Statistics of the detected land cover changes in the basin

Land cover	**Classified maps**						**Relative change (%)**		
	1973		**1986**		**2000**				
	(km^2)	(%)	(km^2)	(%)	(km^2)	(%)	(73-86)	(86-00)	(73-00)
Agriculture	1564.3	44.1	1432.2	40.4	2002.8	56.5	-4	16	12
Grassland	433.4	12.2	999.0	28.2	617.9	17.4	16	-10	5
Forest	1238.5	35.0	750.6	21.2	514.3	14.5	-14	-7	-20
Shrublands	83.7	2.4	159.4	4.5	234.8	6.6	2	2	4
Wetland	69.2	2.0	118.4	3.3	62.0	1.8	1	-2	0
Tea/coffee	151.4	4.2	80.8	2.3	109.3	3.1	-1	1	-1
Water	2.9	0.1	3.1	0.1	3.0	0.1	0	0	0

During the period of 1973-2000, shrublands increased by 4% representing an area of about 151 km^2. This increase was more prevalent in previously forests regions signifying gradual degeneration through selective logging and re-growths. Agricultural lands expanded by 12% over the same period of time, with the highest increase being observed between 1986 and 2000, due to the seasonal conversions between grasslands and agricultural land cover classes. Between 1973 and 2000, the basin was noted to have observed an upsurge of more than 40% in its human population. This population increase could have exerted sufficient economic pressure on the resources within the vicinity. The land cover conversion matrix for 1973-2000 obtained when the classified images were compared across the years is presented in Table 3.

Table 3: Transition matrices of the land covers in km^2 for 1973-2000

2000 [To]	**1973[From]**							**Total [2000]**
	Agriculture	**Grassland**	**Forest**	**Shrubland**	**Wetland**	**Tea/coffee**	**Water**	
Agriculture	1136.5	305.3	385.4	66.5	53.3	55.0	0.8	2002.8
Grassland	335.1	101.7	111.7	13.3	11.3	44.4	0.5	617.9
Forest	13.0	5.0	490.9	0.5	0.8	4.0	0.1	514.3
Shrublands	48.0	10.5	170.8	1.2	1.0	3.2	0.0	234.8
Wetland	11.9	3.2	41.9	1.9	1.8	1.2	0.1	61.9
Tea/coffee	19.4	7.6	37.4	0.4	0.8	43.6	0.0	109.3
Water	0.6	0.1	0.5	0.0	0.2	0.0	1.5	3.0
Total [1973]	1564.5	433.4	1238.6	83.7	69.2	151.4	2.9	3543.8

With this order of priority, it was noted that the majority of the land cover conversions largely resulted into agricultural expansion. About 305 km^2 of grassland area was converted into agriculture between 1973 and 2000. However, about 13 km^2 of agricultural lands and 5 km^2 of grasslands were converted back to forest in the periods of 1973 – 2000. Considering the rapid population increase demanding more land for subsistence agriculture, the conversion of other land covers back to forests may seem unfeasible. However, from our field surveys, a time span of at least 10 years (between the images selected) is sufficiently long enough to accommodate such magnitude of changes. Tree stands that act as wind breakers for the normally fragile tea crops for instance, are occasionally planted and harvested within the uplands and central parts of the basin. It is also a norm for the riparian communities to plant trees for timber and domestic construction purposes, wherever necessary. Nevertheless, it is also acknowledged that classification or data errors could have possibly led to some fluctuation of the results.

Land cover changes within the sub-catchments of the basin were also evaluated. For instance, in sub-catchment no. 14 located in the downstream of the basin, about 228 km^2 was converted into agriculture. And about 98 km^2 of the previously existing forests in 1973 were completely diminished by the year 2000. The area also witnessed significant decline in the wetlands by a margin of about 25 km^2 within the three decades. From the land cover changes; this sub catchment characterizes an area which has been fully converted into an agricultural sub basin. In sub-catchment no. 6 located in the upstream region, deforestation was noted to have affected an area of about 106 km^2. Agriculture and shrublands on the other hand increased by 69 km^2 and 29 km^2 respectively, with the majority of the changes occurring between 1973 and 1986. This sub-catchment exemplifies a forested region which is slowly being converted into a more or less agricultural sub catchment (Figure 2).

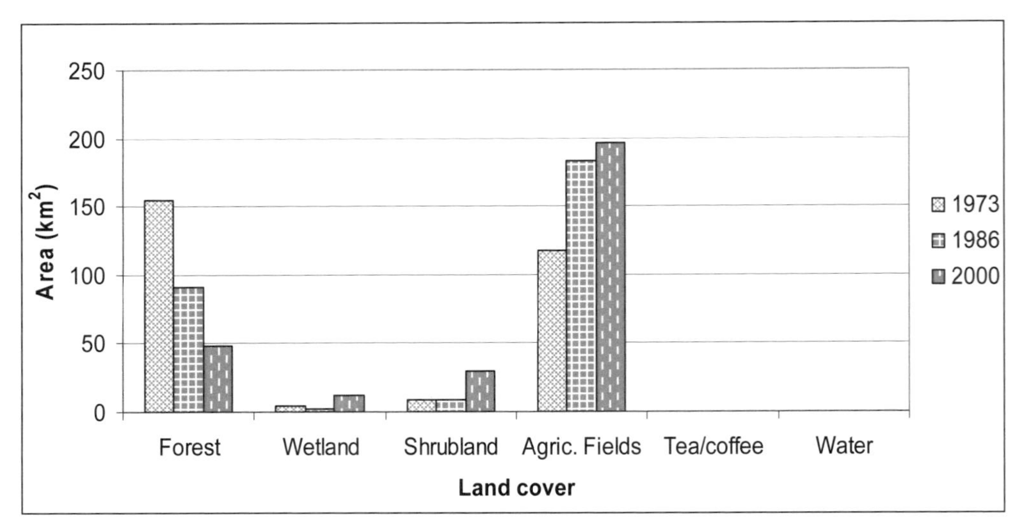

Fig. 2: Land cover changes in sub-catchment No. 6

6 Conclusion and Recommendation

Spatio-temporal changes in the historical land cover states were investigated in this study. The classification results indicated that the basin underwent vast land cover changes over the periods. Apparently clear was the decline in forest coverage with an almost equivalent increase in the agricultural areas. The land cover change results obtained generally depicted similar trends to other studies carried out Kenya. In summary, the study demonstrated the possibility of using multi-temporal Landsat satellite images as a cost effective way of mapping land cover changes. The community based mapping approach used to augment this exercise provided an efficient way to reveal the historical land cover states and trends. Though rigorous in time and cost, such an approach can be used to construct missing information sufficient for mapping of land cover changes in data scarce areas. However, further studies to include other land cover types in future mapping activities are recommended. Such an endeavor will enable extended analysis of the land cover change effects on agricultural and hydrological regimes amongst others. Generally, the agriculture land cover class selected in this study was noted to exhibit immense diversity in terms of their biophysical characteristics. It is hence imperative in future studies to carefully discern these types and assess their spatio-temporal influences on seasonal basis. To achieve this, however, a fully distributed approach involving the application of high resolution spatial datasets may be important.

References

BALDYGA, T. J., MILLER, N. S., DRIESSE, L. K. & GICHABA, N. C. (2007), Assessing land cover change in Kenya's Mau Forest region using remotely sensed data. African Journal of Ecology, 46, pp. 46-54.

CONGALTON, R. G. & GREEN, K. (1999), Assessing the Accuracy of Remotely Sensed Data. CRC Press, Boca Raton, FL, USA.

ERDAS (2002), ERDAS field guide. ERDAS Inc., 6th edition, Atlanta7, USA.

FÜRST, J. (2004), GIS in Hydrologie und Wasserwirtschaft. Herbert Wichmann Verlag, Heidelberg, Germany.

JENSEN, J. R. (2005), Introductory Digital Image processing: A Remote Sensing Perspective, 3rd Edition. Prentice-Hall, Upper Saddle River, NJ, USA.

KUNDU, P. M., CHEMELIL, M. C., ONYANDO, J. O., GICHABA, M. (2008), The use of GIS and remote sensing to evaluate the impact of land cover and land use change on discharges in the River Njoro Watershed, Kenya, Journal of World Association on Soil Water Conservation, J2, pp. 109-120.

OLANG, L. O. & FÜRST, J. (2010), Effects of land cover change on runoff peak discharges and volumes during flood events of the Nyando basin, Kenya. Hydrological Processes (In review).

PELLIKKA, P. K. E., LÖTJÖNEN, M., SILJANDER, M. & LENS, L. (2009), Airborne remote sensing of spatiotemporal change (1955-2004) in indigenous and exotic forest cover in the Taita Hills, Kenya. International Journal of Applied Earth Observations and Geoinformation, 11 (4), pp. 221-232.

RAMBALDI, G., MUCHEMI, J., CRAWHALL, N. & MONACO, L. (2007), Through the Eyes of Hunter-Gatherer: Participatory 3D modelling among Ogiek indigenous peoples in Kenya. Information Development, 23 (2-3), pp. 113-128.

TATEM, A. J., NOOR, A. M. & HAY, S. I. (2005), Assessing the accuracy of satellite derived global and national urban maps in Kenya. Remote Sensing of Environment, 96, pp. 87-97.

TOA, C. V. & HU, Y. (2001), A comprehensive study of the Rational Function Model for photogrammetric processing. Photogrammetric Engineering & Remote Sensing, 67 (12), pp. 1347-1357.

TOTTRUP, C. (2004), Improving tropical forest mapping using multi-date Landsat TM data and pre-classification smoothing. International Journal of Remote Sensing, 25, pp. 717-730.

YUAN, F., SAWAYA, K. E., LOEFFELHOLZ, B. C. & BAUER, M. E. (2005), Land cover classification and change analysis of the twin cities (Minnesota) Metropolitan Area by multitemporal Landsat remote sensing. Journal of Remote Sensing and Environment, 98, pp. 17-328.

GeoNode – A New Approach to Developing SDI

Edward PICKLE

Abstract

Faced with inadequate Spatial Data Infrastructure (SDI) data to execute disaster risk assessments in Central America, the World Bank faced a dilemma: follow the model of major past programs in developed countries (INSPIRE, GOS, etc.) to develop the data, or take an alternate approach. The World Bank has chosen to develop a new approach – GeoNode – working in partnership with the open source geospatial firm OpenGeo. The objective of GeoNode development is to create web-based software that encourages greater SDI data contributions from a wider range of participants through the use of open source software and community building initiatives. This paper provides an overview of the features of GeoNodes, the rationale for the World Bank's decision, and the compelling ideas that are leading other programs to follow a path to SDI building that is informed by key features of open source and Web 2.0 software.

1 CAPRA Program and SDI

The Central America Probabilistic Risk Assessment (CAPRA) program is an ongoing World Bank initiative to enhance the understanding of disaster risk throughout Central America. It encompasses probabilistic risk assessment analysis of hurricane, earthquake, volcano, flood, tsunami and landslide hazards. The platform's architecture has been developed by regional experts to be modular, extensible and open, allowing it to be expanded and improved. The goal is a 'living instrument' where experience is accumulated rather than lost, harnessing the collective work of contributors while minimizing the duplication of their work (CAPRA website 2009).

To meet these requirements, the World Bank concluded that a more participatory facility for building regional SDI information was required. Specifically, the World Bank judged that existing SDI platforms:

- Provided no benefit to user registration
- Had few real users
- Provided no recognition to data contributors
- Offered no reward for the effort
- Used 'stick', not 'carrot' approaches

2 SDI and Risk Management

Government, institution, NGO and community actors rely on and simultaneously create SDI information in the process of multi-hazard risk management. Disaster risk information is a blend of data sources from the national to the local level, with local knowledge updated much faster (SUTANTA et al. 2009). National clearinghouses for SDI information have experienced declines in use and content (CROMPVOETS et al. 2004) even prior to the widespread use of Google Earth and similar geo-applications. Geo-applications, mapping products and location based services have grown into quasi-SDI's themselves due to their benefits for data authors, speed, ease of use and other factors. However, despite their merits, these geo-applications lack critical functionality required for risk assessment. Thus the World Bank partnered with OpenGeo in the GeoNode initiative with the objective to revitalize the SDI domain through the integration of open source software and Web 2.0 principles (McAfee 2006) including search, links, authoring, tagging, etc. The results of this integration are to provide an improved platform for crowd-sourced data, community and professional involvement in risk awareness and cost-effective mitigation. Note that a similar strategy is also being proposed for another prominent global risk assessment initiative (GEM 2009).

3 GeoNode Concepts

At its core, GeoNode is based on open source components GeoServer, GeoNetwork, Django, and GeoExt. These provide a platform for sophisticated web browser spatial visualization and analysis. Atop this stack, the project has built a map composer and viewer, tools for analysis, and reporting tools. GeoNode also supports facilities for styling data as well as collaborative features like ratings, comments, and tagging for data, maps, and styles. GeoNode is built on four key tenets: Collaboration, Distribution, Cartography and Data Collection.

To promote collaboration, GeoNode is designed on Web 2.0 principles to:

- Make it extremely simple to share data
- Provide user statistics
- Easily add comments, ratings, tags
- Allow collaborative filtering
- Provide rankings of best 'views' and data sets contributed
  - Highest rated, most viewed, most shared
- Allow connectivity between several GeoNode instances to augment the collaborative potential of government GIS programs

To secure distribution, GeoNode enables:

- Simple installation and distribution
- Automatic metadata creation
- Versioned metadata
- Search via catalogues and search engines (Google)

To allow for flexible and cost-effective cartography, GeoNode incorporates open source software components:

- GeoExt – The JavaScript toolkit for rich web mapping applications
- OpenLayers – Pure JavaScript library powering the maps of GeoExt
- GeoNetwork – provides Web catalogue service
- GeoWebCache – Cache engine for WMS Tiles
- GeoServer – Standards based server for geospatial information

And, to promote data collection, the GeoNode is aimed to align incentives to create a sustainable Spatial Data Infrastructure to:

- Align efforts so that amateur, commercial, NGO and governmental creators all naturally collaborate
- Figure out workflows, tools and licenses that work to assure data quality
- To promote 'living' data, constantly evolving - authoritative and always up to date

4 GeoNode Objectives and Results

Simply put, the objective of GeoNode is to provide greater ease of use and functionality to data providers of all types, in order to obtain better, more frequent data contributions from as wide a range of data providers as possible. GeoNode development follows standard open source practice: frequent, iterative software releases to enable users to test the system and recommend changes to enhance functionality. For example, early GeoNode releases focused upon both providing the standard ISO metadata that many organizations require, and providing users with a list of the hosted maps of which the data is a part. Any user using GeoNode's built-in cartographic tools can make these maps. The map listing allows users to easily navigate through the GeoNode's cartographic ecosystem. It also tells viewers about the validity of the data itself: widely used data is trusted data.

Figure 1 shows many of the early metadata features of GeoNode.

So far GeoNodes have been deployed by the World Bank in two instances:

1. For the CAPRA program
2. For the dissemination of hazard and damage data in support of Haiti relief

As a result of user feedback from these deployments, new features have been added to GeoNode's underlying open source components including:

- Enhancements to the GeoServer REST API
- Addition of styling components to GeoExt

Further social metadata features are on the GeoNode roadmap. Users will be able to comment and rate data; these ratings will influence search results, much the same way that YouTube uses user feedback to promote quality works. To allow for more institutional oversight, GeoNode will also all participating groups to endorse data. Endorsement provides a "lightweight" way for organizations to provide official guarantees on data without stifling unofficial but productive crowd-sourced data collection, neogeography, and amateur cartography.

Ultimately CAPRA, the World Bank and OpenGeo aim to develop and enhance a set of free and open source GeoNode instances to understand, communicate and support decisions to reduce disaster risk within Central America. The success of GeoNode will be measured by data contributions that result from GeoNode deployments in support of this and similar missions.

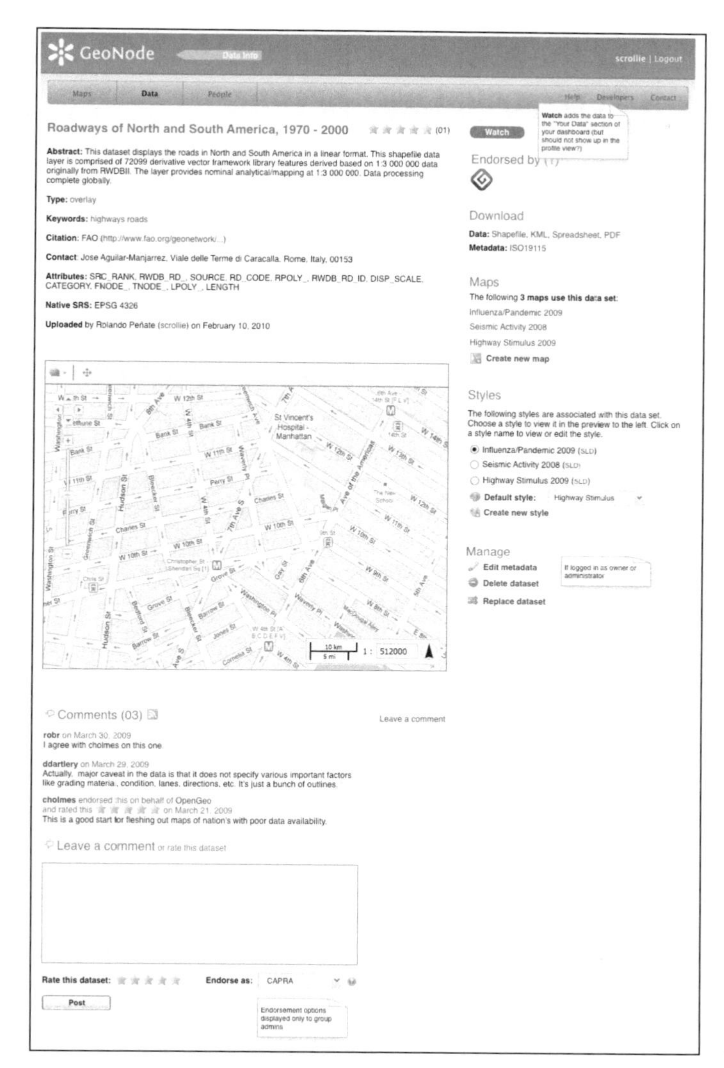

Fig. 1:
Metadata based on through-the-web activity will allow GeoNode to provide new benefits to providers of SDI data.

References

CAPRA WEBSITE (2009) – http://ecapra.org/en/.

CROMPVOETS, A. B., RAJABIFARD, A. & WILLIAMSON, I. (2004), Assessing the worldwide developments of national spatial data clearinghouses. Int. J. Geographic Information Science, (18) 7, pp. 665-689.

GEM FOUNDATION (2009), Global Earthquake Model. – www.globalearthquakemodel.org.

MCAFEE, A. (2006), Enterprise 2.0: The Dawn of Emergent Collaboration. MIT Sloan Management Review, 47 (3), p. 21-28.

SUTANTA, H., RAJABIFARD, A. & BISHOP, I. D. (2009), An Integrated Approach for Disaster Risk Reduction Using Spatial Planning and SDI Platform. In OSTENDORF, B. et al. (Eds.), Proc. of the Surveying & Spatial Sciences Institute Biennial International Conference, Adelaide 2009, Surveying & Spatial Sciences Institute, pp. 341-351.

Measuring the Stem Breast Diameter Using A Terrestrial Laser Scanner

Róbert SMREČEK

The GI_Forum Program Committee accepted this paper as reviewed full paper.

Abstract

An experimental plot of 2500 m^2 was scanned with a terrestrial laser scanner. 72 trees in the sample terrain had a 1.3 m measured diameter. The breast diameter was measured using two methods, by calliper and by laser scanner. The data measured by laser scanner was processed by the programme TLS (version 1.2), where the diameter was calculated using four different methods. The data was analysed using statistical methods, to identify any systematic errors in the measurements and to establish whether there is or isn't any statistically significant divergence between these two measurement methods.

1 Introduction

Terrestrial laser scanners are optical devices, which capture images in their surroundings. Under scanning, the automatic process is understood, where the object is remotely scanned and the information is collected for further computer processing. Scanning is performed in two phases. First, the image is scanned in vertical direction where the laser beam is directed by a mirror. When the vertical scan is finished, the horizontal angle is extended by a constant and next the vertical scan is performed. The situation is captured in a point cloud. Many devices also capture the situation as an image, which can be assigned to the point cloud in post–processing. FRÖHLICH and METTESLEITER (2004) and BIENERT et al. (2007) separated terrestrial laser scanners by many attributes. THIES and SPIECKER (2004), WEZYK et al. (2007) and other authors see the possibility of using terrestrial laser scanners in forestry inventories. Terrestrial laser technology captures forest characteristics. Due to a large number of points in the point cloud, a tree can be automatically identified.

WEZYK et al. (2007) performed measurements in deciduous and coniferous forests in the vicinity of Milicz. The breast height diameter analysis showed a very close ratio ($r^2 > 0.94$) between the scanned stem breast height diameter (next breast diameter) and the reference measurement. HENNING and RADTKE (2006) reached a precision of 2cm by using their tree diameter measurement. They separated the tree stem into 1 m long sections and measured it up to the height of 13m. The result from a tree height of 10m were tree parts represented with only a few points. This was caused by the shadow of tree branches. Measurements were provided in Loblolly Pine (*Pinus taeda* L.) stand. Measurement of THIES and SPIECKER (2004) reached the divergence from -4.1% to 1.3% as opposed to tape measurement. The divergence depends on the applied method of measurement. BIENERT et al. (2007) measured the Sitka Spruce (*Picea sitchensis* (Bong.) Carr.) on a plantation in Ireland. They reached the diameter measurement standard deviation of 2.48cm. In measurements they used 10cm long sections. The results were comparable with the data captured by a harvester.

2 Input data and data processing

The terrestrial laser scanning was performed on a plot of 50 × 50 m in the forest unit no. 451, at the University Forest Enterprise territory of the Zvolen Technical University. The forest stand has two storeys and is 80 years old. The lower storey is 5 years old. The top storey extension is 13.93ha and the lower storey extension is 0.15ha. The stand density of higher storey is 0.9ha and of the lower one is 0.3. Average slope for forest stand is 40 %. There are rocks in the forest and also in the NW part of the plot. European Beech (*Fagus sylvatica* L.) is the major tree species (100% representation) represented in the forest and also in the plot. The Siler Fir (*Abies alba* Mill.), Sycamore Maple (*Acer pseudoplatanus* L.), Norway Maple (*Acer platanoides* L.) and European Hornbeam (*Carpinus betulus* L.) occur only partially.

The scanning was performed by personnel of CEIT s.r.o. with the FARO LS 880 HE terrestrial laser scanner (TLS) in September 2008. 17 scans were obtained and all positions of the scanner and reference sphere were measured with the TOPCON total station. The scans were processed by the personnel of CEIT s.r.o. In October 2008 the terrain measurement established the breast height diameter, the height of tree and the height of crown base, and the parameters of crown projection in four cardinal directions. The breast height diameter was measured with the calliper. The height was measured with the VERTEX ultrasonic hypsometer. There are 168 trees on the plot, due to the low number of points for trees in the marginal parts of scanned area. We used only 72 trees because the area was not scanned from outside.

The data was processed with the TLS version 1.2, a program for processing terrestrial laser data. It is non-commercial program, which was created at the Technical University of Zvolen by Koreň in 2009. The program runs on Microsoft NET Framework 3.5 and XNA Framework 3.0 platform and includes three basic functions (KOREŇ 2009):

- Terrestrial laser scanning data processing: import of point cloud, filtering, conversion to the grid, group creation depending on distance, statistic parameters and distribution of point density calculation, extraction of two-dimensional circle, value assignment.
- Digital elevation model creation and grid processing: new grid creation, import and export of a grid into the exchange text format, gaps filling, overlay.
- Three-dimensional transformation.

For the breast diameter measurement, the extraction of two-dimensional circle was used. The diameter on the cuts can be automatically measured with one of four methods (KOREŇ 2009):

- *Optimal circle* – a method of approximation that searches an optimal position and circle width like the minimum mean square deviation of points. The modified Newton method is used for the calculation. This method comes out from the initial assessment of maximal distance or centroid. Progress of this method is not casual (unlike the *Monte Carlo* method), but the line is specified by the gradient of the optimization function. Calculation will stop after reaching of mean square deviation limit or reaching the specific number of calculation steps.

- *Monte Carlo* – this method searches approximation of the best position and circle diameter, generating many casual changes. At the same time, it results from the initial assessment, calculated by the method of maximal distance or centroid. Step by step, small changes in position and diameter are generated, to assess the best circle. If the minimum mean square deviation is smaller than in the previous assessment, the new circle is considered for the best assessment of the circle. The calculation terminates after reaching the specified number of iterations or after reaching the limit of mean square deviation.
- *Maximal distance* – two maximum distance points are searched for in a group of points. The flow line connecting these two points creates the diameter of the circle. The centre of that circle is calculated as a centre of chord (diameter), which connects the two maximum distance points.
- *Centroid* – calculates a centroid of point group. The coordinates of this centroid are specified by the arithmetic mean value of the point group coordinates. Then the radius is calculated like the arithmetic distance measured from the centre of the circle.

3 Methodology

The aim was to investigate the accuracy of stem breast height diameter measurement using a terrestrial laser scanner (TLS) and to introduce the possibilities to use terrestrial laser scanner technology in forestry. The breast diameter was measured with a calliper (accuracy in cm) and with a terrestrial laser scanner. The difference in the tree number was not analyzed, because the difference was caused by the low number of points for trees in the marginal parts of scanned area. High density of scanner positions guaranteed the good visibility and sufficient number of points for trees. They were not in the marginal parts of scanned area. The breast diameter measured by the TLS was automatically calculated in TLS, program version 1.2. Based on the digital elevation model, which was created from data obtained by the terrestrial laser scanning, two cuts were created, in height of 1.3 m from the ground with different width values:

- Width of 10 cm, cut in height of 125 – 135 cm
- Width of 20 cm, cut in height of 120 – 140 cm

On these cuts the breast diameter was automatically calculated using the TLS program, version 1.2, by applying the following four methods:

- Optimal circle
- Monte Carlo
- Maximal distance
- Centroid

The regression and correlation analysis was conducted to search and confirm the relationship between two quantitative variables and for construction of optimal model which describes their matter. The accuracy of the regression model is described with standard error of measured data ($s_{y.x}$), which describes the variability of measured data (y_i) based on the adjusted data ($\hat{y}_i$) in the current data collection. A meaningful regression model can be constructed only if the independent variable (x) affects the dependent variable (y) significantly. Reliance strength is described with the correlation coefficient (r_{yx}).

3.1 Linear analysis test for the verification of individual parameters correspondence with projected theoretical values

The classic linear analysis test for the verification of individual parameters correspondence with the projected theoretical values H_0: $\beta_j = \beta_0$, is called the WALDOV t test. This test is derived from the interval assessment of individual parameters in the regression model. This test, as well as interval assessment, is based on the theorem of normal incidental sorting of parameters, which is the reasonable assumption in case, when the conditional partitions of residual divergences (e_i) and variables (y_i) are normal (SCHEER 2007).

Null hypothesis H_0: $\beta_j = \beta_0$ is rejected on significance level α in case, if $|t| > t_{\alpha/2(n-p)}$, where $t_{\alpha/2(n-p)}$ is the critical value of Student t distribution by n-p degrees of freedom, where p represents the number of linear regression model parameters.

The formulated null hypothesis is verified by two consequential tests, where the correspondence of regression coefficient (H_0: $\hat{\beta} = \beta_1^0$) is tested first, which insures the equal inclination of comparison lines. If the null hypothesis is rejected on the chosen α significance level, the test of theoretical and empirical model correspondence can be finished, because the regression coefficient disagreement insures the correspondence of checked models. If the null hypothesis (H_0: $\hat{\beta}_1 = \beta_1^0$) is not rejected, we should line test the absolute terms correspondence (H_0: $\hat{\beta}_0 = \beta_0^0$), which insures the position correspondence of lines checked. Then the second hypothesis confirmation means the confirmation of the theoretical and empirical model correspondence, which are not significantly different in their position, neither inclination.

3.2 Paired test

The paired test is used when two values of a variable are measured at every sample unit, using two different processes, methods, workers, etc. The range of both samples are the same and the discovered values x_{Ai} and x_{Bi} of variable X_A and X_B represent corresponding pairs, and the measurements are independent (SCHEER 2007). After verifying the non-deviation, the particular errors of variables calculation are calculated for values measured by formula:

$$e_i = \mu_A - \mu_B$$

Where μ_A represent the values measured by TLS and processed with TLS version 1.2 programme and μ_B represent values measured manually.

Then the null hypothesis H_0: $\mu_A - \mu_B = \mu_\Delta = 0$ against H_1: $\mu_\Delta \neq 0$ is tested. Differences by formula $\bar{\Delta} = \bar{x}_A - \bar{x}_B$ are considered as casually different from zero. Assume, both measurement methods should reach the same result, $\mu_\Delta = 0$. The testing criteria is the quantity t, where for Student t-distribution $f = n - 1$ is a degree of freedom.

After calculating particular errors e_i, the analysis is provided and the non-deviation is tested for the precision and reliability of the method. The analysis relates to the objective errors characteristics, where we distinguish the following terms (ŠMELKO 2000):

- Deviation – indicates the systematic component of an error. The measure of its existence is an arithmetic average of errors regarding the sign.
- Precision – characterises the casual component of an error. It expresses the variability of particular errors ei around their arithmetic average. The measure of precision is represented by the standard error se.
- Accuracy – characterises the total error of measurement, and is as casual as systematic error. Quantitatively, it is expressed by the standard quadratic error me.
- Reliability – defines the probability of the current errors ei occurrence in specified frames of the accuracy.

By using the statistic test, we can confirm, with projected reliability confirmed, or disprove the hypothesis about the non-deviation of the model:

$$t = \frac{\bar{e}}{\frac{s_e}{\sqrt{n-1}}}$$

If $|t| \leq t_{\alpha/2;f}$ by degree of freedom f = n – 1, the null hypothesis about the model non-deviation (H_0: $\bar{e} = 0$) is accepted with 1 – α% reliability and if $|t| \geq t_{\alpha/2;f}$ it is refused on α% significance level.

4 Results

4.1 Optimal circle method

The highest divergence for the cut in height of 125 – 135 cm was +27 cm and it was caused by the branching. For the cut in height of 120 – 140 cm, the biggest divergence was +30 cm, caused also by branching. In both data sets seven values were excluded. These values were defined as extreme values. In both cases, there is a very close correlation. The value of regression coefficient in both cases is higher than 0.99.

Using the WALDOV *t* test, the correspondence of the empirical and theoretical model was tested. In the first stage the correspondence of the regression coefficient H_0: $\hat{\beta}_1 = \beta_1^0$ was tested, which assigns the correspondent inclination of comparison lines. In the table of critical values for Student t-distribution for significance level α = 0.025 and degree of freedom 64 is $t_{0.0125;(64)} = 2.2954$. As you can see in table below $|t| \leq t_{\alpha/2;f}$, the null hypothesis about correspondence of regression coefficient can be accepted with 97.5% of reliability.

Table 1: An overview of statistic characteristics for testing the correspondence regression coefficients by the WALDOV *t* test for the optimal circle method

Cut high (cm)	Regression coefficient ($\hat{\beta}_1$)	Standard error ($s_{y.x}$)	Testing criteria (t)
125 – 135	0.9926	0.0135	0.0553
120 – 140	1.004	0.0137	0.2934

After accepting the null hypothesis for correspondence of the regression coefficient H$_0$: $\hat{\beta}_1 = \beta_1^0$, the null hypothesis about the correspondence of line interception H$_0$: $\hat{\beta}_0 = \beta_0^0$ can be tested, which assigns the position correspondence of checked lines. The critical value is $t_{0.0125;(64)} = 2.2954$. As shown in table 2, we can accept the null hypothesis about the correspondence of line interception with 97.5 % reliability.

Due to these results, we can accept the null hypothesis about the correspondence of empirical and theoretical model with 95% reliability. So we can state that no systematic error was detected in the measurement.

Table 2: An overview of statistic characteristics for testing the line interception correspondence by the WALDOV *t* test for the optimal circle method

Cut high (cm)	Intercept ($\hat{\beta}_0$)	Standard error ($s_{y.x}$)	Testing criteria (t)
125 – 135	0.0582	0.3997	0.1455
120 – 140	-0.4058	0.4078	0.9951

After calculating particular errors, the analysis of non-deviation, precision and accuracy of the method is completed. Thereafter, I tested the null hypothesis about correspondence of the measurement obtained by the calliper and TLS.

Table 3: An overview of errors for Paired test for the optimal circle method

Characteristic	Cut in high (cm)	
	125 – 135	120 – 140
Deviation ($\bar{e}$)	0.1424	0.2970
Precision (s_e)	1.3501	1.3557
Accuracy (m_e)	1.3576	1.3878
Testing criteria (t)	0.8889	1.8458

In the table of critical values for the Student t-distribution for $\alpha = 0.05$ significance level and the degree of freedom of 65, is $t_{0.025;(65)} = 1.9971$. After the comparison of tabular critical values and the calculated testing criteria (see table 3), we can state that $|t| \leq t_{\alpha/2;f}$. With 95% reliability we can accept the null hypothesis about the correspondence of the breast height diameter measurement using the calliper and TLS.

4.2 Monte Carlo method

14 values were marked by cut in height of 125 – 135 cm and 25 values by cut in height of 120 – 140 cm as extreme. The highest divergence of 137 cm occurred by cut in height of 125 – 135 cm and 133 cm by cut in height of 120 – 140 cm. This method confirmed close correlation.

Table 4: An overview of correspondence of statistic characteristics for testing the regression coefficients by the WALDOV t test for the Monte Carlo method

Cut high (cm)	Regression coefficient ($\hat{\beta}_1$)	Standard error ($s_{y.x}$)	Testing criteria (t)	Critical value ($\alpha = 0.025$)
125 – 135	0.9992	0.0234	0.0331	2.3033
120 – 140	0.9857	0.0231	0.62	2.3189

In the WALDOV t test, the correspondence of regression coefficients was tested in the first step. In the table of critical values for Student t-distribution for significance level $\alpha = 0.025$ and degree of freedom 56, the critical value for the cut in height of 125 – 135 cm is $t_{0.0125;(56)} = 2.3033$. For the cut in height of 120 – 140 cm, the critical value for degree of freedom 45 is $t_{0.0125;(45)} = 2.3189$. By comparison of these critical values using testing criteria from table 4, we can accept the null hypothesis H$_0$: $\hat{\beta}_1 = \beta_1^0$ with 97.5% reliability.

Table 5: An overview of statistic characteristics for testing of the line interception correspondence by the WALDOV t test for the Monte Carlo method

Cut high (cm)	Intercept ($\hat{\beta}_0$)	Standard error ($s_{y.x}$)	Testing criteria (t)	Critical value ($\alpha = 0.025$)
125 – 135	-0.475	0.7089	0.67	2.3033
120 – 140	0.5723	0.7161	0.7991	2.3189

In the second step, the correspondence of the line interception was tested. Comparing the critical values with the testing criteria from table 4, we can accept the correspondence with 97.5% reliability. Based on these results we can accept the null hypothesis about the correspondence of the empirical and theoretical model with 95% reliability, and we can also state that no systematic error was detected in the measurement.

The data set was also tested by the Paired test. The statistic characteristics are shown in table 6. The critical values are stated for the significance level α = 0.05 and the degree of freedom 57 for the cut in height of 125 – 135 cm and degree of freedom 46 for the cut in height of 120 – 140 cm.

Table 6: An overview of errors by paired test for the Monte Carlo method

Characteristic	**Cut in height (cm)**	
	125 – 135	**120 – 140**
Deviation ($\bar{e}$)	0.4966	-0.1489
Precision (s_e)	2.0579	1.4554
Accuracy (m_e)	2.1169	1.463
Testing criteria (t)	2.0332	0.8623
Critical value (α=0.05)	2.0025	2.0129

By the cut in height of 125 – 135 cm we found that $|t| \geq t_{\alpha/2;f}$,and therefore we refused the null hypothesis about the correspondence of the two measurement methods at the significance level α = 0.05. We accepted the alternative hypothesis, that between these two methods there are statistically significant differences. Otherwise, for the cut in height of 120 – 140 cm we could accept the null hypothesis about the correspondence of the two breast height diameter measurements methods with 95% reliability.

4.3 Maximal distance method

There are significant differences between the two measurement methods. Only in a case of two values the divergence was less than 10 centimetres. For example, the standard deviation ($\bar{e}$) for the cut in height of 125 – 135 cm was 35.85 cm, and for the cut in height of 120 – 140 cm was 41.93 cm. Based on these results we did not proceed with further analyses.

4.4 Centroid method

As in the method described before, the analysis terminated after investigating the significant differences between the breast height diameters measured by using the two different methods. The standard deviation ($\bar{e}$) for a cut at height of 125 – 135 cm was 16.13 cm and for the cut at height of 120 – 140 cm it was 18.56 cm.

5 Conclusion

Given the test results the Optimal Circle method showed the best results. By the breast height diameter measurement extreme values were identified by every method: 6 by the Optimal circle method, 14 to 25 values by the Monte Carlo method. The differences were too significant by the other two methods.

Considering the significant differences in results obtained with the Maximal Distance and the Centroid method no other analyses were conducted. For both methods the values of the measured breast diameter were overestimated. Only in the case of the Centroids method the diameter values by the cut at the height of 120 – 140 cm were underestimated.

By the Optimal circle method for the cut at the height of 125 – 135 cm the overestimation was about 0.14 cm, and for the cut at the height of 125 – 135 cm it was 0.3 cm. For the Monte Carlo method the overestimation was 0.5 cm in case of the first cut and 0.15 cm in case of the second cut. In both cases the differences were too small and therefore we can consider them negligible. HALAHIJA (2009) reached the mean deviation of 0.39 cm by the manual measurement of the diameter on the same data in ArcGIS software environment. This result is comparable with the results from automated measurements. But the time demands are incomparable.

Based on the results obtained by the WALDOV *t* test we can state that no systematic error was detected in the measurement. The Paired test was used when the two methods of diameter measurement were equivalent. Only in the case of a cut at the height of 125 – 135 cm by Monte Carlo method application was this assessment rejected.

Considering the analysis results, the TLS can substitute the classical methods of diameter measurement and capturing the other forest characteristics. This method is appropriate mainly for forest inventory purposes. WEZYK et al. (2007) mentioned that the data from TLS is hardly used by forestry inventories. Before the concept of precision forestry is applied in the forestry practice, the new technologies like the TLS would be expensive and useless.

The disadvantage for wider TLS use in the practice is the current price, energy and time demanding because of changing the scanner position and terrain conditions in a forest. TLS is not constructed for use in forest conditions. Also the TLS is an optical device, so the laser beam is rebound from the first barrier by measurement. So the data shadow is created behind the barrier. This problem could be solved by susceptible scanner position placement. But in forests with heavy undergrowth, the use of the TLS is practically impossible. Also due to the wind conditions, some data can be useless, because the higher parts are unclear. Therefore, the measurement of a tree height is sometimes impossible. Also the leaf biomass represents a barrier for the laser beam; this problem can be solved by scanning during winter dormancy.

As the advantage of TLS use in forestry science and practice is the permanent registration of the situation, the possibility of repeated and new measurements, as well as the comparison of old and new data. Repeated registration of the same situation from the same place allows for the observation and evaluation of the change of forest ecosystems.

Acknowledgement

The author thanks VEGA for the financial support of the GD 1/0764/10 grant project.

References

BIENERT, A., SCHELLER, S., KEANE, E., MOHAN, F. & NUGET, C. (2007), Tree Detection and Diameter by Analysis of Forest Terrestrial Laser Scanner Point Clouds. International Society for Photogrammetry and Remote Sensing, XXXVI, Part 3/W52, Laser Scanning 2007 and SilviLaser 2007, pp. 50-56. ISSN 1682-1777.

FRÖHLICH, C. & METTENLEITER, M. (2004), Terrestrial Laser Scanning – New Perspectives in 3D Surveying. International Archives of Photogrammetry, Remote Sensing and Spatial Information Sciences, XXXVI, Part 8/W2, 2004, pp. 7-13. ISSN 1682-1750.

HALAHIJA, J. (2009), Stanovenie polohy a hrúbky stojacích stromov z meraní TLS. Diplomová práca, Technická univerzita vo Zvolene, Katedra hospodárskej úpravy lesov a geodézie, 2009, 103 p.

HENNING J. G. & RADTKE P. J. (2006), Detailed Stem Measurements of Standing Trees from Ground-Based Scanning Lidar. Forest Science, 52, pp. 67-80. ISSN 0015-749X.

KOREŇ, M. (2009), Príručka užívateľa k programu TLS verzia 1.2.

SCHEER, Ľ. (2007), Biometria. Technická univerzita vo Zvolene, Zvolen. 333 p. ISBN 978-80-228-1723-3.

ŠMELKO, Š. (2000), Dendrometria. Technická univerzita vo Zvolene, Zvolen. 399 p. ISBN 80-228-0962-4.

THIES, M. & SPIECKER, H. (2004), Evaluation and Future Prospects of Terrestrial Laser Scanning for Standardized Forest Inventories, International Archives of Photogrammetry. In: Remote Sensing and Spatial Information Sciences, XXXVI, Part 8/W2, 2004, pp. 192-197. ISSN 1682-1750.

WEZYK, P., KOZIOL, K., GLISTA, M. & PIERZCHALSKI, M. (2007), Terrestrial Laser Scanning Versus Traditional Forest Inventory First Results from the Polisch Forests. In: International Society for Photogrammetry and Remote Sensing, Volume XXXVI, Part 3/W52, Laser Scanning 2007 and SilviLaser 2007, pp. 424-430. ISSN 1682-1777.

Geocoded Address Point Data and Its Potential for Spatial Modeling of Urban Functional Parameters – A Case Study

Klaus STEINNOCHER, Christoph AUBRECHT and Mario KÖSTL

The GI_Forum Program Committee accepted this paper as reviewed full paper.

Abstract

In this paper the potential of geocoded address data for modeling urban land use patterns is analyzed. The test area covers the city of Dornbirn, located in the most western province of Austria. Starting from the pure address point data, that give a first overview of the basic urban structure, additional data sets are added to the model in order to improve the information content of the model. Linking the point data with real-world geometries in form of a building layer allows for classifying building types. Integration of zoning data gives an indication of functional urban patterns on a large scale. In order to model local land use variations tabular company data are added to the model and linked to the single buildings. Using the synergy of these data sets a functional urban model can be derived to be applied in various fields such as geo-marketing or disaster risk management.

1 Introduction

Handling complex urban information in a spatial context is not a trivial task. Real-world spatial data comprise physical objects and phenomena on the surface of the earth including buildings, infrastructure and vegetation. This general land cover information is mostly acquired through remote sensing analyses or terrestrial surveying and stored in spatial databases. Thematic information such as socio-economic data originating from statistical sources often provide extremely rich information content on a high level of detail (AUBRECHT et al. 2009). Bringing these diverse information sources to fit together and be connected is required in order to reveal functional relations (CHEN 2002) and carry out analyses that go beyond the mere spatial framework and physical parameters. ANSELIN (1999) and WEEKS (2001) explain the role of spatial analysis in social sciences and demographic research and highlight spatial relations of remote sensing data and census data.

Integration of point-based postal data enables the transition of land cover detection to land use identification and leads to a characterization of population-related urban building patterns (MESEV 2005). Additional socio-economic information such as zoning data and company data enable deriving functional patterns on a local scale (KRESSLER & STEINNOCHER 2008). Spatial linking with the georeferenced address information reveals complex urban structures and even facilitates moving to the highest level of detail, i.e. referring to individual inhabitants and employees. In particular the fields of urban planning and environmental management as well as security and risk reduction will benefit from the growing pool of available spatial and space-related information accompanied by new spatial modeling and visualization techniques (AUBRECHT et al. 2008).

The presented paper analyses the potential of address point data for modeling urban land use patterns and the added value that will be gained from integrating real-world and socio-economic information into the model.

2 'Data.Geo' – Geocoded Address Point Data

In the years 2003 and 2004 an area-wide collection of precise geo-coordinates for the existing postal address information was carried out for the entire state of Austria. This extensive compilation of address-based spatial information was enabled by a newly formalized cooperation of the Austrian Post AG and Tele Atlas Austria. These geocoded addresses of all postal-served buildings were mapped as individual point objects. Each object is linked to the official address, typically consisting of street name and number, in certain cases including building descriptions. Starting 2005 this data set was distributed, labeled ACGeo. At the beginning of the year 2008 the product was relaunched and is now available as Data.Geo[1].

The database comprises about 2 million addresses in Austria using a standardized naming convention. As reference system the Austrian Lambert conformal conic projection is used. The extended product also includes information on the number of delivery points per address and the corresponding type of delivery point (private household, company, secondary residence) as well as a quality indicator providing information on spatial accuracy of the point objects (building-accurate, parcel-accurate, or just preliminary localization). Sales and distribution of the data product is carried out centrally by Tele Atlas Austria.

Applications using that kind of geocoded address point information can be found in the fields of emergency and risk management, geomarketing, logistics and transport, telecommunication, insurances and real estate management, tourism, security, navigation, as well as location based services and (Internet-) cartography.

2.1 Assembling the data base

Primary data collection was based on aerial imagery and the Tele Atlas street network. Guided by IT-experts, the postal delivery staff virtually tracked their daily delivery route on screen (postal term: 'order of route') and carried out a precise positioning of the individual addresses. Per definition these point features were digitized in a way that they depict either the center or the entrance of a single building.

Each postal deliverer is responsible for about 300 addresses per day. Bringing together the local know how of the individual postal deliverers proved to be a very efficient way to create and maintain this extensive data compilation. Constant high quality was assured by using a central GIS management.

The first feedback on the concrete use of the data in 2005 revealed strengths and weaknesses, such as positional inaccuracies in urban areas (STEINNOCHER & KÖSTL 2007). This led to a comprehensive revision of the data set in 2007 also associated with a partial restructuring of the address database of the Austrian Post AG. For geographic localization

---

[1] www.datageo.at (last visited on 31 January 2010)

up-to-date and high-quality aerial imagery was used and quality management measurements were further enhanced. That way quality and reliability of the data set increased considerably (compare table 1).

The continuous updating of the Data.Geo data set is performed by personnel of the Austrian Post AG, whereas their local know how is the fundamental factor. New addresses are checked and registered locally day-to-day and the new address records are then archived in the company-wide Postal Address Base Database (PABD) which is in turn directly linked to the spatial data infrastructure of Data.Geo.

Table 1: Quality increase of Data.Geo (2008) compared to ACGeo (2005)

Statistics	Beginning of 2005	Beginning of 2008
Coverage*	95.0%	99.9%
Building-accurate	72.0%	96.0%
Parcel-accurate	20.0%	3.5%
Preliminary localization	8.0%	0.5%

* Austria-wide

3 Cartographic Visualization of Address Point Features Revealing Spatial Patterns of Settlement Areas

Cartographic visualization of the available geo-located postal address point data based on attached thematic attributes (e.g. residential vs. business delivery point) gives a first overview of the basic structure of the analyzed settlement area. Typical spatial patterns expected for a moderately urbanized region are revealed using standardized cartographic methods and design variables (BERTIN 1983) such as point-size-coding (e.g. number of delivery points per address) and point-color-coding (e.g. discrimination of private and business addresses).

Figure 1 shows a rather dense urban area located in the westernmost province of Austria, Vorarlberg. Having approximately 50,000 inhabitants the city of Dornbirn is the largest city in this province and acts as one of the most important regional commercial centers. The city center is characterized by an accumulation of private address points with more than ten delivery points. This indication suggests that the main building structure in this area primarily comprises residential apartment blocks with several floors. Looking just at the spatial distribution of the business address points the main business axis can easily be identified (compare detail Figure 1, 'comparison of designated private [P] and business [B] postal address points').

Individual – somehow "exposed" – business address points with a large number of delivery points located in the suburban area (like the one marked with an arrow in the left part of Figure 1) are clear indicators for typical large commercial centers and/or shopping centers. Another noticeable feature is the clustering of private "address-multi-points" (address points with a larger number of delivery points) at several locations along the railroad, as well as typical residential areas with single address delivery points indicating detached and semi-detached houses.

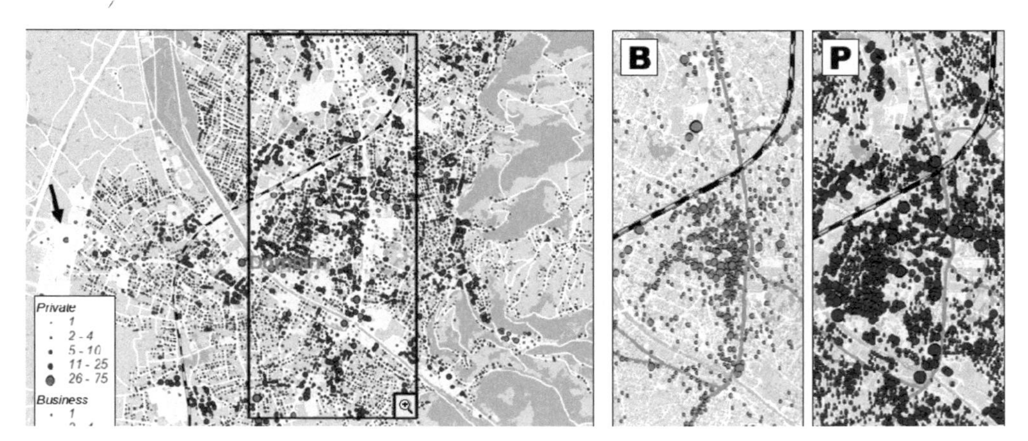

Fig. 1: Postal address points revealing spatial and functional patterns – detail opposing business (B) and private (P) delivery points

A more detailed analysis – e.g. regarding different housing types – is not possible at this stage just considering the point-based address data, but will be discussed in the following sections on integration of ancillary information.

4 Data Revaluation Through Integration of Structural and Thematic Information

The integration of ancillary information, in a both physical and thematic way, implies a considerable added value to the postal address point data. To get a better overall impression of urban structures – e.g. derivation of housing types – it is indispensable to move to a building level and combine the punctual address information with the real-world boundaries of urban development. Using a building data set derived through image interpretation the essential link between real-world geometric features and abstract thematic information can be established. Through intersection with zoning information a first assessment of potential urban land use patterns can be derived. While industrial and commercial areas definitely exclude residential use, economic activities can also take place in residential areas. Mixed use areas are intended to combine residential and commercial use anyway. The introduction of zoning plans gives an indication of the dominant use of buildings and address point clusters, which can also be cross-checked with the information on the type of delivery point implicit in the address point data set. However, local variations of urban land use patterns cannot be taken into account that way. The geo-located address points offer the spatial framework for integrating all kinds of address-based information not featuring an explicit spatial reference. In order to assess local land use variations tabular company data is linked to the address points leading to a highly detailed indication of functional patterns.

4.1 Integrating real-world information to build a spatial framework

A building layer derived through semi-automatic object-oriented interpretation of aerial imagery (acquisition: 2001) is available, forming the real-world spatial framework for working with diverse data sets in an urban data management system. Straightforward automated spatial joining of building objects and address points in the Dornbirn test site results in a 95% success rate, i.e. 8,584 out of 8,987 address points are located within the boundaries of a building polygon. The point location of the postal addresses per definition depicts either the center or the entrance of the building. A small shift can result in address points being located just next to its corresponding building and therefore not being automatically linked. 206 address points are located within a five meter buffer to the next building. Including these points raises the success rate to around 97.5%. These values are pretty much consistent with the data accuracy specifications shown in table 1 (building-accuracy: 96%, parcel-accuracy: 3.5%).

The explanation for the remaining 2.5% of the address points which cannot be related to a distinct building is mostly found in the different recording dates of the data sets – e.g. buildings were built after the image data collection, i.e. the address point already exists in the database, but the corresponding building object is missing in the building layer. Analyzing the linking process the other way round, out of 8,586 building objects 1,451 (approx. 17%) cannot be assigned a postal address. Looking at a publicly available online city map[2] it appears that most of these "anonymous" building objects seem to be garages or adjoining buildings. Another reason is building complexes such as factory premises, where several building objects are related to one single address.

All address points finally linked to the building layer are clearly related to one single building. However, as one building might contain more than one postal address (e.g. apartment buildings with more than one entrance, row houses) more than one point might be related to a building. In addition, one address point can include more than one delivery point. Furthermore private and business delivery points are distinguished.

Considering all the information described above it is possible to derive a preliminary housing type classification. Selecting buildings with address points just including "private" delivery points (6,594 buildings) results in the distribution displayed in table 2. More than half of the dedicated private buildings (59%) are most likely single family detached houses featuring just one unique delivery point. Further 20% can be characterized as semi-detached houses having two delivery points and about the same portion is classified as multi-family (3-4 delivery points) or row houses (5-20 delivery points). Large apartment building blocks just account for 1% of the building objects. In terms of population distribution however these apartment blocks of course include large numbers of persons. For socio-economic related modeling it is therefore essential to know about these types of buildings and geocoded address data give a first idea on spatial patterns. Additional structural information such as building height as derived from Airborne Laser Scanning (ALS) can serve as an important asset to this kind of classification system. Height information enables the calculation of building volume what in contrast to using building footprints represents an additional important factor for housing type classification.

---

[2] www.dornbirn.at/ext/stplan (last visited on 31 January 2010).

Table 2: Preliminary classification of housing types

Number of buildings absolute*	percentage	Delivery points per building	Housing type
3.866	58,6%	1	Single family detached house
1.304	19,8%	2	Semi-detached house
524	8,0%	3-4	Multi-family house
814	12,3%	5-20	Row house
67	1,0%	21-40	Apartment building blocks with several separated sections and entrances
16	0,2%	41-69	
3	0,1%	70-75	

* Total number of buildings including "private" delivery points: 6.594

4.2 Integrating socioeconomic information to identify land use patterns

Zoning plans as defined by the Austrian Land Use Planning Law regulate the process of organizing the use of lands and their resources to best meet people's needs over time, according to the land's capabilities. Its objectives include protection of natural as well as anthropogenic environments, risk reduction in terms of natural hazards, improvement of spatial conditions for economic activities including mining, agriculture and forestry, provision of a reliable infrastructure, economic use of land and landscape protection.

Zoning plans are provided on a municipality level and have to agree with the local development concept including basic statements on building land, traffic aspects and agricultural and natural areas. Building land, transport areas and green areas are differentiated, whereas areas dedicated to building use have to be appropriate for the construction of buildings in terms of infrastructure and natural conditions, excluding areas of natural risk. Areas dedicated to transport should serve the needs of traffic including associated land. All other areas are to be defined as green areas. For application in the presented work the numerous predefined categories were generalized to the following classes:

- Core urban area
- Residential area
- Mixed use
- Industrial and commercial area
- Agricultural area
- Open space

Table 3: Distribution of address points when linked with generalized zoning information

Land use designation	Address points*		Private del. pts.**		Business del. pts.°	
Residential area	6.082	67.7%	12.032	62.0%	260	12.2%
Core urban area	313	3.5%	1.097	5.7%	571	26.9%
Mixed use	2.284	25.4%	5.929	30.6%	953	44.8%
Industrial/commercial	133	1.5%	123	0.6%	302	14.2%
Agricultural area	93	1.0%	108	0.6%	17	0.8%
Open space/misc.	52	0.6%	55	0.3%	13	0.6%

* Total number of address points clearly linked to a building: 8,987
** Total number of related private delivery points: 19,395
° Total number of related business delivery points: 2,126

Linking of the generalized zoning information with the addresses which are clearly related to a building gives insight in the functional structure of the urban area (table 3). 68% of all relevant address points in the Dornbirn test site are located in residential areas, 25.5% in areas of mixed use, and 3.5% in core urban areas. While the private delivery points are distributed as expected with the majority being located in residential areas, core urban areas and areas of mixed use (approx. 98.3%), the business delivery points show special characteristics: Only 14% of these points are located in a predefined industrial or commercial area, 45% are related to areas of mixed use, and still more than 12% lie in residential areas.

As already mentioned, local land use variations cannot be assessed just considering zoning information. In order to progress from describing potential land use patterns to identifying actual building use, tabular company data based on Herold Yellow Pages is linked to the spatial framework consisting of building polygons with attached address information. That way it is possible to derive real-world functional relations and substantially revalue the information content of the urban data management system.

The company data product includes basic information (name, address, economic sector etc.) on about 360,000 companies in Austria and is updated twice a year. The database is not exhaustive but in fact limited to those companies willing to pay for their entry. Information on missing companies is therefore required and can mostly be found on local and regional web pages.

At this point a short note on privacy issues might be appropriate. First of all, the entire information used in this study is publicly available, either from Yellow Pages or from the internet. Second, no personally identifiable information can be retrieved from the data used. In terms of companies the information retrieval was limited to economic sectors and the size of companies. Therefore the authors are convinced that no privacy rights are violated.

Linking company information to address points and building polygons respectively is not a straightforward process. In the first step 72.6% (1,688 out of 2,324) of all available company entries could be directly linked to 1,004 addresses. This can be compared to the information inherent in the postal address data set, where 2,056 delivery points related to 1,119 address points are defined as business points. By slightly adapting the company data base the linking success rate is raised to 93.8% (2,181 out of 2,324 company entries can be linked to 1,203 addresses). Company data corrections basically consist of address adaptations in terms of notation, address appendices and multi-denotations. With eight street name corrections e.g. 247 more addresses could be linked. The majority of the remaining 6% of the Herold company records refer to addresses outside of the study area.

The fact that the number of Herold company entries in some points exceeds the number of business delivery points is easily explained. As an example serves one company which apparently consists of three somehow separate organizations of corporations. All of them appear in the company database, but feature the same address and therefore just one single delivery point.

Linking company data and geometric framework allows precisely locating those buildings in which economic activities take place and identifying buildings which are in residential use only. Furthermore, the integration of company data is not limited to the indication of buildings with economic activities, but also provides information on the type of activity.

Information on the primary branch of business enables a refined functional identification of urban patterns.

To get an idea of the distribution of urban functions within certain zoning classes a comparison with the geo-located company records was carried out. In the residential zone 398 company entries were found referring to a number of 923 employees (information inherent in the Herold database). This would correspond to an average number of 2.3 employees per company. However, if the five entries with the largest number of employees are excluded the average drops to 1 employee per company. That in turn would be in accord with the increasing number of self-employed persons (e.g. software engineers, consultants, or physicians) being located in predominantly residential areas.

As expected, the industrial and commercial zone shows a different picture. 306 company records feature a total number of 5217 employees (average: 17 employees per company). Excluding the four biggest employers the average number of employees still remains rather high at 8.3. Table 4 shows an extensive picture including the values for the areas of mixed use and core urban areas. The biggest employer is the municipality assigning 1,200 employees to one single company record in the core urban zone. In the area of mixed use one hospital employs 520 persons, while in the residential zone 350 people work for a cleaning services company.

Table 4: Joint spatial analysis of company data and zoning information

Zone	Company entries	Employees	Average number of employees per company
Residential area	398	923	2,3
Mixed use	881	3.333	3,8
Core urban area	569	2.702	4,7
Industrial/commercial	306	5.217	17,0
Residential area	393*	411	1,0
Mixed use	875*	2.347	2,7
Core urban area	568*	1.502	2,6
Industrial/commercial	302*	2.499	8,3

* Exclusion of the biggest employers to get a more representative picture of the average company size

5 Conclusion and Outlook

In this paper we presented an approach to link georeferenced postal address data as provided by the Austrian Post AG with real-world geometries and ancillary thematic information. A building layer derived through object-oriented analysis of aerial imagery was used as spatial framework, while generalized zoning plans and tabular company data were used to build up and further enhance the information content of the urban data management system.

Urban structures and functional patterns in high detail could be derived that way, whereas the geocoded address data forms the essential link between real-world geometries and abstract thematic information. Zoning plans give a first impression of the potential use of buildings, but local variations in urban land use patterns cannot be considered that way.

Company data was therefore integrated leading to a detailed functional classification of actual building use. Joint spatial analysis of zoning data and company information attached to building polygons via the geo-referenced address data reveals interesting spatial urban characteristics and shows the complex distribution of diverse branches of business over a settlement area.

This kind of highly detailed address information can be of interest in a lot of different fields including geo-marketing, disaster risk management etc. With this paper we aim at highlighting potential applications and hope that such high-level data will be widely available in the future.

Acknowledgements

The presented work was funded by the Austrian Research Promotion Agency (FFG) in the frame of the Austrian Space Applications Programme (ASAP). Interpretation of the aerial imagery was carried out by GeoVille Information Systems (Innsbruck, Austria). The Data.Geo address data was provided by Tele Atlas Austria.

References

ANSELIN, L. (1999), The future of spatial analysis in the social science. Geographic information sciences 5 (2), pp. 67-76.

AUBRECHT, C., KÖSTL, M. & STEINNOCHER, K. (2008), Visualization of simulated future scenarios in urban and suburban environments. In: CAR, GRIESEBNER, STROBL (eds.), Geospatial Crossroads @ GI_Forum'08. Proceedings of the Second Geoinformatics Forum Salzburg. Herbert Wichmann, Heidelberg, pp 31-41.

AUBRECHT, C., STEINNOCHER, K., HOLLAUS, M. & WAGNER, W. (2009), Integrating earth observation and GIScience for high resolution spatial and functional modeling of urban land use. Computers, Environment and Urban Systems, 33, pp. 15-25.

BERTIN, J. (1983), Semiology of Graphics. English translation of Bertin's Sémiologie Graphique (Paris 1967, 1973).

CHEN, K. (2002), An approach to linking remotely sensed data and areal census data. International Journal of Remote Sensing, 23 (1), pp. 37-48.

KRESSLER, F. & STEINNOCHER, K. (2008), Object-oriented analysis of image and LiDAR data and its potential for a dasymetric mapping application. In: BLASCHKE, LANG & HAY (Eds.), Object-Based Image Analysis. Springer, Heidelberg.

MESEV, V. (2005), Identification and characterization of urban building patterns using IKONOS imagery and point-based postal data. Computers, Environment and Urban Systems, 29, pp. 541-557.

STEINNOCHER, K. & KÖSTL, M. (2007), Zur Qualität österreichischer Geodatensätze. In: SCHRENK (Ed.), CORP 2007, 12th International Conference on Urban Planning and Regional Development in the Information Society. Proceedings, CD-ROM. Vienna, Austria.

WEEKS, J. R. (2001), The role of spatial analysis in demographic research. In: GOODCHILD & JANELLE (Eds.), Spatially integrated social science: examples in best practice. Oxford University Press: New York.

Automated Damage Assessment for Rapid Geospatial Reporting – First Experiences from the Haiti Earthquake 2010

Dirk TIEDE, Christian HOFFMANN, Petra FÜREDER,
Daniel HÖLBLING and Stefan LANG

1 Introduction

The FP7 project G-MOSAIC (*GMES services for Management of Operations, Situation Awareness and Intelligence for regional Crises*, http://www.gmes-gmosaic.eu/) aims at identifying and developing products, methodologies and pilot services for the provision of geo-spatial information in support of EU external relations policies and at contributing to define and demonstrate the sustainability of GMES global security services. Following the Haiti earthquake on January 12th, 2010, the G-MOSAIC Rapid Geospatial Reporting Service has been activated. It was initially requested by the UN cartographic section and the Spanish Red Cross in order to produce geo-spatial products in rush mode to assist relief efforts in Haiti. Satellite imagery immediately acquired after the disaster were processed by G-MOSAIC partners and first geo-spatial information were delivered to the users on January 16th. In this paper we share some experiences resulting from automated damage analysis in this context.

2 Study Area and Data

The Haitian towns Carrefour and Léogâne are located within the Ouest Department of Haiti. The coastal town Carrefour is located about 6 km west of the capital Port-au-Prince (see figure 1).

Fig. 1:
Overview of the study area west of Port-au-Prince

Léogâne lies about 30 km westwards of Port-au-Prince, only 12 km northwest from the epicenter of the earthquake (USGS, 2010). The damage analyses for Carrefour were conducted on GeoEye-1 pre- and post-disaster satellite imagery. For Léogâne GeoEye-1 (pre-images) and WorldView-2 (post-images) data was available. Pre-images for Carrefour were acquired on July 27th 2009 and for Léogâne on October 1st 2009. Post-images for Carrefour were taken on January 13th 2010 and for Léogâne on January 15th 2010. The spatial resolution of the GeoEye-1 and WorldView-2 sensors is 0.5 m in the panchromatic band for commercial use and 1.65 m (1.84 m for WorldView-2) in the multispectral bands. The pre- and post-images for Carrefour consisted of a multispectral image with three optical bands, a NIR band and a panchromatic image. The images for Léogâne were delivered in pan-sharpened format, but only RGB bands were made available.

3 Methods

3.1 Automated Damage Assessment

Rulesets were developed in an object-based image analysis (OBIA) environment to automatically extract relevant information as indications for damaged buildings. Rulesets for information extraction and change detection analysis were written using Cognition Network Language (CNL), a modular programming language implemented in the *Definiens eCognition 8* software environment. Rule-based classifiers are used for knowledge representation, making explicit the required spectral and geometrical properties as well as spatial relationships for advanced class modeling (TIEDE et al. 2010). Because of the tight time frame in this emergency situation, additional challenges, especially concerning the quality of the imagery, had to be tackled:

- Different recording conditions of the timely available images (recording angle, seasons)
- Time constraints in pre-processing of the imagery resulting in limited geometric accuracy and a mismatch between pre- and post-images (geometric shift) and also radiometric differences
- Available imagery for Léogâne was missing the fourth (NIR) band
- GeoEye-1's 0.41 m and WorldView-2's 0.46 m spatial resolution in the panchromatic band was down-sampled by the provider to 0.5 m for commercial use (and also in this humanitarian relief effort)

In the first step the images were orthorectified using the freely available ASTER Global Digital Elevation Map (GDEM), which was especially important in the hilly area in the southern part of Carrefour. The impact from the other limiting factors mentioned above was minimized in the ruleset development process. The damage assessment itself was based on indicators; in this case the shadows casted by buildings before and after the earthquake. To avoid deriving false positives from vegetation shadows a vegetation mask has been created for both images (see also VU et al. 2004). Missing NIR information for vegetation extraction in the town of Léogâne could partly be compensated by using visible greenness instead of the NDVI (Normalized Differenced Vegetation Index).

The geometric inaccuracy of the pre- and post-images and the resulting positional differences of the extracted shadow objects were tackled using object-linking that overcomes strict object hierarchies. Thereby, despite the geometrical shift and divergent overlapping areas between the image objects, we achieved a spatial comparison (size, shape etc.) of the shadow objects resulting in a damage indication class.

3.2 Damage Assessment Maps

The extracted damage indicators for the two areas were analyzed and conditioned applying kernel density methods (SILVERMAN 1986). The kernel density was calculated using point features – in this case the centroids of the extracted damage indicator objects on an output raster cell size of 20 m × 20 m. The resulting maps (damage indicator maps) were supposed to give a fast and easy to grasp overview about the spatial distribution and intensity of damages in the area (see figure 2).

4 Results and Discussion

Figure 2 shows the resulting damage assessment map for the area of Carrefour as it was delivered to the requesting users. An additional map was produced for the area of Léogâne. Distributed computing enabled the rapid processing of the data sets. A multi-core blade server installed at *Definiens* (Definiens AG, Munich) consisting of ~ 20 – 60 cores

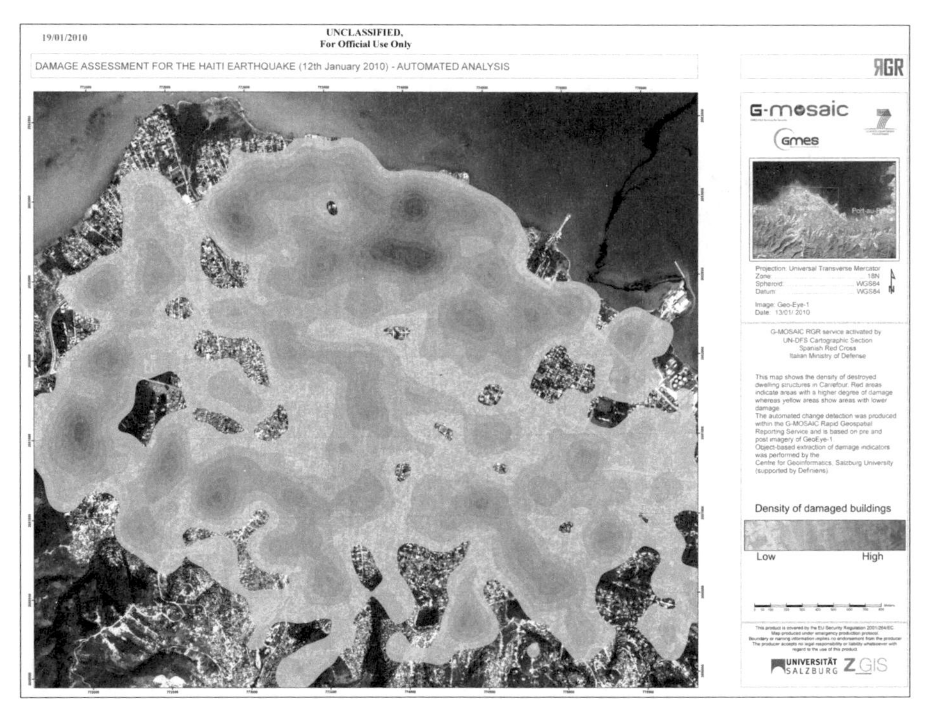

Fig. 2: Damage assessment map based on the automated approach for Carrefour. Darker tones indicate higher damage density.

(dependent on load balancing) could be used for the analysis and reduced the processing time for the area of Carrefour (17 km^2 and 0.5 m GSD) to approximately 4 minutes. Both maps did undergo a project-internal quality check by the service chain leader EUSC (European Union Satellite Centre). The map of Carrefour passed the quality check and was published and delivered to the users. The damage assessment for Léogâne showed good results in the outskirts of the town, but an underestimation of damages in the city centre compared to manual assessments. Two reasons were indentified, which mainly caused problems: (1) The WorldView-2 images for Léogâne were only provided with three spectral bands (RGB), the missing NIR band hampered the differentiation between vegetation / shadow of vegetation and some of the buildings; (2) a lot of construction work in the area led to undesired results (also in the case of visual interpretation). Therefore the map was not delivered to the user but was used internally to support the manual assessment in this area (for cross-checking).

Additional validation of the results was conducted by comparing them with several manually digitized damage assessment maps provided by other institutions. The results confirmed the findings of the project-internal quality check. Additionally, an accuracy assessment for 100 randomly selected damage indicators for each site revealed a user's accuracy of more than 72 % for the Carrefour area and more than 80 % for the area of Léogâne. For the central area of Carrefour some false detected damages were found which resulted from higher buildings and the shadows casted by these causing differences between the pre- and the post-imagery (due to slightly different viewing angles). This could be tackled in the future through threshold modifications in the ruleset. A fairly high user's accuracy in this automated analysis shows the potential of supporting the more time consuming manual analysis.

It has to be stated, that the aim of such an automated approach is not to replace the manual interpretation. It rather helps users and manual interpreters to get a faster impression of the spatial distribution of damages in emergency situations. By purpose the provided maps show no absolute values of detected damages but only tendencies. Absolute figures are very much depending on the resolution of the available imagery and even on aerial images not all damages are visible from the bird's eye view.

References

SILVERMAN, B. W. (1986), Density Estimation for Statistics and Data Analysis. Chapman and Hall, New York.

TIEDE, D., LANG, S., ALBRECHT, F. & HÖLBLING, D. (2010), Object-based class modeling for cadastre-constrained delineation of geo-objects. Photogrammetric Engineering & Remote Sensing, 76 (2), pp. 193-202.

USGS – U.S. GEOLOGICAL SURVEY (2010), Magnitude 7.0 – Haiti Region. – http://earthquake.usgs.gov/earthquakes/eqinthenews/2010/us2010rja6/ (accessed: 22 Apr. 2010).

VU, T. T., MATSUOKA, M. & YAMAZAKI, F. (2004), Shadow analysis in assisting damage detection due to earthquakes from QuickBird imagery. Proceedings of the 10th international society for photogrammetry and remote sensing congress, pp. 607-611.

Residential Location Decisions – A Case of the Bangkok Metropolitan Area

Nij TONTISIRIN and Sutee ANANTSUKSOMSRI

Abstract

Choosing a residential location is indeed a complex problem, influenced by various factors. Many recent studies, in fact, have emphasized that location specific factors such as neighbourhood characteristics and accessibility to local amenities significantly influence residential location choices. Since recent urban planning policies have become increasingly favourable to a more compact or intensified urban development pattern, such an understanding of residential location decisions, therefore, is crucial in formulating effective urban planning policies. In this paper, discrete logit models are used to analyze the effects of household socio-economic characteristics and physical attributes on residential location decisions in the Bangkok Metropolitan area.

1 Introduction

A residential location pattern is one of the predominant factors that shape an urban structure of a city. Residential locations, together with infrastructure networks, may lead to agglomeration or dispersion of a city as well as influence transportation patterns of commuters, thereby affecting environmental pollution and inefficient use of natural resources. As a recent trend in urban planning policies have increasingly promoted a more compact or intensified urban development pattern, an understanding of residential location decisions, therefore, plays a crucial role in formulating effective urban planning policies for controlling and managing sustainable urban development in the future.

Analyses of residential location decisions and urban spatial structure have been done extensively since the 1960s. Early studies were developed by ALONSO (1964) based on much earlier works of VON THÜNEN (1826), the model of agricultural land use. With an assumption that all jobs are located at the Central Business District (CBD), Alonso' mono-centric model has shown that, when choosing a residential location, households trade off a longer commuting distance for larger residential areas. Further studies of household residential choices include the works of MCFADDEN (1978), and ANAS (1985), which are based on the economic random utility theory. Several other empirical studies have demonstrated that demographic and socio-economic factors such as preference, education, income, and household characteristics significantly affect residential location decisions. BHAT & GUO (2007) show that residential location decisions are influenced by location-specific factors such as neighbourhood characteristics. Although empirical research on residential location decisions has been conducted extensively in the developed countries, only few studies have been done in developing country like Thailand.

As the capital of Thailand, Bangkok has long been a socio-economic centre. With recent emergence of rapid transit like the BTS Skytrain and MRT subway in 1999 and 2004 respectively, the city has expanded both vertically and horizontally. However, little has been known concerning residential location decisions of Bangkok residents. As housing developments in Bangkok is mainly market-driven, most studies have focused on a producer side. Only few research has emphasized a consumer side (WISAWEISUAN 2001). This study, therefore, aims to fill in this gap in the literature. Using Geographic Information System (GIS) and logistic regression model, this research analyzes factors influencing residential location decisions in the Bangkok Metropolitan area. To our knowledge, this analysis is one of the first attempts to understand urban spatial structure and residential location decisions of the Bangkok Metropolitan area.

2 Methodology

The logistic regression model, which is used to examine effects of various factors on residential location decisions, is based on random utility theory. For each household *n*, and an alternative location *i*, the linear utility (U_{in}) can be written as follows:

$$U_{in} = V_{in} + \varepsilon_{in}, \qquad \text{location } i \text{ chosen if } U_{in} > \max U_{im},\ n \neq m$$

Where V_{in} denotes a vector of observed explanatory variables; ε_{in} denotes random utility assumed to be independently and identically Gumbel distributed. The probability that a household *n* will choose location *i* ($P_n(i)$) is given by

$$P_n(i) = \Pr(U_{in} > U_{im}), \text{ or } P_n(i) = \Pr(X_{in}\beta - X_{im}\beta < \varepsilon_{in} - \varepsilon_{im}).$$

For this analysis, assuming that observed residential locations are in equilibrium, the explanatory variables are household characteristics, housing attributes, travel-related attributes, and sub-district attributes. The dependent variable reflects residential locations either within the inner ring roads or outside the ring roads (1=within the ring road, 0=otherwise).

3 Data

The study areas cover 316 sub-districts in the Bangkok Metropolitan Area, including the Bangkok Metropolitan Administration and surrounding provinces of Nonthaburi, Pathum Thani, and Samut Prakan. The data used in the analysis is drawn from two major sources: 2008 household socio-economic survey (SES) from the National Statistical Office, and Geographic Information System (GIS) of the Bangkok Metropolitan area from the Department of City Planning, Bangkok Metropolitan Administrative. The SES survey collects individual and household socio-economic characteristics including dwelling types, vehicle ownership, and residential locations at the sub-district level. In order to correctly predict the coefficients, housing units with commercial use (mixed-use units) are excluded in the analysis. Built environment attributes such as numbers of schools and shopping centers in each sub-district are derived from GIS layers, using ArcGIS.

4 Results

Table 1 shows the estimation results of four regression models with different blocks of explanatory variables: household characteristics, housing attributes, travel-related attributes, and sub-district attributes. As can be seen, housing and sub-district attributes are significantly related to residential locations. The results show that households that are owners tend to live outside the ring road. The housings located within the inner ring roads

Table 1: Estimation Results of Residential Location Decisions

	(1)	(2)	(3)	(4)
Household Characteristics				
Sex of the household head	0.230**	0.268***	0.243**	0.014
(1=female)	(0.097)	(0.102)	(0.102)	(0.166)
Number of earners	0.085**	0.127***	0.155***	0.155*
	(0.043)	(0.046)	(0.047)	(0.080)
Age of the household head	0.009***	0.017***	0.017***	0.022***
	(0.003)	(0.003)	(0.003)	(0.005)
Housing Attributes				
Tenure (1=own)		-0.736***	-0.724***	-0.421**
		(0.105)	(0.105)	(0.196)
Total number of room		-0.292***	-0.270***	-0.235***
		(0.051)	(0.052)	(0.072)
Log monthly rent		0.774***	0.807***	0.304**
		(0.105)	(0.107)	(0.145)
Travel-related Attributes				
Own automobile(s)			-0.396***	-0.228
			(0.112)	(0.182)
Own motorcycle(s)			-0.152	0.276*
			(0.098)	(0.163)
Sub-district Attributes				
Population density				0.001***
				(0.000)
Number of transit stations				0.999***
				(0.039)
Number of top schools				0.207***
				(0.052)
Number of shopping mall				0.168***
				(0.045)
Pseudo R-squared	0.0071	0.0435	0.0474	0.6136
Number of observation	4005	4003	4003	4003

Note: Standard errors are shown in parentheses; *, **, and *** indicate significant at 0.10, 0.05, and 0.01 level.

also tend to have less numbers of room but higher monthly rent than those located outside the ring roads. In addition, location specific characteristics such as population density and existence of mass transit stations, schools, and shopping malls in the sub-district significantly associated with residential locations. Travel-related attributes, however, become insignificant when location specific attributes are added. Furthermore, the model goodness-of-fit improves considerably when sub-district attributes are included in the regression, suggesting importance of location specific factors and spatial effects in residential locations.

5 Conclusion

Choosing residential locations are one of the complicated yet important problems faced by households. Understanding factors influencing residential location decisions is, therefore, essential to formulate effective urban planning policies. The results from the logistic regression models show that housing and location specific attributes are significantly associated with residential locations, suggesting that spatial effects may also be contributable to households' decisions. Further studies are still needed to clarify this doubt. Spatial effects on residential location decisions such as spatial weight matrix should be used in further studies of residential location decisions in the Bangkok Metropolitan area.

References

ALONSO, W. (1964), Location and Land Use: Toward a General Theory of Land Rent. Cambridge, Harvard University Press.

Anas, A. (1985), "The Combined Equilibrium of Travel Networks and Residential Location Markets." Regional Science and Urban Economics, 15 (1), pp. 1-21.

BHAT, C. R. & GUO, J. Y. (2007), A comprehensive analysis of built environment characteristics on household residential choice and auto ownership levels. Transportation Research Part B: Methodological, 41(5), pp. 506-526.

MCFADDEN, D. (1978), Modelling the choice of residential location. Spatial Interaction Theory and Planning Models, 25, pp. 75-96.

WISAWEISUAN, N. (2001), An Economic Analysis of Residential Location Patterns: a Case Study of Bangkok, Thailand. Dissertation, University of Cambridge.

Land Use/Cover Changes and Sustainability of Tourism Development in the Bulgarian Black Sea Coastal Zone

Rumiana VATSEVA and Boian KOULOV

1 Introduction

The Black Sea coastal zone is an area of prime concentration of natural resources of vital importance to recreation, biodiversity, and economic development (MARINOV et al. 2002), including international tourism. At the same time, the coast is one of the areas, which has endured the most sizeable and rapid land use/cover change (LUCC) during the post-socialist transition. The dramatic transformation is a direct result of the controversial politico-economic changes related to Eastern Europe's democratization. In an attempt to institute market economy, the Bulgarian state abolished the large-scale cooperative farms and returned the multitude of small parcels of land to their former owners or, rather, to their successors, most of whom already urbanized, without any agricultural skills or the capital to become private farmers. In a similar move to privatize at any cost, the large-scale Black Sea resort complexes (Albena, Golden Sands, Sunny Beach) with hundreds of hotels were broken up and "sold" on a piece-meal basis to individuals, the vast majority of whom without any knowledge or experience in management or the tourism industry. These methods of ownership transfer led to over-construction and over-urbanization in many areas along the Black Sea coast, in contradiction to its sustainable development. Asenova (2007) defines sustainable tourism development as a process, in which tourism-related economic and social changes should diminish the need to protect the environment. The current study strives to analyze the spatio-temporal land use/cover changes and their impact on the sustainability of the tourism industry along the Bulgarian Black Sea coast for the period 1990-2006. It focuses on two aspects: detection of land use/cover change at county level and assessment of the influence of tourism development in the coastal zone.

2 Materials and Methods

The study area is located in East Bulgaria and is approximately bounded by the coordinates: 41.56N to 43.44N and 27.17E to 28.36E (Figure 1a). This land strip with an area of 5829sq. km is 7 to 37 km wide, while the coastline has a length of 378km. Heavily anthropogenized areas dot the well preserved natural landscapes which characterize the region. The northern part features plain and karst relief, while the southern is predominantly lowland with forested hills, lakes, wetlands, and coastal lagoons. The research includes all of the 14 coastal counties in Bulgaria and, at the scale of 1:100000, identified LUCC "hot spots" in seven of them: Kavarna, Balchik, Nesebar, Pomorie, Sozopol, Primorsko, and Tsarevo. The second stage of the investigation focuses on the two-kilometer and one-kilometer coastal bands, since they contain the beach and the internal sea waters, which are exclusive state property and of utmost importance for tourism and recreational sports development.

Computer-aided visual interpretation of orthorectified multispectral satellite imagery applying the methodology and nomenclature of the CORINE Land Cover (CLC) project

(HEYMANN et al., 1994; PERDIGAO & ANNONI, 1997; BOSSARD et al. 2000; FERANEC et al. 2006, EEA 2007, BÜTTNER & KOSZTRA 2007, STEENMANS & BÜTTNER 2007) produced the LUCC data in this study. The CLC changes were registered when the following criteria were met: minimum changed area of 5ha and minimum width of 100m. The remotely sensed data used includes multispectral images from Landsat TM (acquired in 1990), Landsat ETM+ (2000), IRS-P6, SPOT 4 and SPOT 5 (2006). A wide variety of ancillary data was used, such as digital topographic and thematic maps, orthophotos, very high resolution satellite images, statistics, and in-situ data from multi-year field observations. GIS technology assisted the reclassification of the land cover changes into landscape change types. This transformation was performed in conformity with a conversion table (FERANEC et al. 2010) where land cover changes were grouped and classified according to the major land use processes. CORINE land cover changes were reclassified into the following six change types: 1. Urbanization/industrialization; 2. Enlargement/exhaustion of natural resources; 3. Intensification of agriculture; 4. Extensification of agriculture; 5. Afforestation; 6. Deforestation. The spatial change data for the whole coastal zone was overlayed (intersected) with the county boundaries for the extraction and analysis of LUCC for every county.

3 Results and Discussion

The analysis of the LUCC took place according to the land cover state in 1990, 2000 and 2006. The produced map (Figure 1a) shows the structure and spatial distribution of land use/cover in the study area in 2006.

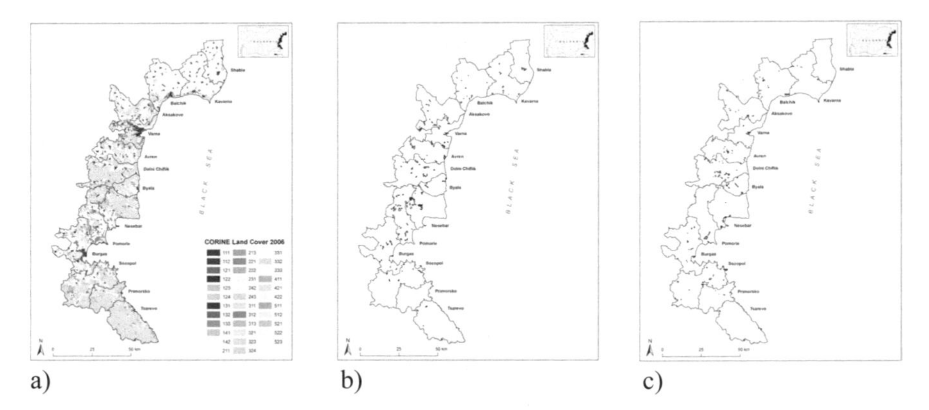

Fig. 1: a) CLC 2006 in the study area; b) LUCC 1990-2000 in the Bulgarian coastal zone; c) LUCC 2000-2006 IN THE BULGARIAN COASTAL ZONE

The total area of LUCC in the 14 Bulgarian Black Sea coast counties was 6682.9ha in 1990-2000 and 3754.7ha for the period 2000 – 2006 (Table 1).

The results indicate that the urban area expanded substantially along the Black Sea coast during the 1990-2006 period, mostly due to the detriment of arable land and woodland. Between 1990 and 2000, the changes consist mostly in abandonment of agricultural land, afforestation, and deforestation. Geographically, they are situated almost exclusively in the

Table 1: Land use/cover change in the Bulgarian Black Sea coastal zone

Land use/cover change type	1990-2000 (ha)	1990-2000 (%)	2000-2006 (ha)	2000-2006 (%)
Enlargement of urban fabric	28.9	0.43	63.7	1.70
Enlargement of construction sites	-	-	417.9	11.13
Enlargement of sport & leisure areas	-	-	289.0	7.70
Enlargement of industrial & transport units	-	-	96.5	2.57
Enlargement of mineral extraction sites	55.9	0.84	18.8	0.50
Intensification of agriculture	1025.1	15.34	228.3	6.08
Extensification of agriculture	2275.8	34.05	377.4	10.05
Afforestation	2005.5	30.01	599.5	15.97
Deforestation	1291.7	19.33	1663.5	44.30
TOTAL:	**6682.9**	**100.00**	**3754.7**	**100.00**

Table 2: Urban changes 2000 – 2006 in the two-kilometre wide band along the Black sea

County	Enlargement of urban fabric (ha)	Enlargement of construction sites (ha)	Enlargement of sport & leisure areas (ha)	Enlargement of transport units (ha)	TOTAL (ha):
Kavarna	-	18.7	-	-	**18.7**
Balchik	-	214.0	-	-	**214.0**
Nesebar	-	164.8	121.8	-	**286.6**
Pomorie	5.1	-	-	-	**5.1**
Sozopol	-	-	119.6	-	**119.6**
Primorsko	21.4	-	-	27.7*	**49.1**
Tsarevo	17.1	-	47.5	-	**64.6**
TOTAL (ha):	**43.6**	**397.5**	**288.9**	**27.7**	**757.7**

* Small airport for charters at a distance of 4 km

interior of the coastal counties and are evenly distributed (Figure 1b). Urban changes are not observed up to two kilometres from the coastline. During the second period, 2000-2006, the LUCC results are due mainly to urbanization "hot spots" in seven counties (Table 2). Most of urban changes are situated in the southern part of the Bulgarian coastal zone (Figure 1c). Urban changes in the two-kilometre band along the coast cover 757.7ha, however, 88.6% (671.6ha) of them are located in the one-kilometre land strip. The areas of the towns of Nesebar (with the Sunny Beach Resort and the village of Ravda), Sozopol, Primorsko and Tsarevo (with the villages of Lozenets and Sinemorets) experience the greatest urbanization pressure. This is in close connection with tourism development and includes construction of hotels, second homes, sport and leisure facilities areas, including golf courses, and a small airport for charter planes.

This research confirms the body of evidence (JABBRA & DWIVEDI 1995, KOULOV 1998b) that the transition to market economy in Eastern Europe caused serious environmental damage. KOULOV (1998a) calls it the "environmental price of the transition".

4 Conclusions

The apparent absence of a working market economy or "rule of law" in the "run" to attract international tourists, as well as the fast multiplication of second homes, some of which also housed tourists, substantially influenced the sustainability of use of natural resources and eco-services production along the Black Sea coast. The study detected a decrease in size of forest and agricultural lands in the coastal zone, in favour of tourism and recreation-related infrastructure even at the 1:100 000 scale.

Acknowledgements

This study was funded by the National Science Fund per contract № DO 02-10/2008 of the Institute of Geography, BAS and Ministry of Education, Youth and Science, Bulgaria.

References

ASENOVA, M. (2007), Monitoring of the Sustainable Tourism Development. Sofia University Year Book, Geography, 99, pp. 279-297.

BOSSARD M., FERANEC, J. & OTAHEL, J. (2000), CORINE Land Cover Technical Guide – Addendum 2000 (European Environmental Agency). – http://terrestrial.eionet.europa.eu/CLC2000.

BÜTTNER, G. & KOSZTRA, B. (2007), CLC2006 Technical Guidelines, 45 p.

EEA. (2007), CLC2006 technical guidelines. EEA Technical report, Copenhagen, 70 p.

FERANEC, J., BÜTTNER, G. & JAFFRAIN G. (2006), CORINE Land Cover Technical Guide – Addendum 2006.

FERANEC, J., JAFFRAIN, G., SOUKUP T. & HAZEU, G. W. (2010), Determining changes and flows in European landscapes 1990 – 2000 using CORINE land cover data. Applied Geography, 30 (1), pp. 19-35. – doi:10.1016/j.apgeog.2009.07.003.

HEYMANN Y., STEENMANS, C., CROISILLE, G. & BOSSARD, M. (1994), CORINE Land Cover. Technical Guide. Office for Official Publications of the European Communities, Luxembourg,136 p.

JABBRA, J. G. & DWIVEDI, O. P. (1995), Managing the Environment: An Eastern European Perspective," Int. Journal of Comparative Sociology and Anthropology, 1 (1), pp. 1-4.

KOULOV, B. (1998a), Political Change and Environmental Policy. Bell J. (ed.) Bulgaria in Transition Politics,Economics, Society and Culture after Communism. Boulder, CO and Oxford, UK: Westview Press, pp. 143-164.

KOULOV, B. (1998b), Post-Communist Change in Ecopolitics and Environmental Management: A Case Study of Bulgaria's Burgas Region. In PASKALEVA et al. (Eds.) Bulgaria in Transition: Environmental Consequences of Economic and Political Transformation. Aldershot: Ashgate, pp. 201-228.

MARINOV, V. & KOULOV, B. (2002), Water Recreation Resources and Environmental Pressures along Bulgaria's Black Sea Coast. In ANDERSON S. & TABB, B (Eds.), Water, Leisure and Culture: European Historical Perspectives. Oxford and New York: Berg Publishers, pp. 165-180.

PERDIGAO V. & ANNONI, A. (1997), Technical and Methodological Guide for Updating CORINE Land Cover Data Base (Joint Research Centre, European Environmental Agency, Luxembourg), 124 p.

STEENMANS, C. & BÜTTNER, G. (2007), Mapping Land Cover of Europe for 2006 under GMES. Proceedings of the 2nd Workshop of the EARSeL SIG on Land Use and Land Cover, pp. 202-207.

Spatio-Temporal Uncertainty in Individual Based Tree Line Modelling

Gudrun WALLENTIN and Adrijana CAR

The GI_Forum Program Committee accepted this paper as reviewed full paper.

Abstract

This paper investigates sources of uncertainty in individual based models. An individual based model (IBM) called TREELIM has been developed to simulate the alpine tree line shift variations and is used here as an example. IBMs can help reveal causal process – pattern relationships by identifying the processes (e.g. landuse or climate change) that are the essential driving factors for the emergence of patterns observed in the real world (e.g. in natural reforestation processes).

A comprehensive overview and understanding of uncertainty is necessary in order to assess the goodness and reliability of simulated results. The steps of conceptualisation, formalisation, parameterisation, analysis and validation in the IBM modelling procedure are investigated as potential sources of uncertainty. The preliminary results contribute to a conceptual view of spatio-temporal uncertainty for assessing the quality of the outputs of an individual based tree line model.

This ongoing work aims at the identification of a set of quantitative, robust, and reproducible methods for the assessment of the most relevant aspects of uncertainty in individual based models that can serve as a standard toolbox for modellers. The major benefit of this approach is that the steps most critical for overall uncertainty can be addressed on a quantifiable and defensible basis rather than be a subject to the modeller's experience or intuition on a trial-and-error basis.

1 Background

In this paper we investigate which sources of uncertainty may arise in the development of a model that simulates the alpine tree line shift. An alpine tree line is not merely a linear feature, but rather a transition zone, where tree density and height diminish over an elevation gradient. Currently, a substantial upward shift of the alpine tree line can be observed due to landuse and climate changes, causing new spatial patterns that change over time.

To gain a better understanding of spatio-temporal patterns at the alpine tree line, the *"TREE LIne Model"* (TREELIM) was developed. In an implementation of TREELIM for the Eastern Alps, the model was validated in a case study at Längenfeld, Ötztal over a period of 52 years (WALLENTIN et al. 2008) (Fig. 1). Conceptually, TREELIM is a simulation model that belongs to the family of individual based models (IBM). The IBM modelling approach

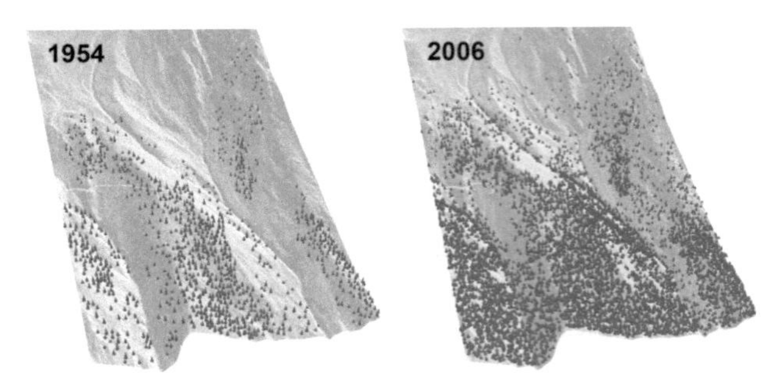

Fig. 1:
TREELIM scenario for the Ötztal case study: Forest patterns in 1954 and 2006

aims at explaining the emergence of system-level patterns from processes of individual animals or plants, their mutual interactions and their local reactions to environmental factors such as climate or elevation gradients. Here, we use TREELIM to discuss issues of uncertainty that heavily impact validation and reliability of individual based models.

The validation of TREELIM was accomplished by comparing patterns of the model outcome to those observed in the real world. In the validation process deviations in space and time from the observed reality were analysed, based on orthophotos available for the modelled period (WALLENTIN et al. 2008). For these points in time deviations between the magnitudes in the model and the reality along the y-axis describe the spatial and those along the x-axis the temporal deviation of the simulated tree distribution (Fig. 2).

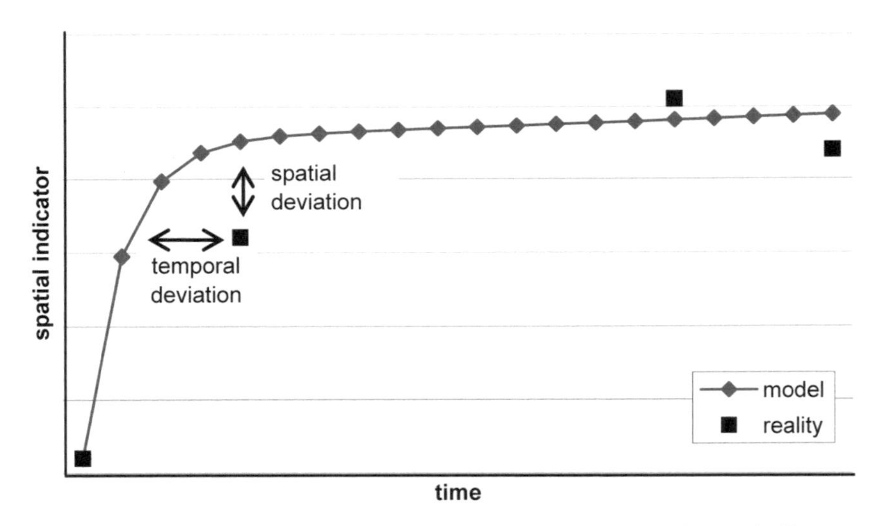

Fig. 2: Conceptual deviation of TREELIM model results, i.e. an indicator, in space and time (square points) from the observed reality (line)

The deviations in the diagram in Fig. 2 can be interpreted in three ways: (1) spatial patterns are captured adequately, meaning trees are distributed correctly in the observed region, but the reforestation happened too slowly or too quickly; (2) the reforestation process is timed correctly, but the spatial distribution does not correspond to reality; and (3) any combination of spatial and temporal deviation is possible. However, there is a lack of understanding what causes these deviations. Such an understanding is essential if we were to accept that model validation can make a statement about the reliability of our model for a particular purpose. The latter is at the centre of the debate in this paper.

Any spatio-temporal model is a more or less imperfect or inaccurate representation of real world phenomena. Thus understanding and communicating the reliability of results produced by such models requires statements of (in)accuracy. *Accuracy* refers to the difference between the reality and our representation of reality. Acknowledging existence of and considering such inaccuracies generally are known as *uncertainty* (LONGLEY et al. 2005). Objectively known inaccuracies are referred to as *errors* (HUNTER & GOODCHILD 1994), and are considered here as a part of the much broader concept of uncertainty. Uncertainty affects and gradually degrades quality of a spatio-temporal representation. It may arise in the very first step of conceptualisation followed by measurement and analysis through to visualisation of a geographical phenomenon, thus ultimately affecting decision making (compare to LONGLEY et al. 2005 Fig. 6.1, p. 129 and Fig. 13.6, p. 296 respectively).

From the GIScience perspective most approaches to conceptualise and formalise geospatial uncertainty acknowledge error components of space, attribute, time, consistency and completeness (VEREGIN 1999). For example, MACEACHREN et al. (2005) focus on uncertainty in visualising spatial data; PLEWE (2002) proposed a concept of uncertainty in temporal geospatial data focusing on a process rather than data itself, and SHOKRI et al. (2006) discuss the same for mobile objects. In the area of ecological modelling using IBM GRIMM et al. (2005) address uncertainty in complex ecological systems (like that of tree line modelling) and propose to use multiple patterns for validation. What remains on our research agenda is a rigorous investigation of sources of uncertainty.

Research achievements in both of these fields – GIScience and ecological modelling – are combined here to investigate potential sources of uncertainty that may arise during the TREELIM modelling cycle. The preliminary results provide a basis for development of a conceptual framework of uncertainty in IBMs. Such a concept is expected to help the modeller optimise an IBM and to quantify uncertainty in order to assess the reliability of model results, i.e. the goodness of the model for the respective purpose.

In section 2 of this paper we discuss the IBM modelling procedure and introduce the framework of uncertainty to be considered for analysis. In section 3 analysis of uncertainty in TREELIM is given based on steps of the IBM modelling procedure. In the final section we summarise our research results so far and identify questions yet to be answered.

2 IBM Modelling Procedure

Individual based models that are structurally realistic model the real-world processes that drive landscape patterns (GRIMM & RAILSBACK 2005). Thus, IBMs can help revealing causal process – pattern relationships by identifying the processes that are the essential driving factors for the emergence of patterns observed in the real world. However, before using a model, sources of uncertainty in the model outcome must be understood to assess the reliability of the model.

The assessment of uncertainty in an IBM is relevant for two model purposes: (1) predictive simulations to forecast future conditions of the system and (2) virtual labs to conduct

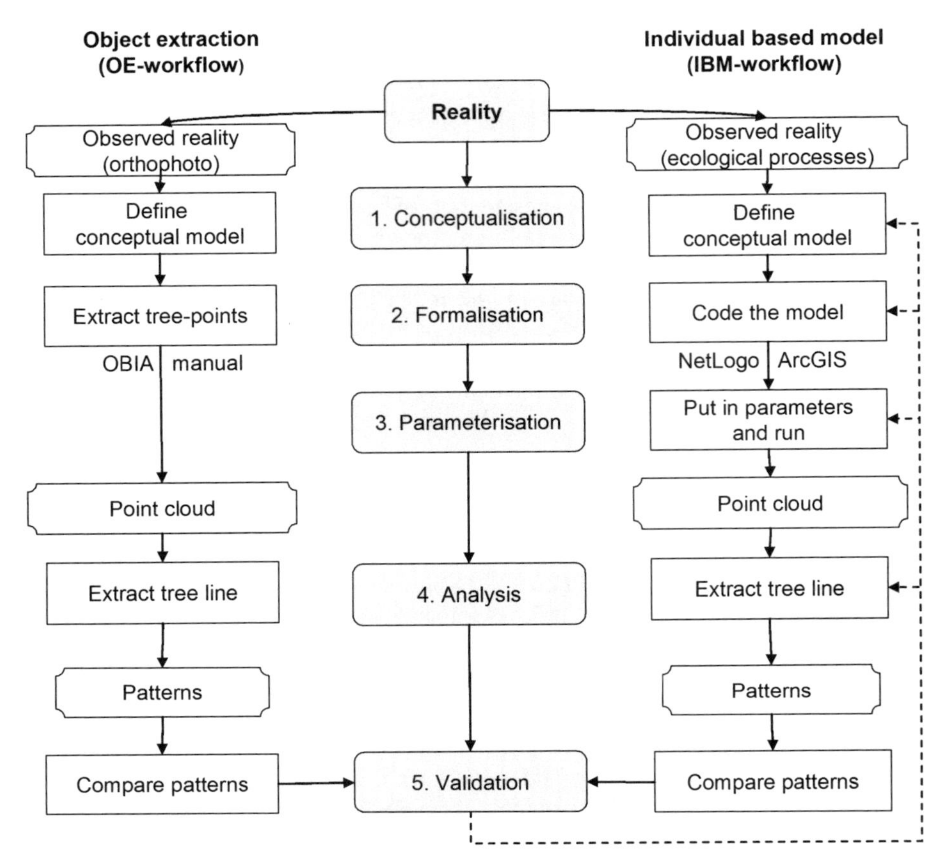

Fig. 3: Two parallel workflows in the TREELIM modelling cycle. Left: object extraction (OE-workflow), and right: individual based model (IBM-workflow).

theoretical experiments for getting a better understanding of the modelled ecosystem itself. We start by analysing the two parallel modelling workflows: the development of the IBM itself ("IBM-workflow"), and the extraction of the objects of interest from the observed reality ("OE-workflow"), where the latter is necessary for model validation. Following the procedure of building and validating IBMs (MULLIGAN & WAINWRIGHT 2004, GRIMM & RAILSBACK 2005), there are five steps in the modelling procedure, in each of which uncertainty may arise: conceptualisation, formalisation, parameterisation, analysis of results, and validation (Fig. 3). This holds for both workflows with exception of parameterisation which occurs only in IBM.

For each step in both workflows, we identify possible sources of uncertainty in the spatial and the temporal domains. We adapt the uncertainty concept defined by LONGLEY et al. (2005; see Section 6 and 13.2 and Figure 13.6) to the domain of spatial simulation models and explicitly include temporal uncertainty aspects. As an IBM ultimately aims at correctly capturing spatial processes, we propose to consider spatio-temporal rather than just spatial patterns at a point in time. *Spatial uncertainty* refers to positional accuracy; here it is a location of an individual tree, and in turn the location of the derived tree line. *Temporal uncertainty* states whether temporal information adequately describes geospatial phenomena; here it refers to temporal dynamics of the tree line shift process, which for example depends on the growth function (growth of an individual tree over time as it only begins producing seeds when reaching height of min 15 m) and the reproduction interval of 7 years. In the following section sources of uncertainty based on the workflow process are discussed in detail.

3 Uncertainty in TREELIM

3.1 Conceptualisation

Defining a concept for a model is an inherently subjective and creative task that involves the "modeller's intuition" (MULLIGAN & WAINWRIGHT 2004). Sources of uncertainty during the model conceptualisation can be manifold as the determination of basic assumptions about objects (e.g. a tree), processes (e.g. seed dispersion), attributes (e.g. tree height, age and species), scale (e.g. spatial extent of study area, time step resolution), modularisation (e.g. aggregating sub-processes into a single process or module), and scheduling are subjective abstractions of the real phenomenon. *Model uncertainty* can arise from

- an incomplete specification of the model i.e. omitting essential processes in a model, or from
- non-uniqueness of a model, i.e. that more than one model concept can lead to the same model output (e.g. climate change vs. landuse change influencing the speed of the tree line shift).

Uncertainty at this level can best be tackled through comparing different concepts, e.g. through progressively raising the level of detail, or testing alternative concepts.

An example of an incomplete specification in TREELIM is its basic version that models tree succession without considering *disturbances* like snow-glides, avalanches or wind storms. From viewing the results of a simulation in an area of the Passeier valley, Italy, it became apparent that snow-glides play an important role in the reforestation at the tree line. A refined concept of the TREELIM is therefore developed that includes snow-gliding tracks. The results of this refined model are yet to be compared to the basic model in order to assess the extent to which uncertainty can be reduced.

3.2 Formalisation

Formalisation refers to turning a concept into a computational model. Uncertainty then can occur due to selection of the software and further in coding within the selected software environment. In our case this holds for both, the IBM- and the OE-workflow.

In the IBM-workflow the magnitude of the *formalisation uncertainty* can be estimated using tests between alternative code structures (TURLEY & FORD 2009); e.g. processes that run simultaneously in nature often must be coded such that they run sequentially thus influencing the outcome by the order of process execution (e.g. aging, maturing, and distributing seeds). Uncertainty of formalisation can further be estimated by comparing model results that are based on the same conceptual model (e.g. of reforestation), but are implemented using different software tools.

An example of formalisation uncertainty in the IBM-workflow is shown by comparing two simulations that result from the implementation of TREELIM in ArcGIS and NetLogo. Conceptually, competition between trees can lead to a higher mortality rate depending on the distance between two trees and the relative tree sizes. In ArcGIS, which is not a true individual modelling framework, the competition-induced mortality rate is modelled based on a calculation of local tree density with a moving window kernel. In NetLogo the same process can be implemented directly as an interaction between individual neighbouring trees. This leads to inherently different results, specifically in the abundance of small trees in dense forests (Fig. 4).

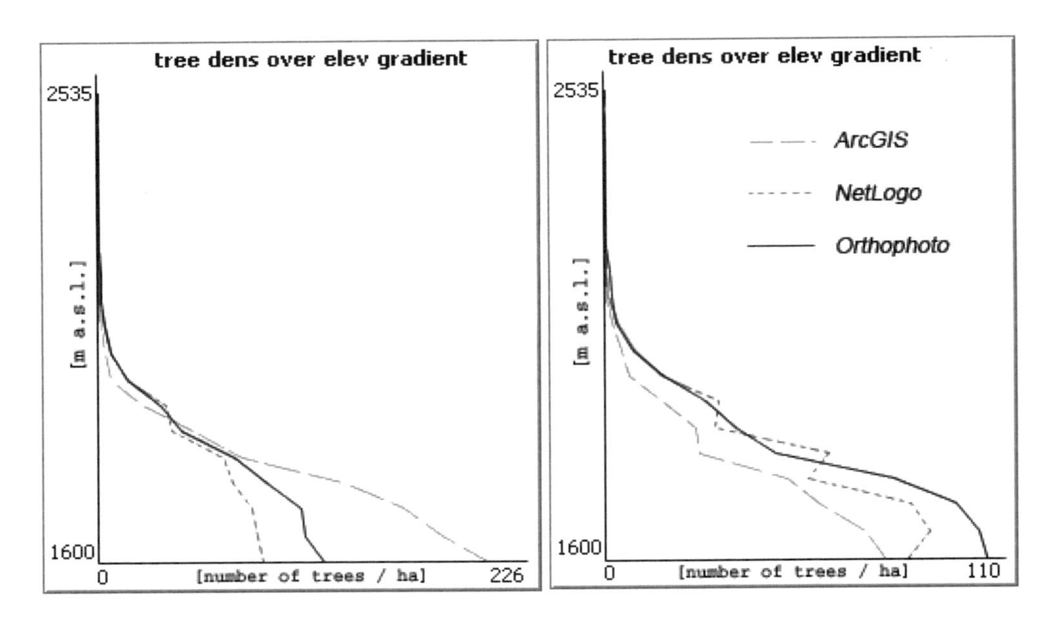

Fig. 4: The tree density over the elevation gradient as modelled in TREELIM (y-axis – meter above sea level; x-axis – number of trees /ha). Left: in dense forests; right: in sparse forests.

Spatial extent of an area of investigation is limited by the capacity of NetLogo, which only can handle roughly 15,000 to 20,000 individuals; otherwise the application crashes. This limits the selection of a study area; in a way it can be seen as a "modifiable area unit problem" (MAUP) and thus cause further uncertainty.

For the OE-workflow an example relates to manual versus automatic (Definiens software) tree extraction from remotely sensed data. The extracted tree points differed in the number of trees and the location. Both of these parameters also depend on the tree height and the distribution patterns; e.g. at near distances the automatic delineation algorithm resulted in an artefact of strong regularity at the local maxima search distance (Fig. 5).

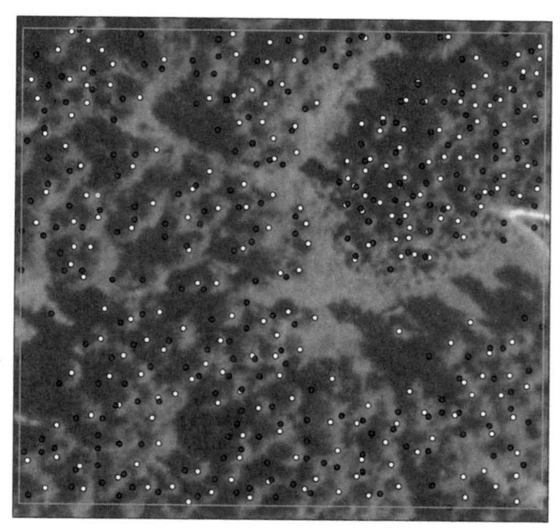

Fig. 5:
The location and number of extracted tree points in a test area: automatic delineation based on LiDAR data (dark points) differs from manually extracted trees, based on the aerial photo (light points)

3.3 Parameterisation

In this step, the magnitude of each *model input parameter* is chosen. Uncertainty may arise if these parameters are false or unrealistic (remark: a missing parameter is already considered in conceptualisation uncertainty). Lacking knowledge regarding the magnitude of input parameters can be addressed through Monte Carlo simulations, i.e. by systematically changing the input parameters.

For the model input parameters of TREELIM, a range of possible values were cited in the literature. To assess the uncertainty resulting from different *parameter values*, scenarios were defined with possible input values for the most essential parameters, as for example the maximum distance of seed dispersal. The patterns emerging from these scenarios were tested and compared to observed patterns. In this way, the most realistic parameters could be identified (WALLENTIN et al. 2008).

3.4 Analysis of model results

The analysis of TREELIM results uses aggregation and interpolation of data to unveil the pattern of interest, which inherently leads to a loss of detail. Here, the analysis of a cloud of tree-points determines a tree line, whereby the point cloud results from a single model run and from extracting trees from an orthophoto, respectively.

Patterns are described through a set of indicators, e.g. frequency of single trees above the forest, the maximum tree line elevation, and the forested area (Table 1). The *choice of indicators* is subjective and thus is a source of uncertainty. This uncertainty can be reduced by using direct indicators (those that clearly correspond to the pattern) but also by gradually increasing the number of indicators for a model at hand.

Table 1: Selection of indicators used in TREELIM (WALLENTIN et al. 2008)

Indicator	1954	2006	TREELIM
forested area [m2]	173311	349236	347444
tree line, maximum elevation [m]	1948	2093	2086
single trees >4 m [frequency]	80	167	127
median distance of single trees ≥4 m to the nearest forest edge (m)	12.4	19.6	22.8

3.5 Validation

In the step of validation, the IBM output is compared to the objects extracted from an orthophoto (Fig. 3). Deviation between the spatio-temporal patterns includes all sources of uncertainty discussed so far. However, the validation process itself can be a source of uncertainty. Uncertainty in this step is mainly due to a lack of observable objects (e.g. small trees hidden in dense forests) and attributes (e.g. tree age) in the real system. To be kept in mind also is the availability of orthophotos with respect to their temporal and spatial resolution.

In the TREELIM example, tree patterns resulting from reforestation simulation and from digitising an orthophoto are inherently different, which makes a comparison difficult. For example, due to the vertical perspective in an orthophoto, small trees may be hidden by taller trees. In dense forests, the tree density is therefore underestimated in the orthophoto-based map. Many attributes cannot be directly derived (like tree age) and are therefore not usable for map comparison. In TREELIM for each individual tree the following attributes can (✓) or cannot (✖) be obtained from reality, aerial photo or a model (Table 2).

Table 2: Possibility of obtaining attributes of an individual tree (WALLENTIN et al. 2008)

tree attribute	attribute value	orthophoto / LiDAR	TREELIM
height	15 m	✓	✓
age	150 years	✖	✓
species	Picea abies	✖	✓
elevation	1900 m	✓	✓
time	1988	✓	✓
location	UTM coords	✓	✓

3.6 Overall model uncertainty

In a complex ecological IBM, the deviation between model results and the observed reality for one single indicator insufficiently describes the overall model uncertainty. Rather, multiple indicators should be used, as advocated by GRIMM et al. (2005) in their pattern oriented modelling approach. Additionally, we suggest to explicitly including temporal aspects in spatial IBMs, as the temporal component is important for describing spatial processes. Only then validation can be used to indicate the goodness and the reliability of a model simulation result.

Building an IBM is an iterative process. In TREELIM the preliminary simulation results are validated and then checked whether they lie within tolerance ranges. If only some spatial and/or temporal indicators describing the patterns of interest (e.g. tree clustering, maximum elevation of single trees) are predicted well by a simulation but others are missed, the reliability of the model is unsatisfactory. At this point we suggest that the proposed framework for sources of uncertainty is applied to structurally search for the main origin(s) of this deviation. Thus, the relevance of single, isolated sources of uncertainty can be understood and also placed in the context of the overall model uncertainty.

4 Conclusion and Outlook

In this paper we explored potential sources of uncertainty in both, the IBM- and the OE-workflows. A comprehensive overview and understanding of uncertainty is necessary in order to assess the goodness and reliability of simulated results. With this information, in the modelling cycle, the steps most affecting the overall uncertainty can be addressed on a quantifiable and defensible basis rather than leaving it subject to the modeller's experience or intuition on a trial-and-error basis. Ultimately we need to know the magnitude of uncertainty of model results with respect to its fitness-for-use rather than just the deviation from the observed world (HUNTER et al. 2009).

However, a comprehensive view on uncertainty of an IBM, including the test of alternative concepts is hard to achieve, as there is probably an infinite number of possibilities in choosing a-priori assumptions and abstractions. The same holds for parameterisation; e.g., the time spent for Monte Carlo simulation is an order of magnitude greater than the basic analysis itself (HUNTER et al. 2009). Thus model development and uncertainty assessment will remain subjective to a certain degree. A trade-off is necessary between time and effort put into model validation and uncertainty assessment and its fitness-for-use.

The specific sources of uncertainty in TREELIM identified here remain to be analysed and discussed more thoroughly in order to contribute to the development of a feasible, yet defensible way of uncertainty analysis in individual based models. If we have sufficiently well identified the potential sources of uncertainty then we expect to answer the following questions:

1. How can we define the goodness of TREELIM in both dimensions: space and time? Ultimately we need to define how trustworthy the model is, when it comes to forecasting.
2. How well is the respective ecosystem represented in the TREELIM model so that it can function as a virtual laboratory?

The major future challenge lies in the identification of a set of quantitative, robust, and reproducible methods for the assessment of the most relevant aspects of uncertainty in individual based models that can serve as a standard toolbox for modellers. The next task is then to find out how to measure the quality of the achieved results, i.e. how to describe the sources constituting overall model uncertainty that is attached to a model result in a scientific and reproducible way.

This would further allow us to suggest methods to assess the magnitude of uncertainty for each step in relation to the overall model uncertainty in order to specifically address steps in the modelling cycle that are disproportionally contributing to uncertainty.

References

GRIMM, V. & RAILSBACK, S. F. (Eds.) (2005), Individual-based Modeling and Ecology. Princeton Series in Theoretical and Computational Biology. Princeton University Press, Princeton and Oxford.

GRIMM, V., REVILLA, E., BERGER, U., JELTSCH, F., MOOIJ, W. M., RAILSBACK, S. F., THULKE, H. H., WEINER, J., WIEGAND, T. & DEANGELIS, D. L. (2005), Pattern-oriented modeling of agent-based complex systems: Lessons from ecology. Science, 310 (5750), pp. 987-991.

HUNTER, G. J., BREGT, A. K., HEUVELINK, G. B. M., DE BRUIN, S. & VIRRANTAUS, K. (2009), Spatial Data Quality: Problems and Prospects. Research Trends in Geographic Information Science. Berlin/Heidelberg, Springer.

HUNTER, G. J. & GOODCHILD, M. F. (1994), Managing uncertainty in spatial databases: putting theory into practice. Journal of the Urban and Regional Information Systems Association, 5 (2), pp. 55-63.

LONGLEY, P. A., GOODCHILD, M. F., RHIND, D. W. & MAGUIRE, D. J. (Eds.) (2005), Geographic Information Systems and Science. J. Wiley & Sons Ltd., Chichester, UK.

MACEACHREN, A. M., ROBINSON, A., HOPPER, S., GARDNER, S., MURRAY, R. & GAHEGAN, M. (2005), Visualizing Geospatial Information Uncertainty: What we know and what we need to know. Cartography and Geographic Information Science, 32 (3), pp. 139-160.

MULLIGAN, M. & WAINWRIGHT, J. (2004), Modelling and Model Building. In WAIN-WRIGHT, J. & MULLIGAN, M. (Eds.). Environmental Modelling: Finding Simplicity in Complexity. London, Wiley.

PLEWE, B. S. (2002), The Nature of Uncertainty in Historical Geographic Information. Transactions in GIS, 6 (4), pp. 431-456.

SHOKRI, T., DELAVAR, M. R., MALEK, M. R. & FRANK, A. U. (2006), Modeling uncertainty in spatiotemporal objects. 7th International Symposium on Spatial Accuracy Assessment in Natural Resources and Environmental Sciences, 5 – 7 July 2006, Lisboa, Instituto Geográfico Português.

TURLEY, M. C. & FORD, E. D. (2009), Definition and calculation of uncertainty in ecological process models. Ecological Modelling, 220 (17), pp. 1968-1983.

VEREGIN, H. (1999), Data Quality Parameters. In: LONGLEY, P. A., GOODCHILD, M. F., RHIND, D. W. & MAGUIRE, D. J. (Eds.), Geographical Information Systems, Principles and Applications. J. Wiley & Sons,p p. 177-189.

WALLENTIN, G., TAPPEINER, U., STROBL, J. & TASSER, E. (2008), Alpine tree line dynamics: an individual based model. Ecological Modelling, 218 (3-4), pp. 235-246.

Three Steps Towards Spatially Explicit Climate Change Analysis

Peter ZALAVARI, Hermann KLUG and Elisabeth WEINKE

The GI_Forum Program Committee accepted this paper as reviewed full paper.

Abstract

For past and present climate change analysis, as well as future projection of climate change, long term primary datasets from meteorological stations are indispensable. However, in raw data format such as *.txt, *.dat, *.csv or any other file format precipitation or temperature datasets are not practical for spatial queries, interpolation, and trend analysis. In three steps we provide users with the facility to make huge amounts of climate data accessible, downloadable, and convertible to a geodatabase. These three steps are available as Python scripts and therefore allow for semi-automated operation. Such a resulting geodatabase can be used for spatial queries or interpolation of data provided as points.

1 Introduction

Analysing the past climate change referring to changes in temperature and precipitation values are an indispensable requirement to understand and to underpin climate change but also to estimate the productivity of agricultural and forestry systems (SOLOMON et al. 2007, EEA 2009, 53). Impacts such as natural hazards are foreseen from the temperature increase which has been observed for the Alps as three times higher than global average since 1980th (BENISTON 2005; OECD 2007, 19). Further global climate change impacts on water resources have been reported by numerous researchers (e.g. OECD 2007, 62). The most well known references on changes at a global scale are the IPCC reports (SOLOMON et al. 2007) and the Millennium Ecosystem Assessment (HASSAN et al. 2005). Even more dramatic are the reported climate induced changes in the European Alps, with increasing temperatures twice the global average (EEA 2009, 9). Local examples from Austria report on decreasing groundwater recharge of 25% within the last 100 years (HARUM et al. 2007). Also in Slovenia measures of incoming and outgoing water still shows decreasing trend (BRANCELJ 2009). These findings clearly show that global climate change assessments are good and necessary but cannot cope with national, regional or even local problems on water shortages. We posit a hypothesis that especially spatio-temporal water availability and water demand need to be considered when analysing water scarcity problems, which are also present in the so called water towers of the Alps. This hypothesis is also underpinned by EEA (2009) that report on limited knowledge on local impacts of climate change on water availability and water demand by different sectors. For instance hydropower energy production will be affected due to seasonal changes in water availability (EEA 2009, 58; LEHNER et al. 2005, SCHAEFLI et al. 2007) and winter tourism will demand a proportional higher share of water compared to today.

To build up a sound basis for politicians, stakeholders, decision makers and water managers, adaptive strategies against climate change in the Alps need to guarantee coordinated action and efficient exchange of information about water availability and water demand across disciplines. To analyse water availability and its past changing trends, climate datasets need to be available, pre- and post-processed and modelled. For the Alps datasets are available e.g. at

- NOAA NCDC (National Oceanic and Atmospheric Administration, National Climatic Data Center),
- ZAMG (Austrian Weather Service),
- hydrographical survey data from each Austrian Federal State,
- data from the Ministry of Life Science,
- HAÖ (Hydrological Atlas of Austria),
- German Weather Service,
- Meteo Swiss Alpe,
- WorldClim (Global Climate Data),
- FAO (United Nations Food and Agriculture Organization,

as well as from several other sources.
To make the datasets available for modelling, pre-processing is necessary. So far a couple of projects and papers report on providing consistent climate parameters:

- HISTALP (Historical Instrumental Climatological Surface Time Series of the greater Alpine Region)
- Adaptalp (Adaptation to Climate Change in the Alpine Space)

Even though for instance HISTALP provide datasets for free of charge, there is still a lack of data available and a strong need to better sharing of information about climate change in the Alps (ALPINE CONFERENCE ACTION PLAN 2009). This lack is due to

- precision of datasets,
- availability of datasets (e.g. access to datasets and property rights),
- black box of pre-processing and knowledge about initial datasets used, and
- updating the existing knowledge and databases.

Additionally, papers report findings on average climate change for the whole of Europe or, as proposed by the Water Framework Directive (WFD) on a catchment basis (Directive 2000-60-EC). Both are unsuitable to use within the Alpine space. Special areas of interest neither can be analysed nor can any conclusions for adaptation measures be framed.

Due to the mentioned shortcomings, the main objective of this paper is to provide a well described procedure ranging from data acquisition from the internet to its geodatabase integration.

2 Methods

From the National Oceanic and Atmospheric Administration's (NOAA) National Climatic Data Center (NCDC, http://gis.ncdc.noaa.gov/geoportal/catalog/main/home.page) geospatial data such as daily temperature and daily precipitation rates are available free of charge. One can either download these datasets using a right mouse click or save the files to the disk or one can use an FTP server to manually download larger parts of the available datasets.

To make the download procedure more straightforward and to enable practitioners to download newly available datasets, we provide the users with a Python script to semi-automate these tasks (Fig. 1). As the first step in Fig. 1 shows, data files are not only downloaded but also unzipped to a predefined destination.

The second step is needed to convert Fahrenheit measurements in degrees Celsius. Moreover, datasets need to be prepared for semi-automated information extraction and to be stored in the geodatabase. The header of the text file needs to be removed and space separations changed to 'comma delimited.'

Finally, a connection will be established between the PostGIS/PorstgreSQL geodatabase and the table necessary for storing information by the station ID, the time of data recording and the values for precipitation and temperature.

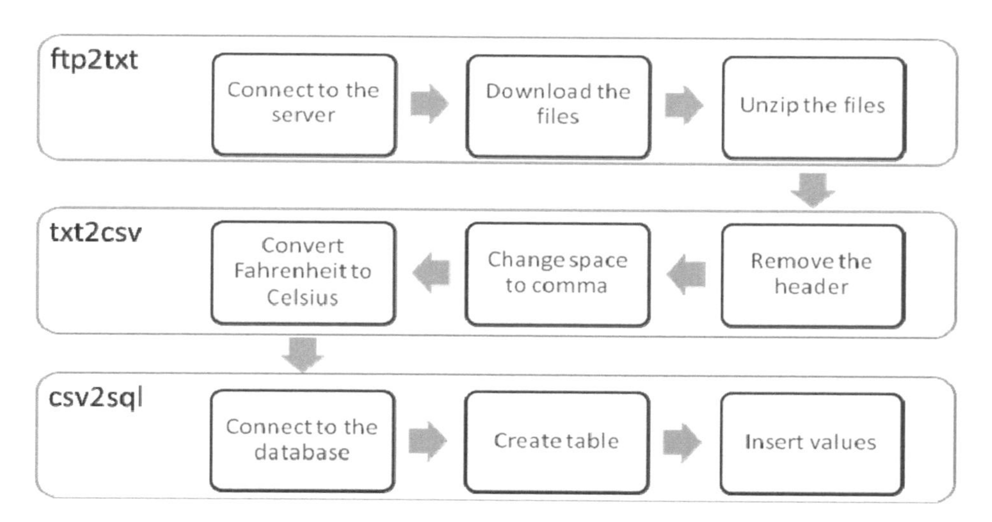

Fig. 1: In three steps from raw data towards a geodatabase

3 Results

3.1 Data acquisition

Datasets from NOAA can be obtained semi-automatically using the Python script in Listing 1. The script connects to the NOAA FTP Server and downloads the worldwide climate datasets. In this example datasets between 1930 and 1932 will be accessed and stored in unzipped format on your hard drive.

Listing 1: Script enabling the semi-operational download and unzip of NOAA datasets

```
#!/usr/bin/env python
import ftplib, tarfile, gzip, os, re

def purge(dir, pattern):
      for f in os.listdir(dir):
            if re.search(pattern, f):
                  os.remove(os.path.join(dir, f))

print 'connecting to the server...'
ftp = ftplib.FTP('ftp.ncdc.noaa.gov')
ftp.login()

year=1930
while year<=1933:
      gsod_year = 'gsod_' + str(year)
      ftp.cwd('/pub/data/gsod/' + str(year))
      print str(year) + ' downloading'

      ftp.retrbinary('RETR gsod_' + str(year) + '.tar',
open(gsod_year + '.tar','wb').write)
      tar =  tarfile.open(gsod_year + '.tar')

      tar.extractall(path=gsod_year)

      ops = os.listdir(gsod_year)
      for i in ops:
            gsod_in = gzip.open(gsod_year + '/' + i)
            gsod_out = open(gsod_year + '/' + i + '.txt',
'w')
            gsod_out.writelines(gsod_in.read())

      purge('./gsod_' + str(year), 'gz$')
      purge('.', 'tar$')

      print str(year) + ' ready'
      year+=1

ftp.quit()
```

```
tar.close()
gsod_out.close()
gsod_in.close()
```

3.2 Pre-processing of unzipped datasets

Pre-processing as mentioned in chapter 2 will be done using Listing 2. After downloading the datasets of the years between 1930 and 1932, this second Python scripts i) removes the first line (header) in all the files; ii) converts temperature data from Fahrenheit to Celsius, and finally iii) changes the file format to CSV (comma-separated values).

Listing 2: Script enabling the semi-operational download and unzip of NOAA datasets

```
#!/usr/bin/env python

import os,sys,re

def rm_line(file,line):
      f = open(file)
      lines = f.readlines()
      f.close
      lines[line-1] = ''
      f = open(file, "w")
      f.writelines(lines)
      f.close()

def txt2csv(file):
      noaa=open(file)
      f = noaa.read().strip()
      noaa.close()
      noaa=open(file, 'w')
      noaa.write(re.sub('[ ]+',',',f))
      noaa.close()

def rm_x(file):
      noaa=open(file)
      f = noaa.read().strip()
      noaa.close()
      noaa=open(file, 'w')
      noaa.write(re.sub('\*','',f))
      noaa.close()

def F2C(row,file):
      noaa = open(file)
      dat=''
      for line in noaa.readlines():
            liste = line.split(",")
            fahrenheit = liste[row]
            celsius = round((float(line.split(',')[row]) -
```

```
32.0) * 5.0 / 9.0, 2)   # [C] = ([F] - 32) x 5/9
            if fahrenheit != '9999.9':
                  liste[row]=celsius
                  new_line=liste[0]
                  for i in range(len(liste)):
                        if i==0:
                              new_line=new_line
                        else:
                              new_line=new_line + ',' +
str(liste[i])
                  dat+=new_line
            else: dat+=line
      noaa.close()
      noaa = open(file,'w')
      noaa.write(dat)
      noaa.close()

a=0
for year in range(1930,1934):
      gsod_year = 'gsod_' + str(year)
      os.chdir(gsod_year)
      for files in os.listdir('.'):
            rm_line(files,1)
            txt2csv(files)
            rm_x(files)
            F2C(3,files)
            F2C(5,files)
            F2C(17,files)
            F2C(18,files)
            noaa=open(files)
            dat=''
            for line in noaa:
                  a+=1
                  dat+=str(a) + ',' + line
            noaa.close()
            noaa = open(files,'w')
            noaa.write(dat)
            noaa.close()
      os.chdir('..')
```

3.3 Transfer of pre-processed text files to the geodatabase

Having downloaded the data files of interest, the script in Listing 3 allows for transferring single text files of precipitation or temperature measures stored in the PostGIS/PostgreSQL geodatabase. This scripts use the psycopg2 Python-PostgreSQL Database Adapter to connect to the database and to create the required table. After the table was established, data from the CSV text files is transferred into the database.

Listing 3: Script enabling the semi-operational download and unzip of NOAA datasets

```
#!/usr/bin/env python

import os

try: import psycopg2
except ImportError:
    raise ImportError, '''The psycopg2 module is required to
run this program.
You can download it from the http://initd.org/pub/software/
psycopg/website.'''

conn = psycopg2.connect('dbname=stations')
curs = conn.cursor()

curs.execute('''
CREATE TABLE data_bbs (
      id    serial PRIMARY KEY,
      stn         char(7),
      wban  char(5),
      year  date,
      temp  float(2),
      count1      char(2),
      dewp  float(2),
      count2      char(2),
      slp         float(2),
      count3      char(2),
      stp         float(2),
      count4      char(2),
      visib float(2),
      count5      char(2),
      wdsp  float(2),
      count6      char(2),
      mxspd float(2),
      gust  float(2),
      MAX         float(2),
      MIN         float(2),
      PRCP  char(5),
      SNDP  float(2),
      FRSHTT      char(7)
);
''')
query = 'INSERT INTO data_bbs VALUES
(%s,%s,%s,%s,%s,%s,%s,%s,%s,%s,%s,%s,%s,%s,%s,%s,%s,%s,%s,%s,
%s,%s,%s)'

for year in range(1930,1934):
```

```
        gsod_year = 'gsod_' + str(year)
        os.chdir(gsod_year)
        for files in os.listdir('.'):
                for line in open(files):
                        fields = line.split(',')
                        curs.execute(query,fields)
        os.chdir('..')

conn.commit()
conn.close()
```

3.4 The spatially explicit results and the structure of the geodatabase

As a result, 2 shows the alpine stations available from NOAA, which has been imported to the geodatabase. The structure of the geodatabase is displayed in Fig 3.

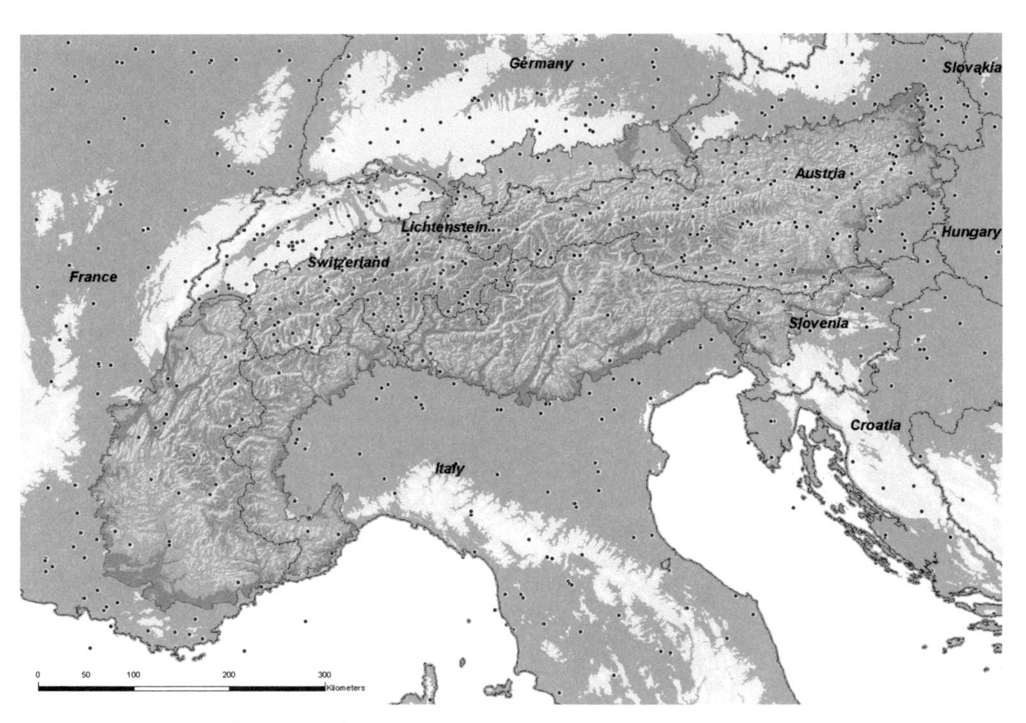

Fig. 2: NOAA climate stations in the Alps

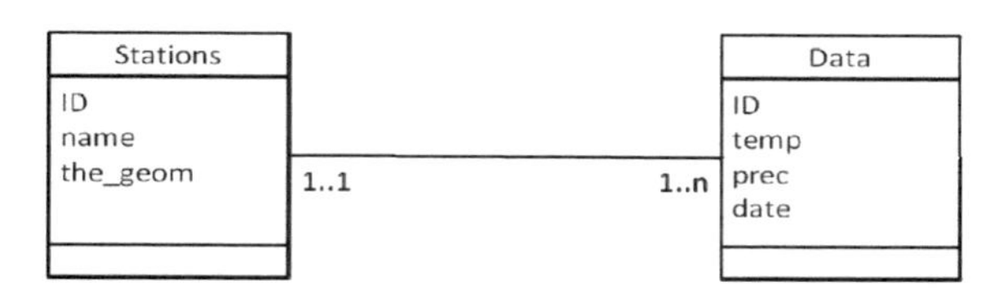

Fig. 3:
Data model for NOAA datasets

4 Discussion

It is clear that many thousands of climate records cannot be manually downloaded and pre-processed for further usage. Thus, this paper gives the users three tools with which they can automate the tasks of downloading of NOAA NCDC datasets and integrating them into a geodatabase.

Since most of the recording of climate parameters are done in a similar way, our experience show that only a few changes need to be done for preparing other datasets with the same scripts, e.g. such as those from the Ministry of Life or the Hydrological Services in Austria.

5 Conclusion and Outlook

As the first step towards regional climate change analysis we managed to integrate climate parameters into a geodatabase in a transparent way. Further steps will provide harmonisation, homogenisation, and analysis especially of temperature and precipitation datasets. This will also be done with Python scripts and thus allow transparent pre-processing for past climate change analysis. Also interpolation of datasets will be implemented in openly accessible software routines, making spatially explicit climate change information available for everyone. Especially the spatially explicit datasets on precipitation and temperature (e.g. to use them as input parameters for calculating evapotranspiration) are required to prove our overall working hypothesis that even in the water towers of the European Alps, spatio-temporal water scarcity problems are still present at the regional or local levels.

Acknowledgements

This research was partially funded by the Interreg IVb "AlpWaterScarce" project.

References

ALPINE CONFERENCE ACTION PLAN (2009), Alpine Conference meeting "Making the Alps an exemplary territory for prevention and adaptation to climate change", 12.03.2009, http://www.alpconv.org/NR/rdonlyres/193D7A9E-0F5E-475D-A48D-E3276F11D292/0/AC_X_B6_en_new_fin.pdf.

BENISTON (2005), Sensitivity Analysis of Snow Cover to Climate Change Scenarios and Their Impact on Plant Habitats in Alpine Terrain, In: Journal of Climatic Change, pp 299-319.

BRANCELJ, A. (2009), Talk in the framework of the 1st annual meeting of the AlpWaterScarce project, April 27, 2009, Vienna, Austria.

DIRECTIVE 2000-60-EC of the European Parliament and of the Council of 23 October 2000. Establishing a framework for Community action in the field of water policy.

EEA (2009), Regional climate change and adaptation. The Alps facing the challenge of changing water resources. EEA Report No 8/2009, ISSN 1725-9177.

HARUM, T., POLTNIG, W., RUCH, C., FREUNDL, G. & SCHLAMBERGER, J. (2007), Variability and trends of groundwater recharge in the last 200 years in a south alpine groundwater system: impact on the water supply. Poster presentation at the International Conference on Managing Alpine Future in Innsbruck, 15 – 17 October 2007.

HASSAN, R., SCHOLES, R. & ASH, N. (Eds) (2005), Ecosystems and Human Well-Being: Current State and Trends. Island Press.

LEHNER, B., CZISCH, G. & VASSOLO, S. (2005), The impact of global change on the hydropower potential of Europe: a model-based analysis'. In: Energy Policy, 33 (7), pp. 839-855.

OECD 2007, Climate Change in the European Alps. Adapting winter tourism and natural hazards management. ISBN: 92-64-03168-5.

SCHAEFLI, B., HINGRAY, B. & MUSY, A. (2007), Climate change and hydro-power production in the Swiss Alps: quantification of potential impacts and related modelling uncertainties. In: Hydrol. Earth Syst. Sci., 11, pp. 1191-1205.

SOLOMON, S., QIN, D., MANNING, M., MARQUIS, M., AVERYT, K., TIGNOR, M. M., MILLER, H. L. & CHEN, Z. (2007), Climate Change 2007. The physical science basis. Contribution of Working Group I to the Fourth Assessment Report of the Intergovernmental Panel on Climate Change, published for the Intergovernmental Panel on Climate Change, ISBN 978-0-521-88009-1.

Collaborative Mapping and Emergency Routing for Disaster Logistics – Case Studies from the Haiti Earthquake and the UN Portal for Afrika

Pascal NEIS, Peter SINGLER and Alexander ZIPF

The GI_Forum Program Committee accepted this paper as reviewed full paper.

Summary

For planning the humanitarian operations of the UN the information about the actual condition of the streets is very important, as well as up-to-date information about hindrances and danger areas. Based on the services behind the OpenRouteService.org (ORS) platform dedicated Web portals for supporting the disaster logistics of the UN have been developed. While ORS uses data from OpenStreetMap in one special portal for Africa the data from the United Nations Spatial Data Infrastructure for Transportation (UN SDI-T) was deployed (ITHACA 2008a, b). This data needed special processing in order to make it usable for routing in the first place. In general recent geographic information is of high importance in disaster management, but in particular in developing countries the availability and access to geodata is still often quite limited. In order to improve the situation and have spatial data ready for planning and managing humanitarian actions the UN SDI has been initiated. It is fed from several sources and has therefore a varying quality. But recently another data source becomes more and more relevant for such activities. These are the user-generated datasets from communities of volunteers that organize themselves through Web 2.0 approaches. This „Volunteered Geographic Information" (GOODCHILD 2007) can play an increasing role in disaster management as the use case of Haiti has showed. Such information – in particular the prominent OpenStreetMap data has been used to realize Emergency Routing Services with highly up-to-data data.

An important feature of ORS for the disaster management operation was to consider blocked areas or streets when routing. The OGC (Open Geospatial Consortium) standard implemented in ORS, the Open Location Services Route Service (OpenLS), defines so called „AvoidAreas", which can be used to realize such functionality – even in an interoperable way. The portal allows UN staff to define and upload spatial data that represent those AvoidAreas into a geodatabase though the Web using the WFS-T (OGC Web Feature Service – Transactional). Further OGC services include a WMS or the OpenLS geocoder. This portal is a follow-up of a first prototype that was actually used during the UN emergency operation in Haiti after hurricane „Ike". Based on those real experiences the UNJLC planned to set up a portal to support in future operations.

Within this paper we present the realized portal based on OGC standards, experiences with the data available for Africa in the UN SDI-T and the functionality that were realized within the emergency routing service based on the technology deployed at OpenRouteService.org. We also introduce the OpenStreetMap (OSM) solution for Haiti and compare the differences and explain the lessons learned. From that we draw conclusions about the need to integrate and synchronize community-based approaches into humanitarian actions carried out by official humanitarian organisations. This includes the need to harmonize the data schemes of e.g. the UN Spatial Data Infrastructure for Transportation and approaches such as OpenStreetMap.

1 The Case of Africa: A Geoportal for the UN Joint Logistics Cluster

The UN Joint Logistics Cluster (UNJLC) is mandated by the UN Inter-Agency Standing Committee (IASC) to complement and co-ordinate the logistics capabilities of co-operating humanitarian agencies during large-scale complex emergencies and natural disasters. The UNJLC defines and implements the UN SDI-T which constitutes the transport-related branch of the United Nations Spatial Data Infrastructure (UNSDI). As explained above the availability of geodata that can be used for planning the logistics of humanitarian actions is often still quite restricted in particular in less developed countries.

The UN tries its best to set up a SDI but because of the general lack of sufficient accessible data the data of the UN-SDI-T consists of very heterogeneous sources and therefore has varying quality (figure 1). In order to make the data usable for route planning in the first place, both automatic procedures and a lot of checking and editing by hand was necessary. The data even included topological errors etc. in spite of being a database for transportation planning. The reason is that the collection and integration of the data only started a few years ago and data sources included (to a few percent) even non-geometric sources such as interviews, reports or measurements even without GPS. For example there are no logical relationship between point-objects such as bridges and obstacles included in the database. This means that those relationships needed to be established through spatial queries. This is problematic because of the heterogeneity of the different data sources with their differing accuracy and scale. There is a strong need for further homogenization and data quality control of this data. Within the work of P. Singler (SINGLER et al. 2009) it was possible to derive a topologically correct, routable street network from this data. This was used as input for the route service.

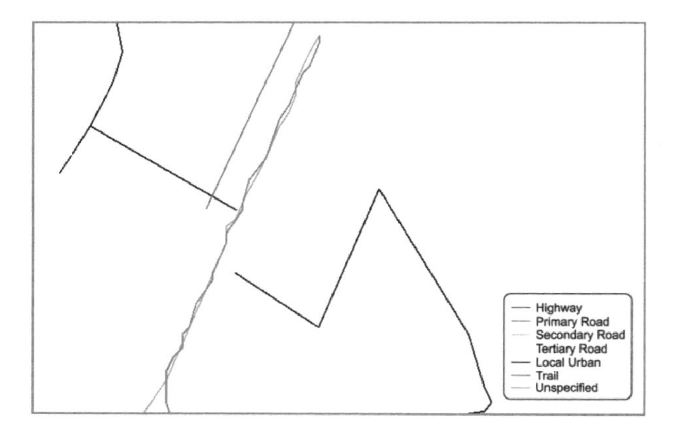

Fig. 1: Example of "heterogeneous" UN-SDI-T data (Africa)

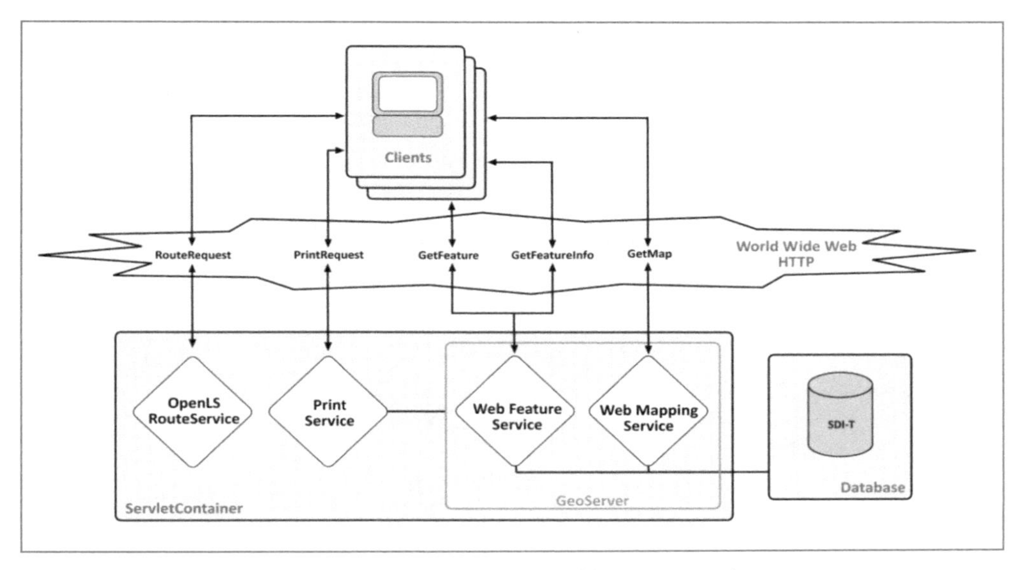

Fig. 2: OGC Service Architecture of the WFP-Africa GeoPortal

The architecture of the realized portal can be seen from figure 2. It is mostly based on the typical OGC web services (WMS, WFS), but includes also the OpenLS Route Service and a new, dedicated Print Service. This is a not-standardized service. It was introduced due to the needs of the humanitarian organisations to get printable planning maps (as high resolution pdf) for the field work with a specific layout and legend etc. This could not easily be realized through WMS with SLD.

The functionality of the Geoportal for Africa includes the possibility to calculate streets (figure 3). The route service implementation has been adapted from OpenRouteService.org to support the new data structures. Earlier versions had even been extended for a 3D context (NEIS et al. 2007). Similar to OpenRouteService.org it is possible to define so called "AvoidAreas" interactively on the map within the portal. But in contrast to ORS, these are being saved to a spatial database for further use by the users of the portal. It is also possible to attach attributes, e.g. in order to specify the grade of the obstacle in order to be able to handle different types of vehicles and different – even temporary – conditions of the roads. For example a non-paved road my become muddy after heavy rain and may only be accessible by All-Wheel Drives of a certain type. Another example is that even after flooding some specialized vehicles may still be able to pass the road if the flooding does not exceed a certain height. This means that it is necessary to be able to add further metadata attributes to the „AvoidAreas“ and specific combinations of those need to be considered for different types of vehicles from bikes to heavy trucks with/without trailers when routing (figure 4).

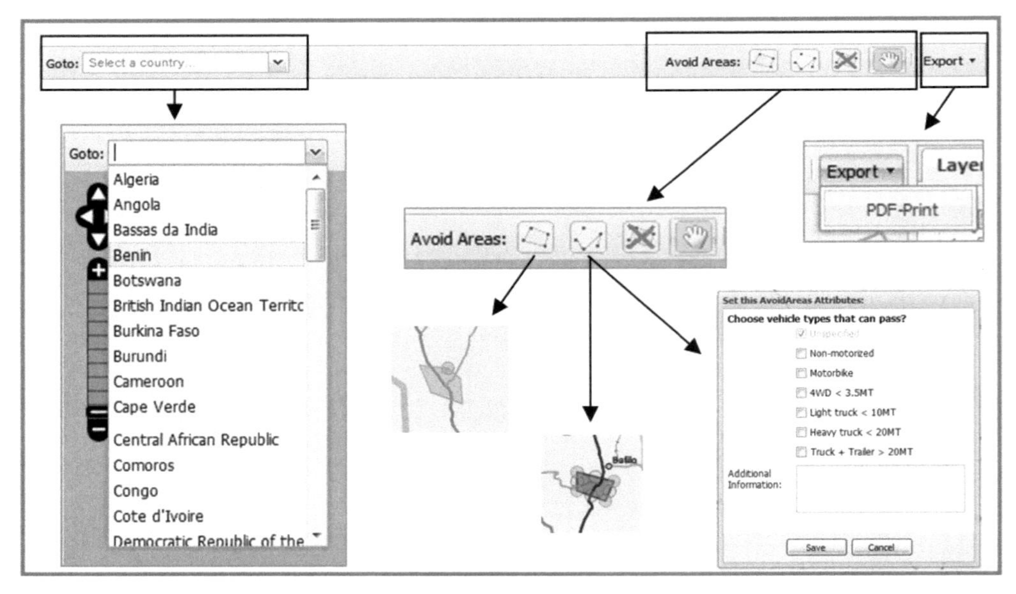

Fig. 3: Some major functionality in the UN WFP Geoportal for Africa

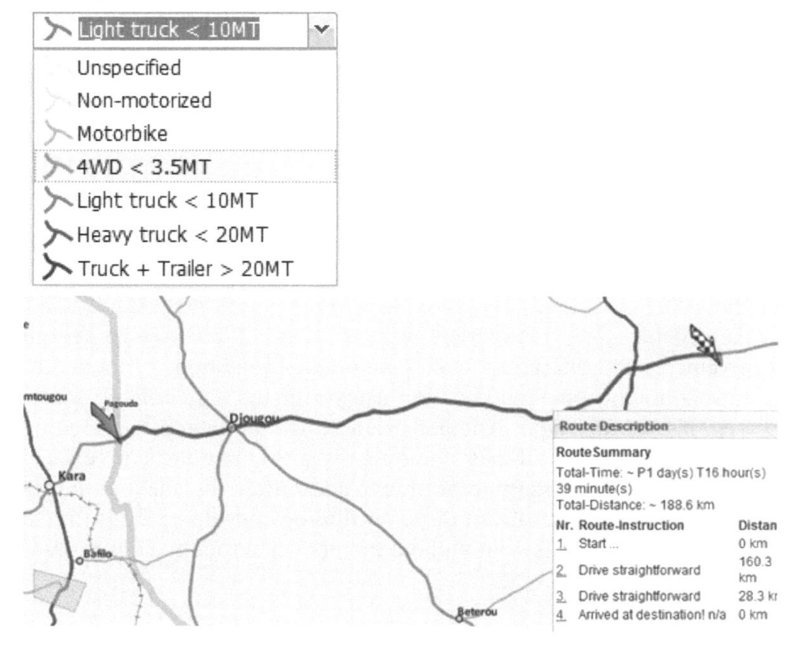

Fig. 4: Emergency Route Service for Africa in the WFP Geoportal based on UN SDI-T data.

2 The Haiti Case: From Hurricane IKE to the Earthquake 2010

The first real application of the new Emergency Route Service was not happening in Africa though. Oddly enough an earthquake hit Haiti in January 2010. Only two days after the Haiti earthquake the team of the Chair of GIScience of the University of Heidelberg published a new Emergency Route Service. This version is online at http://openls.geog.uni-heidelberg.de/osm-haiti/. This service uses the free geodata of OpenStretMap. The Route service is updated every hour in order to reflect the OSM community's great effort to map Haiti as fast as possible. In only a few days the data mapped through a collaborative effort of the OSM community and helpers worldwide has grown by a magnitude. This was possible because a set of recent satellite and aerial images were provided for digitizing the infrastructure and current situation.

Already during the Hurricane IKE disaster an online emergency route planner had been realized (SCHMITZ et al 2009). This was initiated by a request by the UN Logistics Cell that later lead to the development of the UN Africa routing portal. In 2008 OSM was not yet that well known and the data was still very sparse, in particular in less developed countries such as Haiti. Therefore additional data sources for the street network were needed. After some days a quite rough street network was found that was usable for routing between the towns. It did not provide much detailed information about the street network within the cities, though. This means the main difference between the Hurricane Ike version and the Earthquake 2010 version was the data set used. As in the case of the earthquake 2010 OpenStreetMap was used; it was also possible to publish the portal publicly for everybody interested. This was not possible in the Hurricane Ike version due to data license restrictions. This meant that only selected organizations had access to that version. Another difference was related to the functionality offered. In 2010 this functionality was even much more reduced compared to the Hurricane Ike version. The Hurricane Ike version included for example some special functionality for uploading and storing shape-files with already digitized AvoidAreas that should be considered when routing. Additional services such as the Accessibility Analysis Service (AAS) were available that allowed to calculate isochrones based on the street network from a selected location (NEIS & ZIPF 2007). Due to the quite sparse data set this was not really very useful in 2008. In 2010 we did not have the resources to add this service again. In order to increase the usability for non-expert users the web interface and functionally offered remained as simple as possible.

The choice to use OpenStreetMap was based on our experience with OpenRouteService and the speed of the growth of the OpenStreetMap data in general. Also quality assessments of OSM data have been conducted (ZIELSTRA & ZIPF 2010). Already one day after the earthquake it became clear to us that the OSM community would become very active and that there would be a substantial free geodata set in a short time. This meant we did not reactivate the old version from hurricane Ike, but Pascal Neis installed and extended a completely new version based on OpenStreetMap. Next to some efforts to adapt the user interface, the main task was automating the data update process for the route service. This was necessary in order to keep up to date with the fast growing OSM data set. Fortunately the colleagues from Geofabrik Karlsruhe (F. Ramm) provided a very frequent OSM data extract for download, which was more comfortable for handling and sped up the development process as it freed us from caring for this task as well.

The success of the collaborative mapping effort would not have been possible without the institutions owning satellite images providing these images to the public and humanitarian organisations and further giving allowance to use these images for deriving data for OpenStreetMap. The typical way how OSM data is being generated is that volunteers map the area with GPS by walking, biking or driving through the area. This was of course not possible in the Haiti case. Here almost all of the data was derived from satellite images. These are typically not available in other regions for this purpose because of the open licence of the derived vector data, which is not compatible with the licence of the image data. But in case of disasters, individual and relevant images were released to the public and explicit allowance was given to use those images for the purpose of digitizing data for OSM from them. Figures 5 and 6 illustrate the achieved results.

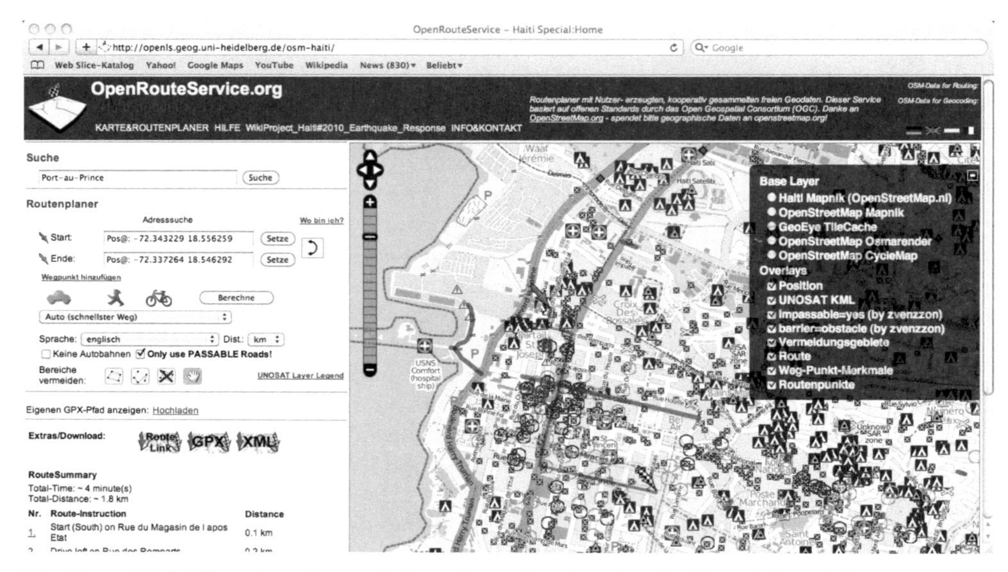

Fig. 5: Web User Interface of OpenRouteService for Haiti and Dominican Republic (the map shows Port-au-Prince with mapped refugee camps, destroyed houses etc.)

One of the activists within the OSM community who initiated the OSM disaster relief effort was Mikel Maron. He sent a mail to the OSM mailing list asking for support for Haiti already on January 12, 2010 and also spread the news on January 14 that the data provider GeoEye has allowed to use their images for mapping for OSM. This was done on Twitter (https://twitter.com/mikel/status/7755676201) – quite typical for the use of new social media and web 2.0 technology for coordinating communities of volunteers. This is in particular true for the case of disasters, where these media often provide relevant information much earlier than the more conventional sources of news and information. The same day another provider – DigitalGlobe – opened their images for OSM mappers. This meant that there were several high-resolution images available as data sources for OSM only after two days and the list of further sources grew continuously – including post-disaster images that allowed mapping the extent of the disaster.

Cmp. http://wiki.openstreetmap.org/wiki/WikiProject_Haiti/Imagery_and_data_sources.

The growth of the OSM data set can be visualized using some maps of the capital Port-Au-Prince before the earthquake, two days after the earthquake and how it looks now (fig. 7).

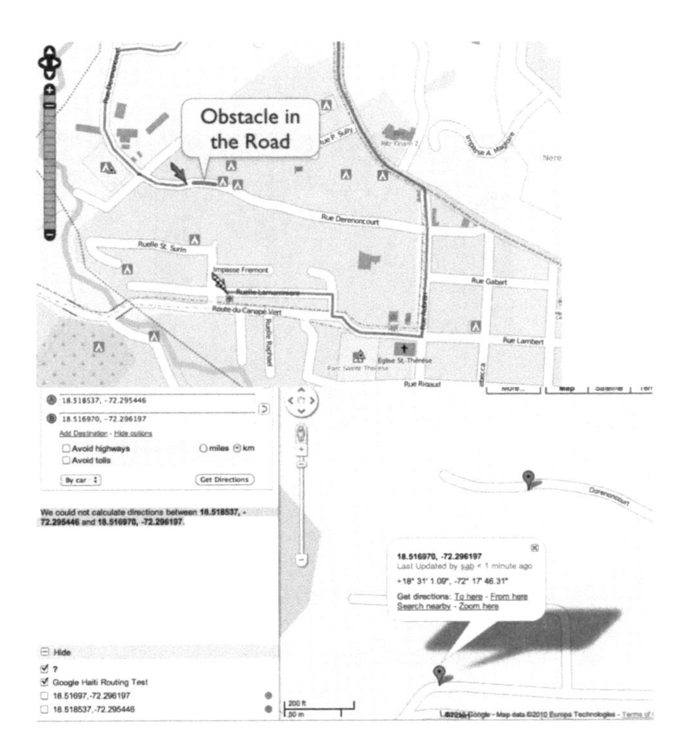

Fig. 6: Result of Routing in Haiti after the earthquake in OpenRouteService (left) ORS supports obstacles and Google (right) (routing failed). (In 2008 after Hurricane Ike Google did not support routing in Haiti at all.)

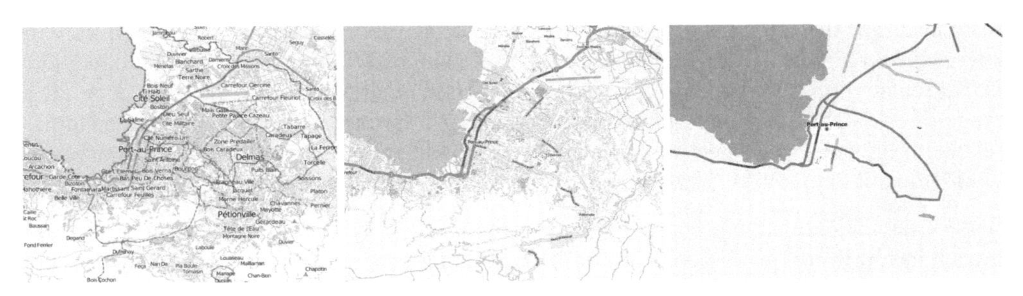

Fig. 7: Maps from OSM data showing the capital Port-Au-Prince before the earthquake, two days after the earthquake, and what it looks like now

Additionally the data became richer through a set of data donations from different sources. This includes maps of destroyed houses from disaster management organisations or data on health infrastructure. A list of data imports can be found here: http://wiki.openstreetmap.org/wiki/WikiProject_Haiti/VectorAndMapData. The following graphs show the increase in the number of places for geocoding as well as route segments within one week after the earthquake (figure 8).

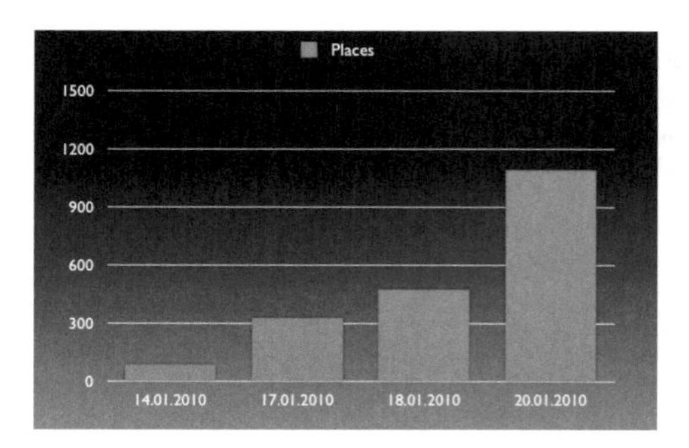

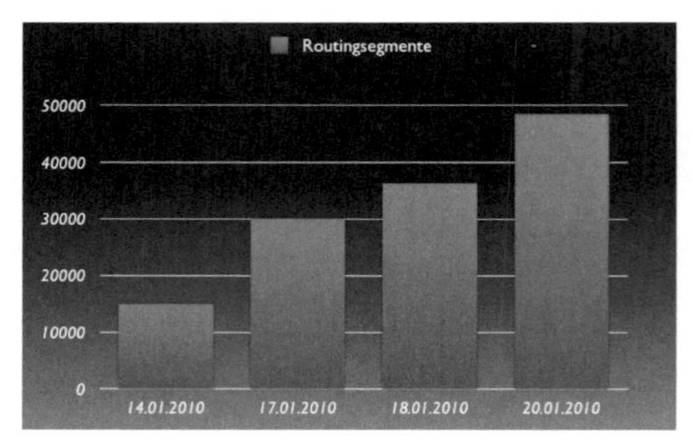

Fig. 8: Increase of OSM data in Haiti after the earthquake http://www.slideshare.net/jokru/crisismapping-in

3 An OSM Tagging Scheme for Humanitarian Actions?

Mapping geometry is obviously not enough. Geodata needs attributes in order to be usable in application in a sensible way. For example in the case of routing applications different street types are needed or attributes describing what types of vehicles can drive on that particular street. We have seen this already in the Africa Geoportal and the UN SDI-T provides an elaborate data schema for this purpose. The goal is to standardize that schema for the UN worldwide in order to increase the interoperability of the data within the UN institutions and other humanitarian organisations. But OSM, on the other hand, did not provide such a standardized schema for attributes related to disasters. In contrast, they use an open list of "tags" that describe attributes. During the mapping effort it became clear that additional tags were needed and mappers started discussing those. Fortunately also people that were aware of the UN SDI-T schema were among the discussants and fed in their experience with such issues. The Humanitarian OSM Team (H.O.T.) needs to be mentioned here in particular: http://wiki.openstreetmap.org/wiki/Humanitarian_OSM_Team. As a result a range of new OSM tags were introduced for adding information on a wide range of topics related to disasters. A respective proposal can be found here: http://wiki.openstreetmap.org/wiki/Humanitarian_OSM_Tags/Humanitarian_Data_Model.

Some examples are shown below. Also here the OSM principle of using the simplest approach has been followed. People do only map what they do understand and what does not mean too much overhead. This avoids sophisticated but overly complex data schemes, as these do not work in crowdsourcing with volunteers. These new tags allowed generating specialized maps (e.g. http://haiti.openstreetmap.nl) and applications. In particular OpenRouteService Haiti has been extended to support the tag „impassable" after it was introduced and used by mappers for Haiti. This allowed introducing a new option in addition to the already mentioned AvoidAreas. This even could be seen as a replacement for the option to save the digitized AvoidAreas (as in the case of the Africa Geoportal or ORS Haiti IKE 2008). Now OSM itself could be the database for this kind of information. But this needs further discussion as it is currently unclear how to handle dynamic situations where attributes change fast and frequently. In the Haiti earthquake case this was easily solved as only the latest version was of interest and the data was updated so frequently. But this is not a general solution. Of course proposals exist, but the question again is, which one is simple enough to work for a Web2.0 community of volunteers.

Examples of relevant tags:

earthquake:damage: collapsed_building ; earthquake:damage: spontaneous_camp;

earthquake:damage: damaged_infrastructure earthquake:damage: landslide

tourism: camp_site ; refugee: yes

Way Finding & Routing: impassable: yes

4 Conclusions

Volunteered information being collected by non-professionals and distributed through social media and Web 2.0 technology has been realized as a new source of information, in particular in the case of disasters. This is also true for geodata as the OpenStreetMap case shows. The Haiti disaster is only the most prominent example for that. This means that this kind of data collection and data source or even technical data infrastructure become an alternative to classical spatial data infrastructures. This leads to the need to discuss how these worlds can be integrated so that we get the best of both: actuality, richness and openness on the one hand, and reliability and semantic structure on the other. Through the experiences with Haiti we see that both players move towards each other and try to work on a solution. Future work will include discussions how to set up future infrastructures and applications for supporting disaster management (such as an emergency routing portal) that can make the best of all relevant data sources. In particular in the field relief workers just need data that works with their equipment – and that is actually true also for OSM, that has been provided in different download formats, e.g. also for use in GPS devices. E.g. a photo showing the Fairfax County Urban Search & Rescue Team using OSM on their GPS devices in Haiti on 22.01.2010 can be found at

http://1.bp.blogspot.com/_GxwMnMY7RgE/S3U_5wBOMSI/AAAAAAAAE8w/ileiANEg1YE/s1600-h/OpenStreetMap_on_a_Garmin_in_Haiti.JPG.
This is gratitude for all mappers that contribute to the project.

Acknowledgements

We thank all volunteers for spending their time deriving geodata for supporting the disaster management. In particular we thank F. Ramm for providing a frequently updated shapefile-download of the Haiti OSM-dataset.

References

GOODCHILD, M. F. (2007), Citizens as sensors: the world of volunteered geography. GeoJournal, 69 (4), pp. 211-221.

ITHACA (2008a), WFP Spatial Data Infrastructure (SDI) in support to a UNSDI. Data model definition, geodatabase implementation. – http://www.ithaca.polito .it/docs/ GDBImplementation_v2_2.pdf.

ITHACA (2008b), WFP Spatial Data Infrastructure (SDI) in support to a UNSDI. Data model definition, geodatabase implementation and WebGIS applications development. www.ithaca.polito.it/docs/WFPSpatialDataInfrastructure_v1_2.pdf.

NEIS, P. & ZIPF, A. (2007), A Web Accessibility Analysis Service based on the OpenLS Route Service. AGILE 2007. Aalborg, Denmark.

NEIS, P., SCHILLING, A. & ZIPF, A. (2007), 3D Emergency Route Service (3D-ERS) based on OpenLS Specifications. GI4DM 2007. 3rd Int. Symp. on Geoinformation for Disaster Management. Toronto, Canada.

OGC (2004), OpenGIS Location Services (OpenLS): Core Services. Part 5 – Route Service. – http://www.opengeospatial.org/standards/ols.

SCHMITZ, S., NEIS, P. & ZIPF, A. (2008), New Applications based on Collaborative Geodata – the Case of Routing. XXVIII INCA International Congress on Collaborative Mapping and SpaceTechnology, Gandhinagar, Gujarat, India.

SINGLER, P., NEIS, P. & ZIPF, A. (2009), OpenLS Emergency Routing Services based on the United Nations Spatial Data Infrastructure for Transportation (UN-SDI-T) – the case of Africa. POSTER at AGIT 2009. Symposium für Angewandte Geoinformatik. Salzburg. Austria.

ZIELSTRA, D. & ZIPF, A. (2010), A Comparative Study of Proprietary Geodata and Volunteered Geographic Information for Germany. AGILE 2010. Guimarães, Portugal.

Acknowledgement

The GI_Forum 2010 programme committee contributed to and shared their expertise in the GIS field with the GI_Forum team. Quality assurance of the GI_Forum programme and publication relies to a great extent on the willingness of these committee members to allocate time for assessing submitted contributions and providing valuable comments for good quality publications.

On behalf of all the participants, and particularly the authors of the GI_Forum 2010, we express our very special *THANK YOU* to the following persons:

Mohamed Aziz	Kuwait University
Euro Beinat	VU Amsterdam
Lars Bernard	University of Technology Dresden
Chris Brunsdon	University of Leicester
George Cho	University of Canberra
Elmar Csaplovics	University of Technology Dresden
Arup R. Dasgupta	GIS Development
Jim Farley	Consultant
Manfred M. Fischer	University of Economics and Business Administration Vienna
Andrew Frank	University of Technology Vienna
Frank Gossette	California State University, Long Beach
Francis Harvey	University of Minnesota
Milan Konecny	Masaryk University Brno
Jacek Kozak	Jagiellonian University Kraków
Stefan Lang	University of Salzburg
Michael Leitner	Louisiana State University
Bela Markus	University of West Hungary
Robert Marschallinger	Austrian Academy of Science – GIScience
Damir Medak	University of Zagreb
Jan-Peter Mund	United Nations University
Marco Painho	New University of Lisbon
Thomas K. Poiker	Simon Fraser University Vancouver
Jüri Roosaare	University of Tartu

Alexander Simonov	Russian State Research Institute on Information Technologies and Telecommunications
Nitin Kumar Tripathi	Asian Institute of Technology
Peter Zeil	University of Salzburg
Alexander Zipf	University of Heidelberg

Adrijana Car
GI_Forum programme committee chair

Josef Strobl
General Chair

Index of Authors

ALBRECHT, Florian (→ p. 1)
University of Salzburg/Austria
florian.albrecht@sbg.ac.at

AMELUNXEN, Christof (→ p. 11)
University of Heidelberg/Germany
christof@amelunxen.net

ANANTSUKSOMSRI, Sutee (→ p. 18, p. 211)
Cornell University, Ithaca/USA
sa457@cornell.edu

ANDERS, Karl-Heinrich (→ p. 69)
Carinthia University of Applied Sciences, Villach/Austria
k.anders@fh-kaernten.at

ASCHE, Hartmut (→ p. 164)
Institut für Geographie, Geoinformatik, Universität Potsdam/Germany

AUBRECHT, Christoph (→ p. 198)
Austrian Institute of Technology, Vienna/Austria
christoph.aubrecht@ait.ac.at

BARBOSA, Marcelo (→ p. 22)
Coffey limited, Brazil
marcelo_barbosa@coffey.com

BASSOUKOS, Anastasios (→ p. 58)
Informatics Apps and Systems Group, Department of Mechanical Engineering, Aristotle University, Thessaloniki/Greece
abas@isag.meng.auth.gr

BAUER, Thomas (→ p. 176)
Institute of Surveying, Remote Sensing and Land Information(IVFL), University of Natural Resources and Applied Life Sciences, Vienna/Austria

BERTERMANN, David (→ p. 160)
Geology, University Erlangen-Nürnberg/Germany
david.bertermann@gzn.uni-erlangen.de

BRUNAUER, Wolfgang (→ p. 87)

BRUNNER, Daniela (→ p. 26)
FH Kärnten/Austria
daniela.brunner@edu.fh-kaernten.ac.at

BURATTO, Mario (→ p. 22)
Hiparc ltda/Brazil
mario@hiparc.com.br

CAR, Adrijana (→ p. 219)
GIScience Institute, Austrian Academy of Sciences, Salzburg/Austria
adrijana.car@oeaw.ac.at

CHIMA, Christopher (→ p. 36)
Coventry University/UK
apy168@coventry.ac.uk

DASGUPTA, Arup (→ p. 40)
Honorary Managing Editor, GIS Development
arup.dasgupta@gisdevelopment.net

DIAS, Leonardo (→ p. 22)
Coffey limited, Brazil
leonardo_santana@coffey.com

DRĂGUŢ, Lucian (→ p. 48)
Department of Geography and Geology, University of Salzburg/Austria
lucian.dragut@sbg.ac.at

EGGER, Gregory (→ p. 26)
EB&P Umweltbüro GmbH, Klagenfurt/Austria
g.egger@umweltbuero-klagenfurt.at

EISANK, Clemens (→ p. 48)
University of Salzburg/Austria
clemens.eisank@sbg.ac.at

EPITROPOU, Victor (→ p. 58)
Informatics Applications and Systems Group, Department of Mechanical Engineering, Aristotle University, Thessaloniki/Greece
vepitrop@isag.meng.auth.gr

ERLACHER, Christoph (→ p. 69)
Carinthia University of Applied Sciences, Villach/Austria
c.erlacher@fh-kaernten.at

FÜREDER, Petra (→ p. 207)
Centre for Geoinformatics, Salzburg/Austria
petra.fuereder@sbg.ac.at

FÜRST, Josef (→ p. 176)
Institute of Water Management, Hydrology and Hydraulic Engineering (IWHW), University of Natural Resources and Applied Life Sciences, Vienna/Austria

GOLDBERG, Valeri (→ p. 101)
Institute of Hydrology and Meteorology, Technische Universität Dresden/Germany
valeri.goldberg@tu-dresden.de

GRÖCHENIG, Simon (→ p. 69)
Carinthia University of Applied Sciences, Villach/Austria
simon.groechenig@edu.fh-kaernten.ac.at

HELBICH, Marco (→ p. 87)
University of Heidelberg/Germany
marco.helbich@gmx.at

HENNERSDORF, Jörg (→ p. 101)
Leibniz Institute of Ecological and Regional Development (IOER), Dresden/Germany
j.hennersdorf@ioer.de

HOCHMAIR, Hartwig H. (→ p. 91)
University of Florida, Gainesville/USA
hhhochmair@ufl.edu

HOECHSTETTER, Sebastian (→ p. 101)
Leibniz Institute of Ecological and Regional Development (IOER), Dresden/Germany
s.hoechstetter@ioer.de

HOFFMANN, Christian (→ p. 207)
Definiens AG, Munich/Germany
choffmann@definiens.com

HÖLBLING, Daniel (→ p. 207)
Centre for Geoinformatics, Salzburg/Austria
daniel.hoelbling@sbg.ac.at

JAUMANN, Ralf (→ p. 164)
Institut für Planetenforschung, Planetengeologie, DLR/Germany

KARAMPOURNIOTIS, Iraklis (→ p. 111)
Rural and Surveying Engineering Department, Aristotle University of Thessaloniki/Greece
iraklis@topo.auth.gr

KARATZAS, Kostas (→ p. 58)
Informatics Apps and Systems Group, Department of Mechanical Engineering, Aristotle University, Thessaloniki/Greece
kkara@eng.auth.gr

KIECHLE, Günter (→ p. 140)
Salzburg Research Forschungsgesellschaft, Salzburg/Austria
guenter.kiechle@salzburgresearch.at

KLINGER, Gernot (→ p. 121)
Austro Control, Vienna/Austria
gernot.klinger@austrocontrol.at

KLUG, Hermann (→ p. 229)
Zentrum für Geoinformatik,
Salzburg/Austria
hermann.klug@sbg.ac.at

KÖSTL, Mario (→ p. 198)
AIT Austrian Institute of Technology,
Vienna/Austria
mario.koestl@ait.ac.at

KOULOV, Boian (→ p. 215)
Institute of Geography,
Bulgarian Academy of Sciences,
Sofia/Bulgaria
bkoulov@yahoo.com

KRÜGER, Tobias (→ p. 101, p. 130)
Leibniz Institute of Ecological and
Regional Development (IOER),
Dresden/Germany
t.krueger@ioer.de

KUNDU, Peter (→ p. 176)
Department of Agricultural Engineering,
Faculty of Engineering and Technology,
Egerton University, Njoro/Kenya

KURBJUHN, Cornelia (→ p. 101)
Institute of Hydrology and Meteorology,
Technische Universität
Dresden/Germany
cornelia.kurbjuhn@
mailbox.tu-dresden.de

LANG, Stefan (→ p. 207)
Centre for Geoinformatics,
Salzburg/Austria
stefan.lang@sbg.ac.at

LEHMANN, Iris (→ p. 101)
Leibniz Institute of Ecological and
Regional Development (IOER),
Dresden/Germany
i.lehmann@ioer.de

LEITNER, Michael (→ S. 26)
Department of Geography and
Anthropology, Louisiana State
University, Baton Rouge/USA
m.leitne@lsu.edu

MANOHAR, Senthanal Sirpi (→ p. 140)
Salzburg Research
Forschungsgesellschaft,
Salzburg/Austria
senthanal.manohar@salzburgresearch.at

MARJANOVIC, Milos (→ p. 150)
Palacky University of
Olomouc/Czech Republic
milosgeomail@yahoo.com

MOELLER, Matthias (→ p. 160)
Austrian Academy of Sciences,
Salzburg/Austria
matthias.moeller@sbg.ac.at

MOSER, Julia (→ p. 1)
Centre for Geoinformatics,
Salzburg/Austria
julia.moser@sbg.ac.at

NASS, Andrea (→ p. 79, p. 164)
Deutsches Zentrum für Luft- und
Raumfahrt, Germany
andrea.nass@dlr.de

NDUWAMUNGU, Jean (→ p. 168)
University of Rwanda,
Kigali/Rwanda
director@cgisnur.org

NEDKOV, Stoyan (→ p. 172)
Institute of Geography, Bulgarian
Academy of Sciences, Sofia/Bulgaria
snedkov@abv.bg

NEIS, Pascal (→ p. 239)
University of Heidelberg/Germany
neis@geographie.uni-bonn.de

OLANG, Luke (→ p. 176)
Department of Agricultural Engineering,
Faculty of Engineering and Technology,
Egerton University, Njoro/Kenya
olanglk@yahoo.com

PARASCHAKIS, Ioannis (→ p. 111)
Rural and Surveying Engineering
Department, Aristotle University of
Thessaloniki/Greece
jpar@topo.auth.gr

PICKLE, Edward (→ p. 184)
OpenGeo, A Division of The Open
Planning Project
epickle@opengeo.org

ROSSNER, Reinhold (→ p. 160)
Geology, University Erlangen-
Nürnberg/Germany
rossner@geol.uni-erlangen.de

SEOANE, José (→ p. 22)
Universidade Federal do
Rio de Janeiro/Brazil
cainho.geo@gmail.com

SINGLER, Peter (→ p. 239)
UN Logistics Cluster, WFP

SMREČEK, Róbert (→ p. 188)
University in Zvolen/Slovak Republic
smrecek@vsld.tuzvo.sk

STEINNOCHER, Klaus (→ p. 198)
Austrian Institute of Technology,
Vienna/Austria
klaus.steinnocher@ait.ac.at

TIEDE, Dirk (→ p. 207)
Univiversity of Salzburg/Austria
dirk.tiede@sbg.ac.at

TONTISIRIN, Nij (→ p. 211)
Cornell University,
Ithaca/USA
nt72@cornell.edu

TRODD, Nigel (→ p. 36)
Coventry University/
United Kingdom
n.trodd@coventry.ac.uk

VAN GASSELT, Stephan (→ p. 79,
p. 164)
Institut für Geologische Wissenschaften,
Planetologie und Fernerkundung,
Freie Universität Berlin/Germany
stephan.vangasselt@fu-berlin.de

VATSEVA, Rumiana (→ p. 215)
Bulgarian Academy of Sciences,
Sofia/Bulgaria
rvatseva@gmail.com

WALLENTIN, Gudrun (→ p. 219)
Austrian Academy of Sciences,
Salzburg/Austria
gudrun.wallentin@oeaw.ac.at

WEINKE, Elisabeth (→ p. 229)
Zentrum für Geoinformatik,
Salzburg/Austria
elisabeth.weinke@sbg.ac.at

ZALAVARI, Peter (→ p. 229)
University of Salzburg/Austria
peter.zalavar@sbg.ac.at

ZIPF, Alexander (→ p. 239)
University of Heidelberg/Germany
zipf@uni-heidelberg.de